MERVIN L. KEEDY, Purdue University

CHARLES W. NELSON, University of Florida

GEOMETRY
A Modern Introduction
SECOND EDITION

ADDISON-WESLEY PUBLISHING COMPANY

Reading, Massachusetts · Menlo Park, California

London · Amsterdam · Don Mills, Ontario · Sydney

This book is in the
ADDISON-WESLEY SERIES IN MATHEMATICS

Richard S. Pieters
Consulting Editor

PREFACE TO THE SECOND EDITION

One of the criticisms of the first edition of this book was that it was too rigorous or too difficult. In preparing this second edition, the authors have been most cognizant of that criticism. We have also received many other kinds of constructive criticism (as well as praise) with respect to the first edition. We hereby thank those who have been so kind as to give us their opinions and we assure all users and prospective users of the book that we have made a vigorous attempt to incorporate as many of their suggested changes as possible into the second edition.

A casual inspection of the following pages will show that we have not eliminated the (optional) material which allows for a rigorous and formal introduction to geometry. We believe that the complete elimination of such material would reduce the level of the course to approximately that of the junior high school, and we feel it would be inappropriate to publish a college book on that level. However, we have reorganized and augmented the exercise sets in such a way as to make the optionality of this material more obvious and to make the book more teachable.

There are essentially three kinds of exercises, and they are clearly separated in the exercise sets. The first exercises ordinarily to be found in an exercise set are those that do not depend upon a rigorous development. They are to be answered on an intuitive basis rather than on the basis of axioms, theorems, and the like, and they do not require rigorous demonstrations. The second kind of exercises are those that do depend upon the rigorous development. They are labeled Rigorous Exercises. These exercises are to be accomplished using axioms, theorems, and the like, and they often require the student to write a proof. In any portion of the book where the instructor wishes to omit or slight the rigorous development, he would omit all, or almost all, of these exercises. The third kind of exercises are preparatory or exploratory, and are labeled Exploratory Exercises. They very often require the student to do some exploring or experimenting, as background for material to be introduced later in that section, or possibly in the following section. These

iii

exercises do not depend upon a rigorous development; hence, they can be assigned even though the rigor is not being followed.

There have been numerous minor changes, additions, and deletions, but there have been two more major changes. The level of the chapter on coordinate geometry has been lowered, and it has been made more intuitive. There is also a completely new chapter on transformations.

It has been our experience in teaching CUPM Level I courses from this text that it is wise to have students do most or all of the exercises in Chapter 1. Of course, we do not do all of these at once. In fact, we spread these exercises over almost a full semester. We then proceed carefully through most of Chapters 2 and 3, introducing the axioms and proving theorems. After this, various choices of topics are possible. For example, if we find that most of our students are already familiar with congruence, we omit or touch lightly, Chapter 4. Most of the material on measures and measurement (Chapters 5, 10, and 11) is most appropriate for elementary teachers, and many instructors may wish to stress it, using the degree of rigor they feel is appropriate. Other selections can then be made from the chapters on parallelism and similarity, non-Euclidean geometry, coordinate geometry, and transformations.

Lafayette, Indiana   M.L.K.
Gainesville, Florida   C.W.N.
August 1972

PREFACE TO THE FIRST EDITION

Undergraduate courses in geometry have in the past been essentially of two kinds. The course usually called *College Geometry*, which is an advanced Euclidean geometry and a continuation of traditional high school geometry, is one of these. The other kind is typically an upper division course in foundations of geometry, usually with a substantial prerequisite. Courses of the second type in fact often carry graduate credit. There is a growing need for a foundational course in geometry at a lower level, and this book was written to fill such a need. Potential users of this text are therefore undergraduates in several categories. They include prospective elementary or secondary school teachers as well as others who may be majoring or minoring in mathematics. At Purdue University the several mimeographed versions of the manuscript have been tested primarily in a freshman course for prospective elementary school teachers, following rather closely the CUPM recommendations for Level I.

Prerequisites for a course based on this text should be two years of high school mathematics, including algebra and geometry, in addition to a course in number systems. It is assumed in this book that the student is already familiar with the systems of natural numbers, integers, rational numbers and, to a lesser extent, real numbers. Set language and notation should already be in the student's repertoire, and he should also have had some experience with mathematical proof beyond that of high school geometry. The prerequisite material beyond that covered in traditional high school courses is outlined in the Appendix. The reader is also referred to the text intended as a companion for this one, *Number Systems: A Modern Introduction*.

The axioms used in the foundational development are those of the authors, designed according to the belief that axioms at this level must necessarily have strong intuitive appeal, yet allow a rigorous development of the subject. Initially only the words *point* and *between* are undefined, and lines, planes, etc. are defined.

This allows the student to participate, in an exploratory fashion with the use of his own pencil, in the development of the abstract concepts and the theory.

It is the philosophy of the authors that for optimum learning of this subject at this level there must be a great deal of student participation and involvement. It is important, for example, that the student explore on his own and, where possible, make his own discoveries or gain his own insights. Chapter 1 is important in this regard. The exercises of this chapter not only provide a quick review of much of high school geometry, but they include activities designed to lead the student to a discovery or an insight. As pointed out in the *Commentary for Instructors*, some of these exercises lead to results already known by some students. Some, on the other hand, lead to results not known by any students. Our experience has shown that most students of a freshman class either will make the desired discoveries or already know the expected result. Either way, or even if neither happens, the student is actively involved in the exercises, and motivation and retention are usually high. It is important that the exercises in Chapter 1 be selected and assigned carefully. We have tried to include enough of them for this kind of activity to be continued, a bit at a time, throughout an entire course. It would be a mistake to use all of these exercises in one block of time. The language used in these exercises is different from that developed later, so as not to confuse the mathematical insights with linguistic problems. Later, at appropriate times, linguistic problems are considered.

In keeping with the philosophy that maximum student involvement pays dividends, we have designed many of the exercises so that they provide experience leading into concepts introduced in the following section of text. Thus, when the student reads text material following such exercises, or hears a lecture or participates in a discussion of it, he already has a background of some experience appropriate to the particular topic. Therefore, although some of these exercises seem fairly trivial, they are important psychologically and should not be slighted. Review Practice Exercise sets are also to be found at strategic places throughout the text. These are designed to provide a quick brush-up review of some topic or skill that will be needed shortly for understanding new material. These exercises should ordinarily be given as a small part of an assignment so that they remain incidental. The instructor will on occasion need to work one or two examples for the class at the time the assignment is made. We have found this kind of quick review to be quite effective. It requires very little time from the course proper, the student does not feel that he is getting a high school course over again, and yet some opportunity is provided for a review of concepts and skills that may have been forgotten. This contributes to a minimizing of both student and instructor frustration.

Both intuition and rigor are important in almost any mathematics course, especially in a course of this kind. In the exercises of Chapter 1 the intuition is given free rein, and it is in Chapter 2 that a rigorous development is begun. According to our philosophy, it is neither desirable nor possible to isolate a rigorous development from the intuitive. Therefore, the text is written so as to encourage the student to use his intuition regularly but, at the same time, to grow and develop in his ability to make a rigorous argument in which intuition plays no formal role. Thus, some exercises call for rigor and some do not. Exercises having questions to be answered rigorously, on the basis of the formal development, are identified by boldface numerals. Exercises not so designated are to be answered on an intuitive basis. If an exercise is marked with an asterisk, then it is generally more difficult and may be assigned accordingly. Certain words in the intuitive development do not occur in the formal development, for example the word "straight." Both kinds of vocabulary are used, and the student is expected gradually to learn not to use "unofficial," or intuitive, language in a formal, or "official," context.

This book is intended to be flexible in order to accommodate students of various backgrounds and to allow for varying course objectives. The rigor in Chapters 2 through 7 is greater than that of the following chapters, yet it is not essential that a class plow through these chapters in full detail. We recommend that the objectives of a course be formulated and then those parts of the text that are appropriate be used. For most students there is much more material here than could be covered in a one-semester course. At Purdue, this text is used in the second of two four-semester-hour courses for prospective elementary school teachers. In this course, foundations, rigor, and ability to do proofs are important but not to the extent that the material on measurement and graphs be excluded. These latter topics are too important for the teacher of elementary school mathematics. Thus, in this course many of the proofs in the formal development are omitted or slighted. The students do some proofs, and are expected to grow in this ability; they work through the formulation of the axiom system to the extent that they gain an appreciation of the importance and nature of axiomatic mathematics; but time is saved for work on later topics. In a course on foundations of geometry for other students, depending on the objectives and the background of the students, it might be more appropriate to concentrate on the details of Chapters 2 through 7, omitting or slighting the later material, which the mathematics major or minor might already know anyway. For further suggestions on ways of constructing courses based on this text, see the *Commentary for Instructors*.

The authors gratefully acknowledge the help they have received from many people. We especially wish to thank Professor Eugene Schenkman, who has taught the course with us and made many valuable suggestions; Mr. William Summers, who helped significantly with the formulation of the axiom system; Mr. Daniel Heisey; and the many patient students, who have been good critics and who were and remain our good friends.

Lafayette, Indiana   M.L.K.
June 1965   C.W.N.

LIST OF SYMBOLS

$\cap$	Set intersection
$\cup$	Set union
$\subset$	Is a subset of
$\in$	Belongs to (a set)
$\bar{S}$	Complement of set S
$\emptyset$	Empty set
$A\text{–}B\text{–}C$	"B is between A and C"
$\overline{AB}$	Segment with endpoints A, B
$\overrightarrow{AB}\!\!\!\circ$	Segment open at B
$\overset{\circ\!-\!\circ}{AB}$	Open segment
$m(\overline{AB})$	Measure of $\overline{AB}$
AB	Measure of $\overline{AB}$
$\overrightarrow{AB}$	Ray, endpoint A and containing point B
$\overset{\circ\!\rightarrow}{AB}$	Half-line, endpoint A and containing point B
$\overleftrightarrow{AB}$	Line, containing A and B
$\angle ABC$	Angle, vertex B and containing A and C on the respective sides
$\angle A$	Angle, vertex A
$m(\angle ABC)$	Measure of $\angle ABC$
$\overleftrightarrow{BC}/A$	The side (on a plane) of line $\overleftrightarrow{BC}$ that contains point A
$\overleftrightarrow{BC}/\sim A$	The side (on a plane) of line $\overleftrightarrow{BC}$ that does not contain A
$\triangle ABC$	The triangle with vertices A, B, C

<	Is less than (for numbers)
≤	Is less than or equal to (for numbers)
>	Is greater than (for numbers)
≥	Is greater than or equal to (for numbers)
≺	Is less than (for segments or angles)
≻	Is greater than (for segments or angles)
=	Is equal to (meaning "is identical with")
≅	Is congruent to
~	Is similar to
‖	Is parallel to
⊥	Is perpendicular to
$\square ABCD$	Parallelogram with vertices A, B, C, D
$\square ABCD$	Square with vertices A, B, C, D
Δa	Error in measurement a
(x, y)	An ordered pair of numbers; coordinates of a point
$P(x, y)$	Point P with coordinates (x, y)
$\lvert x \rvert$	Absolute value of x

CONTENTS

CHAPTER 1

EXPERIMENTAL AND AXIOMATIC GEOMETRY

1. EXPERIMENTAL ORIGIN OF GEOMETRY

Geometry is one of the oldest of the intellectual pursuits. There is evidence that the early Babylonians had developed considerable geometric knowledge dating as far back as 2000 B.C., and the Egyptians' knowledge of geometry dates back to about 1300 B.C. The early quest for geometric knowledge was in an experimental spirit, and that knowledge was of a very practical nature. Geometric knowledge is fundamental to problems of surveying and the construction of buildings, bridges, and the like. The existence of the great pyramids near the Nile is proof that the Egyptians knew their geometry and used it well.

To illustrate how geometric discoveries can be made experimentally, let us consider the following example.

Example. A surveyor wishes to lay out a tract having two parallel sides. He might lay off a line AB and then measure off a distance AC. Using cords of length AB and AC, he then marks arcs, as shown in Illus. 1.1. The arcs meet at a point D, and CD and AB are parallel (as are AC and BD). This could lead to the discovery that if each pair of opposite sides of a quadrilateral are equal, then the opposite sides are parallel.

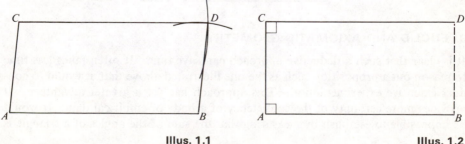

Illus. 1.1   Illus. 1.2

As he experiments further, he might try another method, as shown in Illus. 1.2. In this case he lays out AB and then makes a right angle at A. Then he measures off AC and makes another right angle at C. The lines AB and CD are seen to be parallel in this case also. The discovery here would be that if lines (AB and CD) are perpendicular to the same line (line AC), then they are parallel to each other.

A large collection of geometric facts was in existence in very early times, all or most of them discovered empirically in connection with practical problems of surveying or measurement. It is not strange that this branch of knowledge is called *geometry*, for the word is derived from Greek words meaning "earth measure."

2. DEDUCTIVE GEOMETRY

Beginning about 600 B.C., some Greek philosophers and mathematicians began to systematize the accumulated geometric knowledge. Of note among these were Thales and Pythagoras. Their orientation was no doubt more that of the thinker or philosopher than of the surveyor, and much of their work was deductive. A deductive approach differs from a strictly experimental or *inductive* one in that some facts are *derived*, or *proved*, on the basis of other information already at hand. The following example illustrates this point.

Example. It has been discovered that the sum of the angles of a triangle is 180°. We may now reason out, or prove, that the sum of the angles of a quadrilateral is 360°. No further experimentation will be needed.

Take any quadrilateral and draw a diagonal, thus cutting it into two triangles. The sum of the angles of each triangle is 180°, hence the sum of the angles of the quadrilateral is 360°. (See Illus. 1.3.)

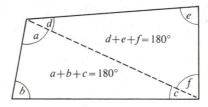

$$d+e+f=180°$$
$$a+b+c=180°$$

Illus. 1.3

3. EUCLID AND AXIOMATIC GEOMETRY

It is clear that such a deductive approach can save time. It often takes less time to reason out a proposition such as the one illustrated above than it would to conduct extensive experimentation. This approach has yet a greater advantage. It provides more certainty of the consistency of a body of empirical data. It would be impossible to establish by measuring that the sum of the angles of a triangle is

180° because measurements are never exact. It would similarly be impossible to establish that the sum of the angles of a quadrilateral is 360° by measurement. By deductive reasoning, however, it is possible to establish beyond doubt that *if* the sum of the angles of any triangle is 180°, then the sum of the angles of any quadrilateral is 360°.

It is sometimes said that geometry before the Greeks was purely inductive, or experimental, and that the Greeks were the first to reason deductively. It is difficult to agree with this, because there are so many closely and easily related facts like those in the above example that intelligent Babylonians or Egyptians must have made some of the connections. It is true, however, that the Greeks first began to systematize and organize the collection of geometric facts known in their time. As they proceeded in this fashion, it apparently became clear that the more facts they could prove, the more certain did the consistency and validity of their discipline become. The preceding example illustrates that *if* we know something about triangles, then we are sure of something about quadrilaterals. The trouble is that the knowledge of triangles may be based on experimental, and hence approximate, evidence. To increase the certainty, then, we might try to *prove* that the sum of the angles of a triangle is 180°, using some simpler or more obvious fact as the starting point. As we proceed in this manner, we find that we have a desire to push things back, as far as possible, to a few unproven but acceptable facts as the starting point. All else is then to be proved from them.

The ultimate work of the Greeks was that of Euclid, who about 300 B.C. wrote a treatise on geometry called *Elements*, containing virtually all that was known of geometry at that time. He began with a few unproven facts, called axioms and postulates,* and proceeded to develop by proof, in a most systematic fashion, all the rest of geometry. This monumental work stood for about 2000 years as the standard work in geometry, and his methods and approach became the model for modern axiomatic mathematics.†

4. INDUCTIVE AND DEDUCTIVE REASONING

It is all too easy for the student of high-school geometry to gain the impression that the subject is purely deductive. This is unfortunate because creative mathematical activity inevitably involves both inductive, or experimental, activity and deductive reasoning. Before a mathematician gets anywhere deductively, he must have something which he can try to prove. Without inductive reasoning or

---

* Today *axiom* and *postulate* are synonyms.

† Later linguistic and logical errors were discovered. For example, it is now clear that Euclid's axioms are not completely adequate. His *Elements* nevertheless remains the cornerstone of modern geometry.

experimentation he does not have it. For example, geometry could not have been organized in the formal deductive manner of Euclid had not a lot of inductively produced knowledge known prior existence. Most mathematical discoveries are made inductively. They are then rigorized or formalized deductively.

If mathematics is taught in such a way that students are not encouraged or allowed to experiment, make discoveries and conjectures, then the nature of mathematical activity is being withheld from them. Even young children can learn much this way. They find it easy as well as exciting, and their interest usually runs high.

As the student matures, however, he should be taught that experimentation and inductive discovery are not the entire answer. In order to gain greater certainty and consistency we must seek to prove that the inductively based discoveries are indeed true. The student who is not taught this is likewise being denied access to the whole nature of mathematical activity.

5. REVIEWING BASIC CONSTRUCTIONS

In the exercises which follow, the reader is given the opportunity to engage in experimentation and inductive discovery. Such experience is considered very important for illustrating this kind of activity, especially for the teacher or prospective teacher.

In later chapters more careful and precise language will be developed, but that need not concern us here. These exercises are, for the most part, written in traditional geometry vocabulary, even though it lacks the precision desired for later development. Certain basic compass and straightedge constructions are necessary in these exercises. We shall review them here. The use of a protractor will also be required.

1. *To bisect a line segment:*

Place compass point at *A* and draw arcs *a* and *b*. Place point at *B* and draw arcs *c* and *d*. Connect the points of intersection. Point *P* is the midpoint of *AB* (Illus. 1.4).

Illus. 1.4

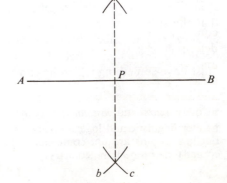

2. *To bisect an angle:*

Place compass point at *A* and draw an
arc, locating *B* and *C*. Place point at
B and draw arc *b*. Place point at *C* and
draw arc *c*. Connect the intersection
with *A* (Illus. 1.5).

Illus. 1.5

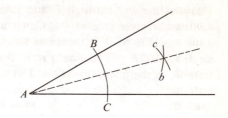

3. *To construct a perpendicular to a line from a point P on the line:*

Place compass point at *P* and draw an
arc, locating points *A* and *B*. Place
point at *A* and draw arc *a*. Place point
at *B* and draw arc *b*. Connect the
intersection with *P* (Illus. 1.6).

Illus. 1.6

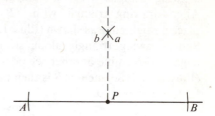

4. *To construct a perpendicular to a line from a point P off the line:*

Place compass point at *P* and draw an
arc, locating points *A* and *B*. Place
point at *A* and draw arc *a*. Place point
at *B* and draw arc *b*. Connect the intersection with *P* (Illus. 1.7).

Illus. 1.7

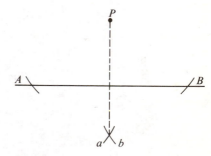

5. *To construct an angle the same size as a given angle:*

If angle *A* is the given angle, place
compass point at *A* and draw an arc to
determine *B* and *C*. Draw a line at *A'*
and with the same compass setting draw
an arc, determining *P*. Place point at *B*
and open compass so that the pencil is
at *C*. With the same setting, place point
at *P* and draw arc *p*. Connect the intersection with *A'* (Illus. 1.8).

Illus. 1.8

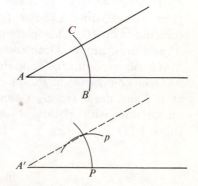

6. *To construct a line parallel to a given line through a point P off the line:*

Draw a line *PQ* through *P* and intersecting the given line at *Q*. Construct
angle ∠*QPR* so it is the same size as
angle ∠*MQP*. The line *PR* is then
parallel to the given line (Illus. 1.9).

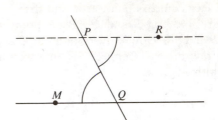

Illus. 1.9

It is also easy to draw a pair of parallel
segments, using a straightedge and
drawing triangle. Draw a segment *AB*.
Align drawing triangle with *AB* and
place straightedge as shown (Illus. 1.10).
Slide drawing triangle along straightedge to determine another segment *CD*
(Illus. 1.11). Segment *AB* is then parallel to segment *CD*.

Illus. 1.10

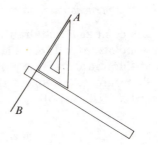

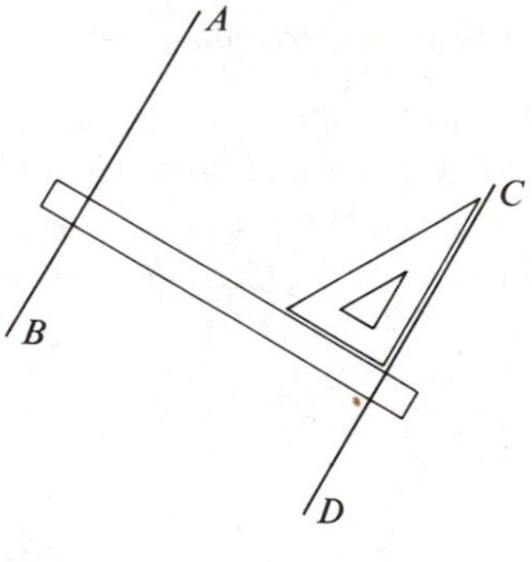

Illus. 1.11

7. *To construct a triangle the same size and shape as a given triangle:*

First method (SSS). If △*ABC* is the
given triangle, draw a line at *A'*. Open
compass to fit *AC*. Then place point at
A' and draw arc *a*. Open compass to
fit *AB*, place point at *A'*, and draw arc
b. Open compass to fit *BC*, place point
at *C'*, and draw arc *c*. Connect the
intersection with points *A'* and *C'*
(Illus. 1.12).

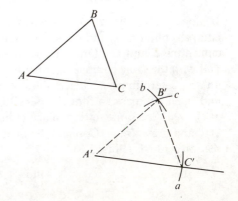

Illus. 1.12

Second method (SAS). Draw a line at A' and find C' as in method 1. At A' construct an angle the same size as $\angle A$. Open compass to fit AB, place point at A' and draw arc a. Connect B' and C' (Illus. 1.13).

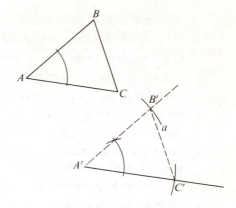

Illus. 1.13

Third method (ASA). Draw a line at A' and find C' as in methods 1 and 2. At A' construct an angle the same size as $\angle A$. At C' construct an angle the same size as $\angle C$ (Illus. 1.14).

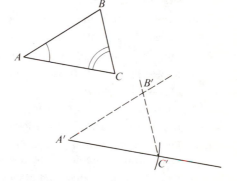

Illus. 1.14

EXERCISES

The primary objective in these exercises is *to look for patterns and to make discoveries.* It is important that drawings be made rather large, usually only one to a page. They should be made carefully and precisely, with a sharp pencil.

It is not intended that all of these exercises should be completed now. The best procedure is to complete a few of them periodically as one proceeds through the book; in addition to providing a review of some basic geometric concepts, it will continue to emphasize the importance of searching for patterns and making independent discoveries.

1. a) Draw four points, no three of which are on a line.

   b) Draw a quadrilateral with these four points as vertices.

   c) Draw a different quadrilateral (dashed or in a different color), using the same four points. This quadrilateral may cross itself.

   d) How many quadrilaterals are there having four given points as vertices?

2. a) Draw five points and see how many pentagons you can find having these points as vertices.

   b) Can you generalize this to n points?

3. a) Draw two angles with their sides parallel.

   b) Measure the angles and generalize.

4. a) Draw two angles with their respective sides perpendicular.

   b) Measure the angles and generalize.

5. a) Construct a right triangle, using compass and straightedge.

   b) Measure the two acute angles.

6. a) Draw an obtuse triangle.

   b) Construct another triangle, larger than this one, having an obtuse angle the same size as the first and one acute angle the same size as one of the angles of the first.

   c) Measure the angles of the first triangle. Measure the angles of the second triangle.

7. a) Draw an acute triangle.

   b) In three ways, construct a triangle congruent to this one, using compass and straightedge only.

8. a) Draw a triangle and construct the perpendicular bisectors of its sides.

   b) Repeat for several other triangles. Generalize if you can.

9. a) Draw a triangle and construct the altitudes.

   b) Repeat for several other triangles. Generalize if you can. (Do not fail to include obtuse triangles.)

10. a) Draw a triangle and construct the angle bisectors.

    b) Repeat for several other triangles. Generalize if you can.

11. a) Draw an obtuse triangle and draw its medians. (A median is a segment from a vertex to the midpoint of the opposite side.)

    b) Draw an acute triangle and draw its medians.

    c) Draw a right triangle and draw its medians.

    d) Repeat. See if you can generalize.

12. a) Draw a triangle and find the orthocenter, centroid, and circumcenter.

    b) Repeat for some other triangles. See if you can generalize.

13. a) Draw a triangle ABC and its medians. Call the centroid P.

    b) Measure the medians.

    c) Measure AP, BP, and CP. Generalize and discuss.

14. a) Draw a large right triangle.

    b) Draw the altitude from the right angle. This forms two triangles within the large one.

    c) Measure the angles of all three triangles.

    d) Repeat (a), (b), and (c) for several other right triangles. Generalize.

15. a) Draw a fairly large triangle on cardboard.

    b) Draw its medians.

    c) Cut out the triangle, tie a thread to a pin, and stick the pin in at the centroid. Suspend the triangle by the thread. Discuss what you observe.

16. a) Draw an isosceles triangle.

    b) Construct a triangle congruent to it. Cut out the second one and place it over the first one.

    c) Now flip the second one over and see if it still fits. If it does, what does this tell you about the angles of the triangle?

17. a) Draw an equilateral triangle.

    b) Construct a triangle congruent to it. Cut out the second one and place it over the first one.

    c) Now rotate the second one (do not flip it over) and see how many positions there are for which it fits the first one. What does this tell you about the angles of the triangle?

18. a) Draw a triangle and bisect one side. Call the midpoint P.

    b) Through P construct a line parallel to one side of the triangle. Note where it meets the third side.

    c) Repeat for some other triangles.

19. a) Draw a scalene triangle ABC. Denote the sides opposite A, B, and C by a, b, and c, respectively.

    b) Measure all the sides and all the angles. List the angles in order of size. List the sides in order of size.

    c) Repeat for some other triangles.

20. a) Draw a triangle and one of its exterior angles.

    b) Measure the exterior angle. Measure the two angles of the triangle that are not adjacent to the exterior angle. Add the latter two measures.

    c) Repeat for some other triangles.

21. a) Draw a triangle ABC and bisect angles A and B.

    b) Denote the points where the bisectors meet BC and AC by X and Y, respectively.

    c) Consider the exterior angle at C which is on the opposite side of BC from A. Bisect this angle. Call the point where the bisector meets line AB point Z.

    d) Repeat for several other triangles.

22. a) Draw a quadrilateral that does not cross itself.

    b) Bisect all four sides. Call the midpoints A, B, C, and D.

    c) Connect the midpoints of the sides in order to form a quadrilateral $ABCD$.

    d) Repeat for several other quadrilaterals.

23. a) Draw a quadrilateral that does not cross itself.

    b) Bisect all four sides. Call the midpoints A, B, C, and D.

    c) Connect opposite midpoints to form two intersecting segments $\overline{AC}$ and $\overline{BD}$. Call the point of intersection O.

    d) Compare the lengths of $\overline{AO}$ and $\overline{OC}$.

    e) Repeat for several other quadrilaterals.

24. a) Draw a parallelogram $ABCD$ and its diagonals.
    b) Call the intersection of the diagonals point P. Compare the lengths of $\overline{BP}$ and $\overline{PD}$.
    c) Compare the lengths of $\overline{AP}$ and $\overline{PC}$.
    d) Repeat for several other parallelograms.

25. a) Draw a rhombus and its diagonals.
    b) At the point of intersection of the diagonals, four angles are formed. Measure them.
    c) Repeat for some other rhombi.

26. a) Draw a rectangle and its diagonals.
    b) Compare the lengths of the diagonals.
    c) Draw a parallelogram that is not a rectangle and its diagonals. Compare the lengths of the diagonals.
    d) Repeat for several other rectangles and parallelograms.

27. a) Draw a rectangle and bisect all of its sides.
    b) Connect the midpoints of the sides to form a quadrilateral.
    c) Repeat for some other rectangles.

28. a) Draw a triangle ABC and the medians from A and B. Call the midpoints of $\overline{AC}$ and $\overline{BC}$ points E and D, respectively. Call the intersection of the medians point O.
    b) Bisect $\overline{AO}$ at F and bisect $\overline{OB}$ at G. Draw quadrilateral $EFGD$.
    c) Repeat for some other triangles.

29. a) Draw a parallelogram $ABCD$ and its diagonals intersecting at O.
    b) Bisect $\overline{AO}$ at E, $\overline{BO}$ at F, $\overline{CO}$ at G, and $\overline{DO}$ at H. Draw quadrilateral $EFGH$.
    c) Repeat for some other parallelograms.

30. a) Draw a circle. Without changing the compass setting, step off radius lengths around the circle.
    b) How many radius lengths are required to go around the circle once?
    c) Using the several points on the circle as vertices, draw a polygon.
    d) Repeat for some other circles, describe and generalize.

31. A regular polygon is one which has all of its sides congruent and can be inscribed in a circle. It must not cross itself.
    a) By experimentally determining a compass opening, draw a regular pentagon; that is, draw a circle and find a compass setting that will step around the circle "exactly" five times.
    b) Measure the angles of the pentagon.
    c) Draw several other regular polygons, some with more than five sides. Generalize and discuss.

32. a) Draw a regular hexagon.
    b) By measuring, drawing in other lines, etc., see what you can find out about it. You might wish to consider such things as angle size, perimeter, area, relation with triangles, circles, etc.

33. a) Draw a circle and mark two points, A and B, on it that are not opposite ends of a diameter. The two points thus determine a major arc and a minor arc.

   b) Inscribe an angle ACB in the minor arc and measure it. Inscribe several other angles in the same arc and measure them.

   c) Inscribe several angles in the major arc and measure them.

34. a) Draw a circle and a diameter $\overline{AB}$.

   b) Choose a point C on the circle, different from A or B. Draw segments $\overline{AC}$ and $\overline{BC}$.

   c) Measure angle C.

   d) Repeat for several other points such as C.

35. a) Draw an acute triangle.

   b) Construct the angle bisector of angle A and the perpendicular bisector of side BC.

   c) Do the same for angle B and side AC.

   d) Do the same for angle C and side AB.

   e) Circumscribe a circle about triangle ABC.

   f) Repeat (b) through (e) for several other acute triangles.

36. Follow the directions in Exercise 35 for a right triangle.

37. Follow the directions in Exercise 35 for an obtuse triangle.

38. a) Draw a circle and mark a point P on it. Call the center O.

   b) Draw a tangent to the circle at P. (A tangent is a line that touches the circle at just one point.) Use a straightedge, but use no other construction.

   c) Draw the segment $\overline{OP}$. Measure the angles that it makes with the tangent line.

   d) Repeat for several other circles. ·

39. a) Draw a circle and a point P outside the circle. Call the center of the circle O.

   b) Draw a line containing P and tangent to the circle. Call the point of tangency T.

   c) Draw segment $\overline{OP}$ and bisect it. Call the midpoint M.

   d) Draw a circle with center M and $\overline{MP}$ as a radius.

   e) Where is T with respect to this circle?

40. a) Draw a circle and in it a chord $\overline{AB}$. Call the center O.

   b) Bisect $\overline{AB}$ at C and draw a radius containing point C.

   c) Measure the angles made by the radius and the chord.

   d) Repeat for some other circles and chords.

   e) Draw a circle and a chord $\overline{AB}$. Construct the perpendicular bisector of $\overline{AB}$. Repeat for some other chords.

41. a) Draw a segment $\overline{AB}$ and construct the perpendicular bisector.

   b) Select any point P on the bisector and draw $\overline{PA}$ and $\overline{PB}$.

   c) Compare the lengths of $\overline{PA}$ and $\overline{PB}$.

   d) Choose any point Q not on the perpendicular bisector. Compare the lengths of $\overline{QA}$ and $\overline{QB}$.

e) Repeat for some other points on the bisector and some other points not on the bisector.

42. a) Draw three noncollinear points, A, B, and C, and draw segments $\overline{AB}$ and $\overline{BC}$.

   b) Construct the perpendicular bisectors of the two segments. Call the intersection of the bisectors O.

   c) Draw a circle, with center at O and with radius $\overline{OA}$.

   d) Repeat for some other points.

43. a) Draw a triangle ABC.

   b) On $\overline{AB}$ choose a point R, on $\overline{BC}$ choose a point S, and on $\overline{CA}$ choose a point T.

   c) Construct the circle containing points A, R, and T.

   d) Construct the circle containing points B, R, and S.

   e) Construct the circle containing points C, S, and T.

   f) Repeat for some other triangles.

44. a) Draw a circle and in it a chord $\overline{AB}$.

   b) Bisect the chord, calling the midpoint M.

   c) Draw two other chords both containing the point M.

   d) Connect the endpoints of the latter two chords in such a way that the connecting segments cross $\overline{AB}$. Call the points at which they intersect $\overline{AB}$ points S and T, respectively.

   e) Compare the lengths of $\overline{MS}$ and $\overline{MT}$.

   f) Repeat for several other circles and with chords in different positions.

45. a) Draw a circle and construct a tangent at a point P on the circle. Call the center O.

   b) Select some other point of the circle, Q, and draw ray $\overrightarrow{PQ}$. Draw $\overline{OP}$ and $\overline{OQ}$ to form a central angle.

   c) Measure the central angle and compare it with the angle formed by the tangent and the ray $\overrightarrow{PQ}$.

   d) Repeat for some other circles.

46. a) Draw a circle. Circumscribe about this circle a quadrilateral $ABCD$ by making four tangents.

   b) Measure the sides of the quadrilateral. Add the measures of $\overline{AB}$ and $\overline{CD}$. Add the measures of $\overline{BC}$ and $\overline{DA}$.

   c) Repeat for some other quadrilaterals circumscribed about circles.

47. a) Draw a circle and a point P outside the circle. Construct the two tangents (Exercise 36) from P to the circle. Call the points of tangency R and S. Call the center of the circle O.

   b) Draw $\overline{OR}$, $\overline{OS}$, and $\overline{RS}$. Compare the sizes of angles OSR, ORS, and SPR.

48. a) Draw a circle and from a point P outside the circle construct a tangent. Call the point of tangency T.

b) From P draw a secant, meeting the circle at points R and S (where R is between P and S).

c) Represent the measure of $\overline{PT}$ by "a," the measure of $\overline{PR}$ by "b," and the measure of $\overline{PS}$ by "c." By measuring, find a, b, and c.

d) Find the ratio b/a. Find the ratio a/c.

e) Repeat for some other circles and points P.

49. a) Draw a circle and a point P outside the circle.

b) Draw two secants from P, meeting the circle at points Q and R (with Q between R and P) and points S and T (with S between P and T), respectively.

c) Represent the measure of $\overline{PQ}$ by "a," the measure of $\overline{PR}$ by "b," the measure of $\overline{PS}$ by "c," and the measure of $\overline{PT}$ by "d." Find these measures.

d) What relationships between a, b, c, and d can you find?

e) Repeat for some other circles.

50. a) Draw a circle and inscribe a hexagon. No two sides of the hexagon should be parallel. The hexagon need not be a simple hexagon. Label the vertices, in order, A, B, C, D, E, and F.

b) Draw lines $\overleftrightarrow{AB}$ and $\overleftrightarrow{ED}$ and find their point of intersection.

c) Do the same thing for lines $\overleftrightarrow{CB}$ and $\overleftrightarrow{EF}$.

d) Do the same thing for lines $\overleftrightarrow{AF}$ and $\overleftrightarrow{CD}$. [*Note:* If the intersection points of parts b, c, and d are off the paper, you will need to draw a different hexagon (or find some large paper).]

e) Repeat for some other hexagons inscribed in circles.

REVIEW PRACTICE EXERCISES

In Exercises 1 through 10, find the intersections.

Example. $\{4, 5, 6, 7\} \cap \{1, 2, 3, 4, 5\} = \{4, 5\}$

1. $\{A, B, C\} \cap \{C, D, E\}$
2. $\{X, Y, Z\} \cap \{V, W, X\}$

3. $\{2, 4, 6, 8\} \cap \{1, 3, 5, 7\}$
4. $\{a, d, e, f\} \cap \{g, a, e, f\}$

5. $\{3, 2, 1, 0\} \cap \{0, 1, 2, 3\}$
6. $\{1, 2, 3\} \cap \{1, 2, 3, 4, 5\}$

7. $\{1, 2, 3, 4, 5\} \cap \emptyset$
8. $\{A, B, C\} \cap \{C, B, D\} \cap \{A, D, B\}$

9. $\{x \mid x \text{ is a whole number and } x < 5\} \cap \{x \mid x \text{ is a whole number and } x > 2\}$

10. $\{x \mid x \text{ is a whole number and } x < 9\} \cap \{x \mid x \text{ is a whole number and } x > 9\}$

11 through 20. Find the unions of the sets in Exercises 1 through 10.

Example. $\{A, E, F\} \cup \{B, E, C\} = \{A, B, E, F, C\}$

CHAPTER 2

POINTS, LINES, AND PLANES

1. INTRODUCTION

In the preceding chapter we have seen the importance not only of inductive activity, but also of deductively establishing as many of the experimentally determined facts as possible. This is precisely what Euclid did.

The work of Euclid, thought for many centuries to be infallible, has been shown to lack logical rigor and to have some linguistic weaknesses. Our approach to geometry will be somewhat different from that of Euclid, although the basic objective, and many of the deduced results, will be ultimately the same.

In this chapter we shall begin the deductive development. One of the first tasks will be to formulate some axioms. These will consist of plausible assumptions about the most basic nature of geometry. Then on the basis of these assumptions we shall prove as much about the rest of geometry as space and time permit. Space here does not permit a thorough and rigorous deductive development of elementary geometry in all details. We therefore shall attempt only to develop an adequate set of axioms, formulate definitions, and then accomplish enough proofs to illustrate clearly the spirit of modern formal geometry.

As we begin to look at basic notions of geometry, we shall consider certain configurations because they remind us of objects we see in the physical world. This is important, but it should be remembered that geometric configurations, such as points, lines, planes, polygons, etc., are merely idealizations of the real configurations that exist in the physical world and that the geometry we study here is one of the many mathematical systems man has created in an attempt to describe the world around him. The basic geometric concepts are therefore abstractions and exist only in our minds. This is of course true of numbers also. They too are abstract ideas which we cannot see or feel. We can only contemplate them.

In this chapter, as well as in some to follow, we shall consider only properties that do not involve the concept of *size* or *distance*. We shall not speak of the "length" of a line segment, for example, or the "area" of a square. This may seem strange at first, for one may feel there are no properties that can be discussed to any extent without bringing in the ideas of size or measurement. We shall soon see that this is not the case. The most fundamental ideas of geometry are precisely those that do not involve the ideas of size or measurement. Geometry in which such notions are considered is called "metric" geometry, and geometry in which there is no concept of size or measurement is called "nonmetric" geometry.

2. POINTS

As in developing most mathematical systems, we begin by considering a set of elements. The fundamental elements in our geometry are called *points*. Such things as lines, circles, triangles, etc., will later be defined as certain sets of points.

How should we think of a point? Since a point is the smallest entity in geometry, we might assume it to be an abstraction of the smallest thing in our physical world. At first, one might choose an atom as the smallest thing in the world; but atoms can be split into smaller particles, the smallest of which is an electron. Should we think of a point as an abstraction of an electron? Probably not, since electrons have some size and mass and we have no assurance that an electron cannot be split. We therefore would need to look for something smaller than an electron.

How, then, shall we define, or describe, a point as we proceed with a formal development of geometry? If we say it is that which has no length, breadth, or thickness, or if we say it is "that which has no part," as Euclid did, we would then need to say what we mean by *breadth*, *length*, *thickness*, or *part*, and we would be no better off than before. In fact, we would then have several words to define rather than one. The way out of this dilemma is simple. We simply leave the word *point* undefined. This is common procedure in contemporary mathematics. Since we cannot, in any situation, define everything without being circular,* we always begin with one or more terms that are not defined.

The word point is undefined.

We have a fairly good intuitive notion about what a point is. This is essential. A point, in the intuitive sense, is the smallest thing in our geometry. So far as the formal sense is concerned, however, we have placed ourselves in the position of not knowing what a point is. We may say that we "draw a point," but what we mean is that we make a mark which is *a picture* representing our intuitive notion of what a point is. "Officially" we cannot draw a point or even a picture of it, because we do not know what it is. It is undefined. This poses no particular

---

* A definition using the word being defined is circular. Several definitions taken together may be circular. For example, if we define a bush as a small tree and a tree as a tall bush, we have circularity. Dictionaries employ such circularity as a matter of necessity.

difficulty, except perhaps that of distinguishing when we are speaking intuitively from when we are speaking formally or "officially." Later we shall discuss this at greater length.

REVIEW PRACTICE EXERCISES

Set A *is a subset* of set B ($A \subset B$) if and only if *every* element of A is also an element of B.

Example. If $A = \{2, 5\}$ and $B = \{1, 2, 3, 4, 5\}$, then $A \subset B$. Similarly, if $C = \{3\}$, then $C \subset B$.

In Exercises 1 through 8, decide whether A is a subset of B.

1. $A = \{6, 3\}$ $B = \{1, 2, 3, 4, 5, 6\}$
2. $A = \{a, d\}$ $B = \{d, c, b, e\}$
3. $A = \{10, 11, 12\}$ $B = \{0, 1, 2, \dots\}$ (i.e., the set of whole numbers)
4. $A = \{c\}$ $B = \{a, b, c, d\}$
5. $A = \{2, 4, 6, 8\}$ $B = \{8, 6, 4, 2\}$
6. $A = \emptyset$ $B = \{1, 2, 3\}$
7. $A = \{0\}$ $B = \{1, 2, 3\}$
8. $A = \{a, b, c, d\}$ $B = \{a, b, c\}$
9. a) In Exercise 1, what is $A \cap B$? $A \cup B$?
   b) In Exercise 3, what is $A \cap B$? $A \cup B$?
   c) In Exercise 4, what is $A \cap B$? $A \cup B$?
   d) If $A \subset B$, what is $A \cap B$? $A \cup B$?
10. In Exercise 5 is it true that $A \subset B$ and $B \subset A$? $A = B$?
11. All of the subsets of $\{1, 2, 3\}$ are given below:

    $\emptyset$, $\{1\}$, $\{2\}$, $\{3\}$, $\{1, 2\}$, $\{1, 3\}$, $\{2, 3\}$, $\{1, 2, 3\}$
    a) Name all the subsets of $\{a, b\}$.
    b) Name all the subsets of $\{1, 2, 3, 4\}$.
12. Consider the universal set $U = \{1, 2, 3, 4, 5, 6, 7, 8, 9, 10\}$. Find the complement of each set.
    a) $S = \{8, 9, 10\}$ b) $S = \{2, 4, 6, 8, 10\}$
    c) $S = \{1, 2, 3, 4, 5, 6, 7, 8, 9\}$ d) $S = \emptyset$
    e) $S = \{1, 2, 3, 4, 5, 6, 7, 8, 9, 10\}$
13. If U is the universal set and $S \subset U$, what is
    a) $S \cup \bar{S}$?   b) $S \cap \bar{S}$?

3. GEOMETRIC FIGURES

There are many and varied kinds of geometric figures, familiar to anyone who has studied even a modest amount of geometry. We shall now make a definition of *geometric figure* as a second step in our formal development of geometry. Often,

in traditional treatments, no definition of *figure* is made, but we shall find it necessary here in order that our development be complete. In making such a definition,
we find it advisable to look at many of the different kinds of things we call *figures*
to see, if we can, what they have in common. As this is done, we find that all
figures seem to contain points but that they have nothing else in common. We
are therefore led to say that a figure is a set of points. If we make such a definition,
then of course *any* set of points will be a figure, including a set containing just one
point, just three points, and sets which look like those in Illus. 2.1.* This may
seem a bit strange at first, but such a general definition would include not only
strange kinds of sets of points but also those usually studied. Such general definitions are almost always more useful than those that tend to be restrictive. Let us
now make the definition formally.

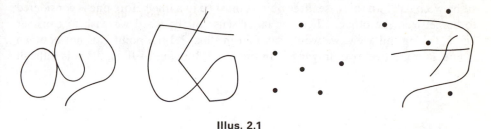

Illus. 2.1

Definition 2.1 A *geometric figure* (or simply *figure*) is a nonempty set of points.

The word "set" appears in Definition 2.1. This word has not been defined in
our formal development, and the reader may wonder why it has been neither
defined nor listed as undefined. The answer to this question is that the word
"set" is regarded as being a word in our regular language and not a technical
word in our formal geometric development. Thus the word "set" is one whose
meaning and use are understood by English-speaking persons as are the words
"is" and "of" that also appear in Definition 2.1.

On the other hand, it is assumed that the reader is familiar with the fundamental
notions of sets and set language, as well as set notation, even though this cannot
be expected of every English-speaking person. In particular, the reader is expected
to be familiar with the concepts of *union, intersection, complement, subset, empty
set*, and *universal set*.† It may be noted at this point that the union of two figures
is also a figure and that the intersection of two figures is a figure, provided it is
not empty.

It should also be noted that in defining geometric figures we used the undefined
concept of *point*.

---

* In this book, the word "Illustration" is used, rather than the usual "Figure" because we
are about to define, and will use, the latter in a technical geometric sense.
† See Appendix for a review of these concepts.

4. BETWEENNESS

What does it mean for one object to be between two others, and what does it mean for a point to be between two others? In the examples of Illus. 2.2, one should have no trouble deciding whether point C is between points A and B. In the examples of Illus. 2.3 it may not be quite so easy to decide. The word "between" in ordinary usage has ambiguous meaning, as do most words used nontechnically. One house may be said to be between two others, even though it is set back from the road 50 ft farther than the others. In this case we allow that it is *between* the others even though the three objects are not in a line. On the other hand, we often use the word in another sense. In a theater, for example, two people may be sitting in the same row with a vacant seat between them. A person sitting in the seat directly behind the vacant seat might be asked, "Won't you come sit between us?" In this case three objects must be in a line before one is considered to be between the others. It is the latter sense of the word we wish to consider here. Taking this view, we would say that in Illus. 2.3 the point C is not between A and B in the first two drawings. In the other drawing of Illus. 2.3 it is difficult to tell.

Illus. 2.2

Illus. 2.3

Then how should we define *between*? We may be tempted to include in such a definition something about the points being "on a line." But let us consider for a moment what this would mean. We would have to either define *line* first or leave *line* undefined. Geometry can be developed in many different ways, and it is possible to leave *line* undefined. In this development, however, we prefer not even to speak of lines yet, so that we can define them later. Since we have a fairly good intuitive idea of betweenness, we prefer to leave it undefined.

The word *between* is undefined.

We now have two undefined terms, *point* and *between*. We shall be using the notion of betweenness so often that it will be helpful to adopt some compact notation, as follows:

The sentence "*B* is between *A* and *C*." may be abbreviated "*A-B-C*."

We shall almost always name points by upper case letters and shall most often use the above abbreviation. The sentence "*A-B-C*" may also be read simply as "*A, B, C*."

We have not yet discussed how many points we shall want to have in our geometry nor just what properties we shall want to assume about the undefined relation *between*. This will be done after the following exercises, in which the reader is asked to develop his own intuitive feelings in these respects.

EXERCISES

1. One dictionary defines a *point* as "something thought of as having definite position in space but no size or shape." What other definitions would be needed if we used this definition? Why would we eventually need some undefined words?

2. Find two definitions from a dictionary that illustrate circularity.

3. We have not said how many points are to be in our geometry.
   a) If there were just one point, how many figures would there be in the geometry?
   b) If there were just two points, how many figures would there be?
   c) If there were just four points, how many figures would there be?
   d) If we allow a figure to be empty, instead of defining it as we did, what would then be the answers to parts (a) through (c)?

4. a) Draw a figure consisting of three points. Label them *A*, *B*, and *C*.
   b) Draw a point *D* between *A* and *B*.
   c) Draw a point *E* between *B* and *C*.
   d) Draw a point *F* between *C* and *A*.

5. For Exercise 3, generalize to *n* points (that is, if there were *n* points, how many figures would there be?).

6. a) Draw three points *A*, *B*, and *C* such that *A-B-C*.
   b) Is it possible to have the relationship *A-B-C* if the points *A*, *B*, and *C* are not distinct, that is, if *A* = *B*, or *B* = *C*, or *A* = *C*?
   c) When the relation *A-B-C* holds, does the relation *C-B-A* necessarily hold?

7. a) Draw two points, *A* and *B*, about 2 in. apart.
   b) Draw a point *C* such that *A-C-B*.
   c) Draw *D* such that *C-D-B*. Is *D* between *A* and *B*?
   d) Draw *E* such that *D-E-B*. Is *E* between *A* and *B*?

e) Draw F such that E-F-B. Is F between A and B?

f) How many points are there between A and B?

8. a) Draw three points A, B, and C such that A-B-C.

b) Draw D such that A-D-B.

c) What betweenness relations hold among A, C, and D?

d) What betweenness relations hold among B, C, and D?

9. Draw two points, X and Y, and the figure consisting of X, Y, and all the points between them.

REVIEW PRACTICE EXERCISES

In Exercises 1 through 5, rewrite the conditional sentence in *if...*, *then* form.

1. The game will be postponed if it rains.

2. Triangles are not rectangles.

3. Numbers greater than 10 are greater than 5.

4. A natural number is even if it is not odd.

5. Tomorrow is Wednesday because today is Tuesday.

In Exercises 6 through 10, write the converse and contrapositive of the conditional sentence (see Appendix).

6. If $a + y = a + x$, then $y = x$.

7. If today is Saturday, then yesterday was Friday.

8. If Bill lives in Los Angeles, then he lives in California.

9. If a number is even, then it is a multiple of 4.

10. If $x < 5$, then $x < 9$.

11. A conditional sentence $A \to B$ is false whenever an interpretation can be found in which A is true while B is false. Otherwise, the conditional sentence is true.

Example. If x is a prime number, then x is an odd number. This conditional sentence is false, since there is an interpretation, namely $x = 2$, for which the antecedent A is true, while the consequent B is false.

a) For Exercises 6 through 10, find the truth value of the conditional sentence, its converse, and its contrapositive.

b) How do the truth values of the conditional sentence and its contrapositive compare? The conditional sentence and its converse?

5. ASSUMPTIONS ABOUT POINTS AND BETWEENNESS

Since *point* and *between* are both undefined, we "officially" know nothing about them. In the preceding exercises, however, the reader was asked to use his imagination and to some extent organize his intuitive (or "unofficial") ideas about these undefined notions. This is important because in order to draw any conclusions about points, betweenness, and related concepts, we must know something about the relation *between*. Since it is undefined, all we know about it is that it involves triples of points. In order to proceed with our formal development of geometry, we must next make some assumptions about the nature of betweenness. Actually, we are free to assume whatever we wish, and the kind of geometry we get will depend on what kind of assumptions we make, but our purposes will be served better if we make assumptions that are in accord with our intuitive feelings and our experience. The assumptions we make will be formally stated as *axioms*, after which they become a part of our "official" or formal vocabulary, and can be used in the rest of the formal development.

How many points shall we assume our geometry to contain? A geometry with only one point would be of practically no interest, so we should certainly insist that there be at least two points. A geometry with just two points would likewise be of little interest, so, unless we can find some way to *prove* that there are more, we should assume that there are more than two points. Let us assume that there are at least two points and, as we shall see, this, together with some of our other assumptions about betweenness, implies the existence of a great many more points.

Axiom 2.1 There exist at least two points.

It is easy to "see," as in the preceding exercises, that the idea expressed by the sentence "*A* is between *B* and *C*" does not conform to our intuitive notion unless the three points are distinct. We shall make this an assumption also.

Axiom 2.2 For any points *A*, *B*, *C*, if *A-B-C*, then *A*, *B*, and *C* are three distinct (different) points.

It is also easy to "see" that given any two distinct points, there are always points between them. To simplify this, we can say that given any two distinct points, there is *at least* one point between them. We shall make this our next assumption.

Axiom 2.3 For any distinct points *A* and *B*, there is at least one point between them.

We should note here that Axiom 2.3, together with Axiom 2.1, guarantees that there are *at least* three points in our geometry. Axiom 2.1 says there are at least two, and Axiom 2.3 says there must be another point between them. Actually, we need Axiom 2.2 here also, to guarantee that the point *between* the other two cannot be one of them. We shall not stop to prove a theorem here, because presently we shall prove a more general theorem.

We should also note that Axiom 2.3 says nothing to the effect that we "can *draw* a point between two others." This would not be true. For example, if we have "drawn two points" a billionth of an inch apart, it might be impossible to draw another point between them. What we are assuming in 2.3 is that at least one point *exists*. Actually, the emphasis in our entire geometric development will be on what *exists*, rather than what *can be drawn*. The latter notion was of paramount importance to Euclid, and is also fundamental in traditional high-school geometry courses, but it leads to logical difficulties. Hence we take the present view.

If a point is between A and B, then our intuition leads us to say that the point is also between B and A. This will be our next assumption.

Axiom 2.4 If a point B is between points A and C, then it is also between C and A. (If A-B-C, then also C-B-A.)

We shall need one more assumption about betweenness for the ensuing development. This one is also very easy to accept intuitively and is as follows:

Axiom 2.5 For any distinct points A, B, C, and D, if A-B-C and A-D-B, then A-D-C and D-B-C.

This axiom refers to the situation pictured in Illus. 2.4. We have B between A and C. Then if there is a point D between A and B, the point D must also be between A and C; furthermore, B must be between D and C.

A D B C **Illus. 2.4**

Now that we have made the assumptions in Axioms 2.1 through 2.5 we have given quite a lot of structure to our geometry. Although *point* and *between* remain undefined, we now have some machinery with which to work in the further development.

6. SEGMENTS

We now consider a special kind of geometric figure. Suppose we have two points, A and B. We know from Axiom 2.3 that there is at least one point between them. In fact, there are many points between A and B. We shall prove presently that the number of points between A and B is infinite.* The figure consisting of points A and B and *all* of the points between them we shall call a *segment*. Let us make this our second definition.

Definition 2.2 For any two distinct points, A and B, the set consisting of A, B, and all the points between A and B is called a *segment*. It may be named "$\overline{AB}$."

It will be instructive to state this definition in set notation. The set consisting of A and B may thus be named "$\{A, B\}$." The set of all points between A and B may be named "$\{X \mid A\text{-}X\text{-}B\}$," read "the set of all points X such that A-X-B." The segment $\overline{AB}$ consists of the union of these sets and hence may be named "$\{A, B\} \cup \{X \mid A\text{-}X\text{-}B\}$."

To "draw segment $\overline{AB}$" we make a drawing like that below (Illus. 2.5).

---

* The word *finite* means "ending" and the word *infinite* means "unending."

$A \rule{6cm}{0.4pt} B$ **Illus. 2.5**

Actually, we do not draw the segment, but a picture which is satisfactory from an intuitive standpoint.

Suppose we consider a segment $\overline{AB}$. This is a certain set of points as defined in Definition 2.2. How does this compare with the segment $\overline{BA}$? By looking at the definition we can see that these are the same set. This amounts to an easy theorem —an appropriate one to be the first in our development.

Theorem 2.1 For any two distinct points, A and B, $\overline{AB} = \overline{BA}$.*

Proof. By Definition 2.2, $\overline{AB} = \{A, B\} \cup \{X \mid A\text{-}X\text{-}B\}$, and $\overline{BA} = \{B, A\} \cup \{X \mid B\text{-}X\text{-}A\}$. We know that when a set is named by listing its elements, order is not important. Thus, $\{A, B\} = \{B, A\}$. Also, $\{X \mid A\text{-}X\text{-}B\} = \{X \mid B\text{-}X\text{-}A\}$ since, by Axiom 2.4, any X that is between A and B is also between B and A. Thus, the two sets $\overline{AB}$ and $\overline{BA}$ are the same, which was to be shown.

From the definition of segment and Theorem 2.1 we now see that any two distinct points determine a unique segment. We are thus justified in speaking of the "segment determined by two points" or the segment "joining two points." The points A and B of a segment $\overline{AB}$ are called its *endpoints* and the other points are called *interior points*. The set of all interior points is called the *interior* of the segment. The interior of $\overline{AB}$ is thus $\{X \mid A\text{-}X\text{-}B\}$, consisting of all of the segment except its endpoints.

We have stated, and our intuition tells us, that every segment is an infinite set of points. Let us now make this a theorem.

Theorem 2.2 Every segment is an infinite set of points.

Outline of proof.† Consider any segment $\overline{AB}$. By Axiom 2.3 there is a point C_1 between A and B. By Definition 2.2, C_1 belongs to the segment and, by Axiom 2.2, C_1 is distinct from A and B. Hence the segment contains at least three points.

By Axiom 2.3 there is a point C_2 between A and C_1. By Axiom 2.5, C_2 is between A and B and hence belongs to the segment. By Axioms 2.2 and 2.5, C_2 is distinct from A, B, and C_1. Thus the segment contains at least four points. Similarly, there is a point C_3 between A and C_2 which is distinct from the four

---

* In this book = is always used in the sense of *logical identity*. That is, any sentence with the verb = asserts that the symbols on the right and the left represent (or name) the same thing.

† By an *outline of proof* we mean the following: A series of steps in an argument that shows the essential sequence of a proof. There may be minor steps omitted, and the supporting reasons may be omitted in order not to complicate the presentation too much.

points just mentioned. This point likewise belongs to the segment; hence the segment contains at least five points.

This argument may be repeated endlessly; hence there is an infinite number of points between A and B.*

7. INTUITION, DRAWINGS, AND FORMALISM

The preceding development has illustrated the role of intuition in building a formal mathematical system. The importance of allowing one's intuition to guide him is difficult to overstate.

In our formal development we choose some undefined terms, state some axioms, make some definitions, and, most importantly, prove theorems. In proving theorems we may not introduce anything of an intuitive nature, even though intuition may provide us with hunches or insight as to how to proceed. In proving a theorem, it is often helpful to the intuition to make a drawing, but it is imperative to remember that the drawing is only a guide to the intuition and is *never* a part of the proof. We have purposely avoided making a drawing in connection with Theorem 2.2 to illustrate that the drawing is not essential.

The use of definitions in proving theorems is almost always essential. It is extremely important to refer to definitions wherever needed in a proof. It is also important to use the definitions given in the formal development but never to use any intuitive notions that may accompany them. It is thus advisable that the student study the definitions carefully and remember to use them when needed in proofs.

It is not possible, apparently, to make completely separate developments of geometry on an intuitive and a formal basis. The two kinds of thinking usually go on concurrently, paralleling each other. For this reason, from here on, exercises have been divided into three groups. The first group of exercises in each exercise set will, in general, be less formal, and the student will often look to his intuition for an answer. Such exercises may involve drawing figures, describing relationships, and, in general, applying the concepts introduced in the preceding section. The second group, labeled "Rigorous exercises," will usually require the student to look to the axioms, definitions, and theorems for a formal answer. Proofs or outlines of proofs will often be required. Occasionally, a third group, labeled "Exploratory exercises," will appear as a part or all of an exercise set. These exercises are designed for exploring ideas and concepts which will be introduced in a following section. They provide an opportunity to discover and experience a new concept at an intuitive level before it is introduced formally later on.

---

* To finish this proof in the most rigorous fashion one would need to use a technique of proof called *mathematical induction*. The student who is familiar with this technique should use it.

EXERCISES

1. a) Restate Axiom 2.2, using the letters P, Q, and R instead of A, B, and C.

   b) Restate Axiom 2.3, using the letters R and S instead of A and B.

   c) Restate Axiom 2.4, using the letters X, Y, and Z instead of A, B, and C.

   d) Restate Definition 2.2, using the letters P and Q instead of A and B.

   e) Restate Axiom 2.5, using the letters S, T, U, and V instead of A, B, C, and D, respectively.

   f) Restate Axiom 2.5, using the letters B, A, C, and D instead of A, D, B, and C, respectively.

   g) Restate Axiom 2.5, using the letters C, B, D, and A instead of A, D, B, and C, respectively.

2. a) Draw two points, A and B.   b) Draw a point C_1 between A and B.

   c) Draw a point C_2 between A and C_1.   d) Draw a point C_3 between A and C_2.

   e) Draw a point C_4 between A and C_3.

   f) Theoretically, could one obtain a segment by continuing to add points in this fashion?

3. Find five objects in the physical world, each of which bring to mind the geometric abstraction called a *segment*.

4. a) Draw a segment $\overline{AB}$ and mark an interior point C.

   b) Your drawing now shows two segments besides $\overline{AB}$. Name them.

   c) What is the union of the two segments named in part b? What is the intersection?

   d) Consider the segment $\overline{AB}$ as a universal set. Describe the complement of $\overline{AC}$. Describe the complement of the set which contains just the point C.

5. Consider a segment $\overline{AB}$ and an interior point C. Which of the following are true?

   a) $C \in \overline{AB}$ b) $\overline{AB} \subset \overline{AB}$ c) $B \in \overline{AC}$

   d) $\overline{AC} \cup \overline{AB} = \overline{AB}$ e) $\overline{AC} \cap \overline{BC} = \emptyset$ f) $\overline{AB} \cap \overline{CB} = \overline{BC}$

Rigorous exercise

6. For each part of Exercise 5 that is true, give a proof.

Exploratory exercises

7. a) Draw a segment $\overline{AB}$ with a point C_1 in its interior.

   b) Draw C_2 between B and C_1. Draw C_3 between B and C_2.

   c) Draw a point C_4 between B and C_3.

   d) Theoretically, how long can this process be continued?

   e) In the segment $\overline{AB}$, is there a point *next* to B? That is, can there be a point Y, near B, such that there is no point between Y and B?

f) If we take a segment and "remove" its endpoints, does the resulting set of points contain endpoints?

g) If we take a string of beads and remove the end beads, does the resulting set of beads have end beads?

8. a) Draw a segment $\overline{AB}$.   b) Draw a point C_1 such that A-B-C_1.

c) Draw a point C_2 such that B-C_1-C_2.   d) Draw a point C_3 such that B-C_2-C_3.

e) How far can this process be continued?

f) Draw the set of points which consists of the union, $\overline{AB} \cup \{X \mid A$-$B$-$X\}$.

9. a) Draw a segment $\overline{AB}$.   b) Draw a point P_1 such that P_1-A-B.

c) Draw a point P_2 such that P_2-P_1-A.   d) Draw a point P_3 such that P_3-P_2-A.

e) Draw the set of points $\overline{AB} \cup \{Y \mid Y$-$A$-$B\}$.

10. a) Draw a segment $\overline{AB}$.   b) Draw $\{X \mid A$-B-$X\}$.

c) Draw $\{Y \mid Y$-A-$B\}$.   d) Draw $\{X \mid A$-B-$X\} \cup \{Y \mid Y$-A-$B\}$.

e) Draw $\{X \mid A$-B-$X\} \cup \{Y \mid Y$-A-$B\} \cup \overline{AB}$.

8. OPEN SEGMENTS

Let us consider a kind of geometric figure similar to a segment. A figure which consists of all points of a segment except its endpoints is the interior of that segment. If the endpoints of a segment are A and B, respectively, then the interior is $\{X \mid A$-X-$B\}$. This figure is called an *open segment* as well as the *interior* of the segment. A figure consisting of all points of a segment except one of the endpoints is called a *half-open segment*.

Definition 2.3 For any two distinct points, A and B, the figure $\{X \mid A$-X-$B\}$ is called an *open segment* and may be named "$\overset{\circ\!-\!\circ}{AB}$." The figure $\{A\} \cup \{X \mid A$-X-$B\}$ is called a *half-open* segment. It may be named "$\overset{-\!\circ}{AB}$."

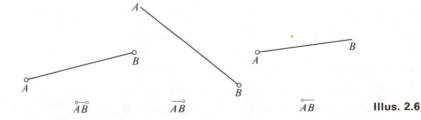

$\overset{\circ\!-\!\circ}{A\,B}$ $\overset{-\!\circ}{A\,B}$ $\overset{\circ\!-}{A\,B}$ **Illus. 2.6**

Open segments and half-open segments may be drawn as in Illus. 2.6. The drawing on the left represents an open segment and the other two drawings, half-open segments. The center drawing represents a segment open at B and the other drawing represents a segment open at A.

                                                              Illus. 2.7

It is of some importance that an open segment does not *contain* endpoints. This may be seen as follows: Suppose open segment $\overset{\circ\!-\!\circ}{AB}$ did have an endpoint P as shown in Illus. 2.7. This would mean that there would be no point of the open segment "beyond" P. In other words, there would be no point between P and B. Now P and B are distinct points, so by Axioms 2.3, 2.4, and 2.5 there must be a point between them which is on $\overset{\circ\!-\!\circ}{AB}$. Our assumption that there is an endpoint P therefore cannot be true, and the open segment contains no endpoints.*

The fact that an open segment contains no endpoints (and of course that a half-open segment contains just one endpoint) may seem strange to one's intuition. The reason for this is probably the fact that in the physical world we do not observe infinite sets. If we think of the points of a segment like the beads in a string of pearls, for example, our intuition tells us that when an endpoint (pearl) is removed, the one next to it will then serve as an endpoint. But a string of pearls is not an infinite set and hence may not serve our intuition well. This example illustrates how our intuition may lead us to contradict ourselves. All of the axioms we have assumed so far seem easy to accept intuitively. Once we have accepted them, we are forced to accept the fact that open segments do not contain endpoints or else abandon any attempt at consistency. This is so, in spite of the fact that the deduced result may not appeal to our intuition.

An open segment $\overset{\circ\!-\!\circ}{AB}$ is of course the same as the open segment $\overset{\circ\!-\!\circ}{BA}$. A similar thing is true about half-open segments. Thus we can name an open segment $\overset{\circ\!-\!\circ}{AB}$ as "$\overset{\circ\!-\!\circ}{BA}$" whenever we choose, and similarly for half-open segments.

9. RAYS AND LINES

When we formulated Axiom 2.3, which says that between any two points there is at least one other, we followed our intuition. In some of the preceding exercises it must have appeared that there is another, similar assumption that would be easy to accept. If we consider any two distinct points A and B, as in Illus. 2.8, it would seem that there should always be a point P "beyond" B (such that A-B-P). Let us make this assumption as our next axiom.

Axiom 2.6 For any two distinct points, A and B, there exists at least one point P such that A-B-P.

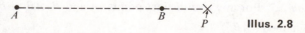

                                                              Illus. 2.8

---

* Although this argument is not a formal one, it illustrates *proof by contradiction*, in which we assume the opposite of what we are trying to prove and then deduce from that assumption a contradiction (something known to be false). The conclusion is thus that this assumption was false and thus the original statement is true.

It now follows that there are many points "beyond" point B of a segment $\overline{AB}$. This can be seen as follows: As in Illus. 2.8, there is a point P as shown. Then A and P are distinct points (why?), and thus there is another point, P_1, such that A-P-P_1. We may similarly show the existence of a point P_2 such that A-P_1-P_2, and so on, endlessly. It also follows that there are many points "beyond" point A. This can be seen by finding P such that P-A-B (combining Axioms 2.4 and 2.6), and proceeding as before.

If we consider a segment $\overline{AB}$ and add points to it as described in the preceding paragraph, we obtain a figure with an infinite set of points "beyond" one end of the segment. If we should take *all* of the points X such that A-B-X, we would obtain a figure like the one shown in Illus. 2.9. The arrowhead indicates that there is no endpoint on the right, and intuitively we think of this figure as being of unlimited extent in that direction. This kind of figure is called a *ray*. Let us now define it precisely.

Illus. 2.9

Definition 2.4 For any two distinct points, A and B, $\overline{AB} \cup \{X \mid A\text{-}B\text{-}X\}$ is called a *ray*. It may be named "$\overrightarrow{AB}$."

It is now easy to see that any two distinct points determine a ray. In fact two rays are determined by each pair of points. In Illus. 2.9 one of the rays determined by points A and B is shown, and in Illus. 2.10 the other ray determined by these points is shown. It is the ray $\overrightarrow{BA}$. Note that the endpoint of the ray (B in this case) is named first.

Illus. 2.10

Let us now consider the union of rays $\overrightarrow{AB}$ and $\overrightarrow{BA}$, as shown in Illus. 2.11. This is a most important kind of figure, which we shall call a *line*. Let us make this our next definition.

Illus. 2.11

Definition 2.5 For any two distinct points, A and B, $\overrightarrow{AB} \cup \overrightarrow{BA}$ is called a *line*. It may be named "$\overleftrightarrow{AB}$."

It is easy to show that line $\overleftrightarrow{AB}$ is the same as line $\overleftrightarrow{BA}$. We shall state this as a theorem, leaving the proof as an exercise.

Theorem 2.3 For any two distinct points, A and B, $\overleftrightarrow{AB} = \overleftrightarrow{BA}$.

A line is, of course, a familiar geometric figure. We think of it intuitively as being of unlimited extent in both directions and also as being "straight." The words "unlimited extent," "directions," and "straight" are not in our official vocabulary, however, so we should not say "straight line," but simply "line," as in Definition 2.5. To represent a line we make a drawing like that of Illus. 2.11, the arrowheads indicating our intuitive notion that the line is of unlimited extent. Note that the line of Illus. 2.12 has been named "*l*." This is another usual way of naming lines, by a single lower-case letter.

Another figure similar to a ray will be of importance in the sequel. It is like a ray except that it contains no endpoint. Such a figure is shown in Illus. 2.13 and is called a *half-line*. Let us now define *half-line*.

Illus. 2.13

Definition 2.6 For any two distinct points, A and B, $\overset{\circ\!-}{AB} \cup \{X \mid A\text{-}B\text{-}X\}$ is called a *half-line*. It may be named "$\overset{\circ\!\rightarrow}{AB}$."

It is important to remember that the emphasis in this development is on what *exists*, rather than what *can be drawn*. One can contemplate, for example, a line which contains the top of the Empire State Building and the top of the Taj Mahal. Such a line would exist but could scarcely be drawn! When we make a drawing, we should remember that it is a picture to represent in some fashion our (abstract) geometric notions. What we show on a drawing is not everything that exists but only those things to which we wish to pay attention at the moment.

Illus. 2.14

Example. Illustration 2.14 shows a line with two named points. The drawing also shows a segment $\overline{PQ}$, in the sense that it is a marked part of the drawing. There are many other points on the line, and hence many other segments, but only two points and, hence, one segment are *shown*. In the same sense, the drawing also shows several rays, half-open segments, half-lines, etc. For example, there are rays $\overset{\rightarrow}{PQ}$ and $\overset{\rightarrow}{QP}$ visible, as well as a ray extending to the left from P and a ray extending to the right from Q. Likewise, open segment $\overset{\circ\!-\!\circ}{PQ}$ and half-line $\overset{\circ\!-}{QP}$ are visible, as well as other sets of points.

Many other sets of points *exist* on this figure, but only a few of them are *shown* in the drawing.

It is apparently helpful to think of geometric figures as motionless sets of points. In traditional treatments of geometry the idea of motion plays an important role.

A line might be thought to be generated by a moving point, for example. The traditional language encourages this. It is common, for instance, to say that a line "passes through" a point or that a locus is the "path of a moving point."* Some preliminary pedagogical investigations have shown that it is better for the student to think of geometric figures as static, or motionless, sets of points. It has also been noted that some students intuitively regard geometric figures as having properties like those of a stretched rubber band or a coiled spring. Such an intuitive notion gives rise to some bewilderment. One student, who regarded a segment as being like a coiled spring, thought that if an endpoint of a segment were removed, then there would be nothing to hold the interior points in and they would fly outward, transforming the figure into a ray. In summary, the reader will do well to regard geometric figures as sets of points, static, and with no internal forces to encourage movement.

10. A POINT OF LANGUAGE

Since figures are sets of points, we can of course speak of a point as being *a member of* a figure, or as being *in* a figure. For example we can write "$P \in l$" to say that point P belongs to line l, or "$Q \in \overline{AB}$" to say that point Q belongs to segment $\overline{AB}$. It is convenient in some instances to speak of a point as being *on* a figure with the same meaning. For example, we can say "point A is *on* line m," meaning that it is a member of the set of points that we call line m.

It is also convenient to use the word "on" in a symmetric fashion, as is often done by mathematicians. We can say that a line *contains* point P (as a member), but we shall also agree that "the line m is *on* point P" means the same thing. Thus we can say that a line is on a point or that a point is on a line.

Points that are on the same line are said to be *collinear*. If a point C is not on line $\overleftrightarrow{AB}$, points A, B, and C are said to be noncollinear. Since two points determine a line, two points are of course always collinear.

EXERCISES

1. a) Draw two points, A and B.

    b) Draw $\overrightarrow{AB}$.

    c) In a different color, draw $\overrightarrow{BA}$.

2. a) Draw three noncollinear points, P, Q, and R.

    b) Draw $\overrightarrow{PQ}$, $\overrightarrow{QR}$, and $\overrightarrow{RP}$.

    c) In a different color, draw $\overrightarrow{QP}$, $\overrightarrow{PR}$, and $\overrightarrow{RQ}$.

---

* The concept of a *locus* is not treated *per se* in this book. A locus would be defined here merely to be a *set of points*, and the concept of motion would not be pertinent. *Locus* and *figure* are synonyms here.

3. a) Draw a line l and on it mark points A, B, and C.

    b) How many segments are shown on your drawing? Name them.

    c) How many rays are shown on your drawing? Name them.

    d) How many open segments are shown on your drawing? Name them.

    e) How many half-lines are shown on your drawing? Name them.

4. a) Draw a ray $\overrightarrow{AB}$ and points P, Q, such that A-P-B and A-B-Q.

    b) Is $\overrightarrow{AP}$ the same as $\overrightarrow{AB}$?

    c) From your drawing, how many ways can you name $\overrightarrow{AB}$?

5. Is the union of two half-lines necessarily a line? Support your answer with drawings.

6. Is the union of two rays necessarily a line? Support your answer with drawings.

7. Draw a line m, and on it mark points P, Q, and R such that P-Q-R. Describe the following figures or sets of points.

    a) $\overrightarrow{PQ} \cap \overrightarrow{QP}$ b) $\overrightarrow{PR} \cup \overrightarrow{QR}$ c) $\overrightarrow{QP} \cap \overrightarrow{QR}$

    d) $\overset{\circ\quad\circ}{QR} \cup \overset{\circ\quad\circ}{RQ}$ e) $\overset{\circ}{QP} \cup \overset{\circ}{RQ}$ f) $\overset{\circ}{QP} \cap \overrightarrow{PQ}$

    g) $\overset{\circ}{QP} \cap \overrightarrow{QR}$ h) $\overset{\circ\quad\circ}{PQ} \cup \overset{\circ\quad\circ}{QR}$ i) $\overline{PQ} \cup \overline{QR}$

    j) $\overline{PQ} \cap \overline{QR}$

8. Definition 2.4 uses the concept of a segment and hence relies on Definition 2.2. Restate Definition 2.4, using only the concept of betweenness and without using Definition 2.2.

9. Definition 2.5 uses the concept of a ray and hence Definition 2.4. Restate Definition 2.5, using only the concept of betweenness and without using Definition 2.4 or 2.2.

10. a) Restate Axiom 2.6, using the letters P, A, and X instead of A, B, and P, respectively.

     b) Restate Definition 2.4, using the letters B, T, and Y instead of A, B, and X, respectively.

11. In the figure shown in Illus. 2.15:

    a) Which points are on line m?

    b) Which points are members of line l?

    c) Which lines are on point D?

    d) Which lines are on point A?

    e) Segment $\overline{AB}$ is a subset of which line(s)?

    f) Segment $\overline{CE}$ is a subset of which line(s)?

    g) Segment $\overline{AC}$ is a member of which line(s)?

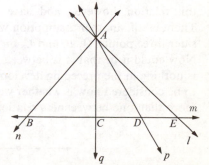

Illus. 2.15

Rigorous exercises

12. Prove that for any two distinct points A and B, $\overset{\circ\!-\!\circ}{AB} = \overset{\circ\!-\!\circ}{BA}$.

13. Prove Theorem 2.3.

14. Prove that for any ray $\overrightarrow{AB}$, there is no endpoint X such that A-B-X. (That is, prove that for any such X there is always another point Y such that A-X-Y and A-B-Y.)

15. Prove that an open segment contains no endpoints.

REVIEW PRACTICE EXERCISES

Write the negation of each of these sentences (see Appendix).

1. The meeting is scheduled for next week.

2. Tomorrow is Friday.

3. Tom did not pass the test.

4. The job will take at least three days.

5. The job will take no more than two days.

6. $2 + 5 \neq 6$

7. The letter was sent last week and it was not lost in the mail.

8. Triangles are squares or triangles are circles.

9. Two is an odd number and two is prime.

10. Today is not Tuesday or tomorrow is Wednesday.

11. If he has time on Saturday mornings, he studies.

12. We'll get a raise if sales increase this year.

13. If n is even, then $n/2$ is odd.

14. If x is a prime number, then $x = 2$ or x is an odd number.

11. MORE ABOUT LINES AND BETWEENNESS

Let us recall that *between* is undefined. We have made some assumptions about this relation, however, and have used them to proceed with the development. There is still another assumption we shall wish to make about betweenness. Consider three points, A, B, and C, on a line such that A-B-C, as shown in Illus. 2.16. Now could it be that A is between B and C or that C is between A and B? There is nothing in the preceding development to preclude this possibility. The question to be considered now is whether we wish to have a geometry in which we can have more than one betweenness relation among three points of a line.

A B C **Illus. 2.16**

There is nothing to prevent us from making any kind of axioms we please, except that we must be consistent. The kind of geometry we get will vary according to the kinds of assumptions we make. Before we decide, let us bring our intuition to bear a bit further. Suppose two airplanes take off at some field on the equator and fly in opposite directions, one east and one west. Some time later, while both are traveling along what would seem to be a "straight line," they would meet. When they begin their flight, we would say that they are flying away from each other, yet since they meet, they must also be flying toward each other.

If we abstract this situation to our geometric development, we might then wish to have lines that are "closed," as shown in Illus. 2.17. Three points on such a line would be situated something like three people seated at a circular table, as in Illus. 2.18. From these examples it is apparent that we have not only A-B-C but also B-A-C and A-C-B. Each of the three points is between the other two.

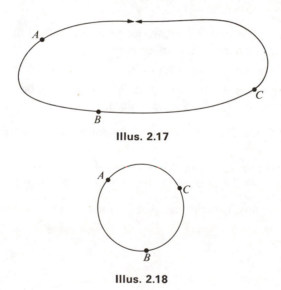

Illus. 2.17

Illus. 2.18

As we have said, we are at liberty to make an axiom, saying that for any three points of a line, each is between the other two. It is more in accord with our intuitive notion of "straightness," however, not to say this. We can ensure that our lines will not be closed as in Illus. 2.17 by making a different assumption, as follows:

Axiom 2.7 For any three distinct points of a line exactly one is between the other two.

Intuitively it would seem that if we have a ray $\overrightarrow{AB}$ and any point P of that ray except the endpoint, we could also name the ray $\overrightarrow{AP}$. We are justified in doing this, however, only if we can be sure that $\overrightarrow{AB}$ and $\overrightarrow{AP}$ are the same set of points. This requires proof. Contrary to what one might guess, the proof is not simple. We shall outline it by stating a sequence of four theorems, the last of which is the desired result. The first three are also of interest in themselves. We omit the proofs.

Theorem 2.4 For any distinct points A and B, and for any two distinct points X and Y between them, one and only one of the following holds: A-X-Y or A-Y-X.

Theorem 2.5 For any points A, X, and B, if A-X-B, then $\overline{AB} = \overline{AX} \cup \overline{XB}$.

Theorem 2.6 For any points A, B, X, and Y, if A-X-B and X-B-Y, then A-X-Y and also A-B-Y.

Theorem 2.7 For any ray $\overrightarrow{AB}$ and any point P on the ray, except the endpoint, $\overrightarrow{AB} = \overrightarrow{AP}$.

Now that we have Theorem 2.7 we are justified in naming any ray, using a name of the endpoint and a name of any other point whatsoever of the ray. It is equally plausible that if we have a line $\overleftrightarrow{AB}$ and were to choose any two points on it, say P and Q, then we could also name the line $\overleftrightarrow{PQ}$. In order to be sure of this, we need the following theorem, which we state without proof.

Theorem 2.8 If C and D are any two distinct points on a line $\overleftrightarrow{AB}$, then $\overleftrightarrow{CD} = \overleftrightarrow{AB}$.

Another important theorem which follows from Theorem 2.8, but which we shall not prove here, states that any two points are contained in a unique line.

Theorem 2.9 For any two distinct points, there is one and only one line that contains them.

EXERCISES

1. a) Draw a segment $\overline{AB}$ and place points on it in two different ways to illustrate Theorem 2.4.

   b) Make a circular diagram, similar to that of Illus. 2.18, to show that Theorem 2.4 would not hold if we did not have Axiom 2.7.

2. Draw a diagram to illustrate Theorem 2.5.

3. Illustrate Theorem 2.6 by one or more diagrams. Use accompanying descriptions or explanations if necessary.

4. a) Name the ray $\overrightarrow{AB}$, shown in Illus. 2.19, in two different ways. What theorem guarantees that you are correct?

Illus. 2.19

b) Name line *l* in four different ways. What theorem guarantees that you are correct?

5. a) Suppose Theorem 2.7 were false. Make a drawing to illustrate a situation that would then be possible but is not possible if it is true.

　　b) Suppose Theorem 2.8 were false. Make a drawing to illustrate a situation that would then be possible but is not possible if it is true.

Rigorous exercises

6. Prove that if *A-X-B*, and a point is in $\overrightarrow{AX}$, then it is also in $\overrightarrow{AB}$.

7. Complete the proof of Theorem 2.5.

8. Prove that for any three points *A*, *B*, and *C*, if *A-B-C*, then the points are on the same line.

Exploratory exercises

9. a) Draw a line $\overleftrightarrow{AB}$ and a point *C* not on the line.

　　b) Draw $\overleftrightarrow{AC}$ and $\overleftrightarrow{BC}$.

　　c) How many lines are determined by three different noncollinear points?

10. a) Draw three noncollinear points, *A*, *B*, and *C*, and the segments joining them.

　　b) How many segments are determined by three such points?

　　c) What kind of figure is $\overline{AB} \cup \overline{BC} \cup \overline{CA}$?

11. a) Draw a line $\overleftrightarrow{AB}$ and a point *C* not on the line.

　　b) Do $\overleftrightarrow{AB}$ and $\overleftrightarrow{AC}$ have a common point?

　　c) Do $\overleftrightarrow{AB}$ and $\overleftrightarrow{AC}$ have *every* point in common?

　　d) How many points do $\overleftrightarrow{AB}$ and $\overleftrightarrow{AC}$ have in common?

　　e) What is the maximum number of points that two different lines may have in common?

12. Suppose lines *l* and *m* have two different points *P* and *Q* in common.

　　a) Then *l* contains *P* and *Q*. How does *l* compare with $\overleftrightarrow{PQ}$? Why?

　　b) Now also *m* contains *P* and *Q*. How does *m* compare with $\overleftrightarrow{PQ}$? Why?

　　c) How does *l* compare with *m*?

　　d) You have just proved something. State what you have proved.

Experience leads us toward the intuitive belief that if two distinct lines meet (meaning that their intersection is not empty), then that intersection contains just one point. This is actually true in the geometry we have been developing, and we shall make it our next theorem.

Theorem 2.10 Any two distinct lines meet in at most one point.

Proof. Let us assume, as hypothesis, that we have two lines, *m* and *n*, and that two different points *P* and *Q*, are contained in their intersection. Now *P* and *Q* are on line *m*; thus by Theorem 2.9, $\overleftrightarrow{PQ} = m$. Similarly, *P* and *Q* are also on line n; thus by Theorem 2.9, $\overleftrightarrow{PQ} = n$. Therefore $m = n$. We have now shown that if lines m and n have two different points in common, they are the same line. It follows (the contrapositive) that if two lines m and n are not the same line, then there are not two different points in their intersection, which is what we were to prove.

In some of the previous exercises, the student was instructed to find a point not on a given line. Intuitively this presents no problem, but so far as our formal development is concerned we do not know that such points exist. We do know many things about points and betweenness, of course. In particular we know that every line, and indeed every segment (even the smallest one), contains an infinite number of points. There is nothing in our axioms or theorems, however, that guarantees that our geometry has points off a given line. It might happen that all the points of our geometry lie in a single line. If that were true, then Theorems 2.7, 2.8, 2.9, and 2.10 would not be important.

Let us recall that the kind of geometric system we build will depend on the assumptions we make. If we should wish to have a geometry all of whose points are on a line, we could do so. Such a geometry would not be the kind we have in mind, however. We shall thus wish to postulate the existence of enough points so that they are not all on one line. Shall we make it an axiom that there are an infinite number of points off a line? Perhaps we should assume that there are 2 or 3 points off a line. It turns out to be sufficient to assume that for any line there is *at least one* point not on that line. It will follow that there is an infinite number.

Axiom 2.8 For any line l there is at least one point P such that $P \notin l$.

EXERCISES

Exploratory exercises

1. a) Draw a segment $\overline{AB}$.
   b) Draw a point C not on $\overleftrightarrow{AB}$.
   c) Draw $\overline{AC}$ and $\overline{BC}$.
   d) Find a point D between A and B and a point E between A and C. Draw $\overline{DE}$.
   e) Find a point F between D and E. Find a point G between B and C. Draw $\overline{FG}$.

2. a) Draw a line l and a point P not on l. Choose any point on l, and name it Q.
   b) Draw $\overleftrightarrow{PQ}$.
   c) Choose another point of l, naming it R.
   d) Draw $\overleftrightarrow{PR}$.
   e) How many lines are there that contain P and some point of l? Why?

3. a) Draw three noncollinear points, A, B, and C.
   b) Draw $\overline{BC}$.

   c) Choose some point X on $\overline{BC}$, and draw $\overleftrightarrow{AX}$.

   d) Choose another point Y on $\overline{BC}$ and draw $\overleftrightarrow{AY}$.

   e) Repeat the instructions in parts (c) and (d) for 6 more points.

   f) How many lines are there that contain A and some point of $\overline{BC}$? Why?

4. On your drawing for Exercise 3, using a different color:

   a) Choose a point P on $\overline{AC}$ and draw $\overleftrightarrow{BP}$.

   b) Choose another point Q on $\overline{AC}$ and draw $\overleftrightarrow{BQ}$.

   c) Repeat parts (a) and (b) for 5 more points of $\overline{AC}$.

5. On your drawing for Exercises 3 and 4, using a third color:

   a) Choose a point S of $\overline{AB}$ and draw $\overleftrightarrow{CS}$.

   b) Choose another point T of $\overline{AB}$ and draw $\overleftrightarrow{CT}$.

   c) Repeat parts (a) and (b) for 5 more points of $\overline{AB}$.

6. Describe the figure consisting of the union of all possible lines like those you drew for Exercises 3, 4, and 5.

12. PLANES

A figure consisting of the union of many lines, as developed in the preceding exercise, might be described as a "flat sheet" of points. It is also, intuitively speaking, of unlimited extent because there are many lines that are subsets of it. This is the kind of figure we ordinarily call a *plane*, such as is shown in Illus. 2.20. If

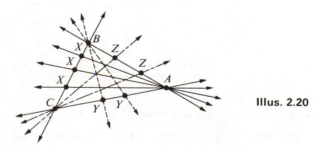

Illus. 2.20

we start with any three distinct and noncollinear points, A, B, and C, we then consider all lines $\overleftrightarrow{AX}$, where X is *any* point of $\overline{BC}$. Similarly, we consider all lines $\overleftrightarrow{BY}$, where Y is any point of $\overline{AC}$, and all lines $\overleftrightarrow{CZ}$, where Z is any point of $\overline{AB}$. The union of all these lines will be called the *plane ABC*. We now make the definition.

Definition 2.7 For any three distinct noncollinear points, A, B, and C, the union $\{P \mid P \in \overleftrightarrow{AX}$, where $X \in \overline{BC}\} \cup \{P \mid P \in \overleftrightarrow{BY}$, where $Y \in \overline{AC}\} \cup \{P \mid P \in \overleftrightarrow{CZ}$, where $Z \in \overline{AB}\}$ is called a *plane*.

It should be noted that according to this definition a plane is a set of points, as we expect it will be. Although we define a plane in terms of lines, a plane is not a set of lines, but a set of points. From the definition it does follow immediately, however, that the number of lines that are subsets of any plane is infinite. Thus, the plane itself is an infinite set of points.

It is clear from the definition of a plane that, given any three noncollinear points, there is a plane that contains them. It is not so clear that there is *only one* plane that contains them. Since it fits our intuition and is otherwise useful to have a geometry in which three noncollinear points determine a unique plane, we wish to either prove this fact or assume it. We shall do the latter, since the fact does not follow from our previous axioms.

Axiom 2.9 For any three distinct, noncollinear points, there is no more than one plane that contains them.

Axiom 2.9, together with Definition 2.7, tells us that for any three distinct, non-collinear points, there is *one and only one* plane that contains them. In other words, three noncollinear points determine a unique plane. Thus if we consider a plane *ABC*, as shown in Illus. 2.21, containing other points such as *P*, *Q*, *R*, and *S*, we may name the plane in several ways, for example, *PQR*, *QRS*, and *PRS*. Let us note also that planes are often named by using lower-case Greek letters. The plane in Illus. 2.21, is called α.

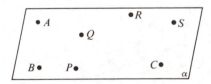

Illus. 2.21

Suppose we choose any two points, *P* and *Q*, on a plane β and consider the line containing them. Does the line $\overleftrightarrow{PQ}$ lie entirely within (is it a subset of) the plane β? Is it possible that part of the line "sticks up out of" β, as shown in Illus. 2.22? It is not difficult to show that the line must be a subset of the plane. Let us make this a theorem.

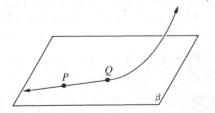

Illus. 2.22

Theorem 2.11 If P and Q are any two points on a plane α, then $\overleftrightarrow{PQ} \subset \alpha$.

Outline of proof. Choose a point R on α but not on $\overleftrightarrow{PQ}$ and consider plane PQR, which must be the same plane as α. Use Definition 2.7 to show that $\overleftrightarrow{PQ} \subset \alpha$.

Another easy but useful theorem now follows at once. This theorem says that any line and a point off the line determine (and lie in) a unique plane. The proof of this theorem is left as an exercise.

Theorem 2.12 For any line and any point not on that line, there is one and only one plane that contains them.

The next theorem is similar to 2.12 and is likewise easy to prove but useful. It says that two intersecting lines determine, and lie within, a unique plane.

Theorem 2.13 For any two distinct lines, l and m, if $l \cap m \neq \emptyset$, then the two lines are contained in one and only one plane.

13. SPACE

So far we have made enough assumptions to know that there are many points in our geometry and that they do not all lie on a single line. In fact, we know that the number of lines is infinite. However, we do not know whether all of our points lie on a single plane. We do not wish for this to be the case, so we make an axiom analogous to 2.8 to guarantee that our geometry contains points off a plane. Thus we have what may be called "three dimensions." The set of all the points in our geometry is called *space*.

Axiom 2.10 For any plane α there is at least one point P not on α.

It now follows easily that the number of planes in our geometry is infinite. This can be shown in much the same way that we show that the number of lines is infinite, following Axiom 2.8.

In a few sections of the book we shall consider space, but for the most part we shall be concerned primarily with geometry in a plane.

Drawings of Planes

Drawings of course are merely representations, or pictures, of abstract concepts or relationships. It is particularly important to remember this when drawing planes. If we attempt to draw a flat surface of unlimited extent, we find it impossible. Instead, we ordinarily try to draw something which looks like a rectangular piece of cardboard, viewed in perspective. In other words, we try to make a picture that looks like a portion of a plane. The following illustrations show various ways in which a plane might be represented.

If we consider a rectangular portion of a plane and imagine it lying flat, as on a table, such that it is viewed from above and to one side, it would appear something like Illus. 2.23. Note that the drawing is actually a parallelogram. To show a vertical plane, we make the same kind of drawing but oriented differently, as in Illus. 2.24. A plane that is neither vertical nor horizontal is shown in Illus. 2.25. Two intersecting planes α and β could be drawn as in Illus. 2.26. Note that in drawings like this, "hidden" parts are represented by dashed lines; also, all opposite segments are drawn parallel. Illustration 2.27 shows a plane α with a line l in it.

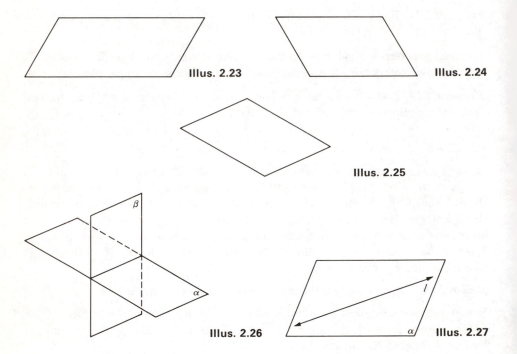

Illus. 2.23

Illus. 2.24

Illus. 2.25

Illus. 2.26

Illus. 2.27

EXERCISES

1. We have used the term "three distinct, noncollinear points." Actually this is redundant terminology, the word *distinct* being superfluous. Why?

2. Why are tripods used to hold things like cameras, telescopes, etc.?

3. If a line l and a point P are "contained in" a plane α, in what sense are they "contained"? That is, are they members or subsets of α?

4. a) Draw a picture to represent two horizontal planes.

    b) Draw two vertical planes.

5. a) Draw a horizontal plane with a line segment on it.

    b) Draw a vertical plane with three noncollinear points A, B, and C on it. Also draw lines $\overleftrightarrow{AB}$, $\overleftrightarrow{BC}$, and $\overleftrightarrow{AC}$.

6. a) Draw a plane that is neither vertical nor horizontal, and a line that meets the plane but is not in the plane.

   b) If a line not on a plane meets the plane, how many points will be in the inter-section?

7. a) Use a full sheet of paper and think of it as representing part of a plane. Draw a line *l* on the plane.

   b) Draw a point *P* on the plane not on *l*.

   c) Draw a line *m* that contains *P* and also meets *l*.

   d) Draw a line *n* that contains *P* but does not meet *l*. (Remember, the plane must be thought of as being unlimited in every direction; likewise, the lines are of un-limited extent.)

   e) How many lines are there on the plane which contain *P* but do not meet *l* (that is, their intersection with *l* is empty)?

8. a) Draw two intersecting planes, one vertical and one horizontal.

   b) Describe the intersection of any two planes.

Rigorous exercises

9. Prove Theorem 2.11.

10. Prove Theorem 2.12.

11. Prove Theorem 2.13.

12. Prove that space contains at least three planes.

14. INTERSECTIONS OF LINES AND PLANES

We have proved that two distinct lines meet in at most one point. Let us now consider similar questions about planes and lines. In what ways may a line and a plane intersect, for example? We already know that it is possible for a line to be a subset of a plane. In fact, by Theorem 2.11 if at least two points of the line are in the plane, then every point of the line is in the plane. What are the other possibilities? Suppose there is at least one point of the line on the plane and at least one point off the plane. Then all the points of the line, except one of them, must be off the plane, as shown in Illus. 2.28.

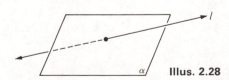

Illus. 2.28

In effect we have just outlined a proof of a theorem.

Theorem 2.14 For any line *l* and any plane α, if *l* is not on α, then $l \cap \alpha$ contains at most one point.

Note that when we say *at most one point*, we allow the possibility that the line and the plane have an empty intersection. This theorem does not assert that such a situation is possible, but merely that more than one point under these conditions is impossible.

We now consider the possible intersections of two planes. From the preceding exercises the reader may feel that the intersection of two distinct planes is a line. Although this may appear obvious, it cannot be proved until we add another axiom to our list, as follows:

Axiom 2.11 If two planes intersect, their intersection contains at least two distinct points.

This axiom is intuitively clear and easy to accept, and it allows us to prove the next theorem. Axiom 2.11 was used by Euclid but was not stated as an axiom or postulate. This was one of the weaknesses in his development of geometry.

Theorem 2.15 If two distinct planes have a nonempty intersection, then that intersection is a line.

Outline of proof. To prove this theorem we must show two things: (1) that the intersection contains a line and (2) that the intersection contains nothing else (that is, does not contain three noncollinear points). Consider any two distinct planes, α and β, with a nonempty intersection. (1) follows from Axiom 2.11, Theorem 2.11, and the definition of set intersection. Now assume the intersection contains a point R not on this line. This leads to the conclusion that $\alpha = \beta$, contrary to hypothesis.

EXERCISES

1. If α, β, and γ are three planes, describe the possible intersections $\alpha \cap \beta \cap \gamma$.

2. If two distinct planes intersect, their intersection is a line. Consider two such intersecting planes and a line that does not lie on either plane. Describe the possible intersections of this line and either or both planes.

Rigorous exercises

3. Prove Theorem 2.14.   4. Prove Theorem 2.15.

15. PARALLELS

We have considered two lines on a plane. In fact, we have proved that if two lines are different, their intersection contains at most one point. If two lines do meet, then they both lie in the same plane. We might now ask whether it is possible for two different lines to lie in the same plane and not meet (that is, have an empty intersection). We have not proved that such a relationship is possible, but our intuition tells us this is reasonable, and we shall therefore give it a name. If two distinct lines lie in a plane and do not meet, we will agree to call them *parallel*. We shall make our definition accordingly.

Definition 2.8 Two lines are said to be *parallel* if and only if they are in the same plane and have an empty intersection.

We shall often wish to speak of parallel segments, rays, or half-lines. We will therefore have to define what we mean by parallelism for such figures. Illustration 2.29 may suggest a definition. A partial definition is given below. Its completion is left to the reader.

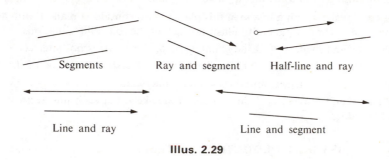

Segments   Ray and segment   Half-line and ray

Line and ray   Line and segment

Illus. 2.29

Definition 2.9 Any two of the following kinds of figures (segments, rays, half-lines, lines, and open segments) are said to be *parallel* if and only if . . .

It is possible for two lines to have an empty intersection and not be parallel. In such a case the lines would not be coplanar (that is, would not lie in the same plane). If two lines are not coplanar, then they cannot meet, and we say that they are *skew lines*. If two planes do not meet (have an empty intersection), we say they are parallel.

Definition 2.10 Two lines are said to be *skew lines* if and only if they are not in the same plane.

Definition 2.11 Two planes are said to be *parallel* if and only if their intersection is empty.

Similarly, we can make a definition of parallelism for a line and a plane.

Definition 2.12 A line and plane are *parallel* if and only if their intersection is empty.

EXERCISES

1. Complete Definition 2.9.

2. Write a definition of parallelism for a plane and a ray. A plane and a segment.

3. Find an example of a physical object to illustrate each of the following.
   - a) parallel lines
   - b) parallel segments
   - c) parallel planes
   - d) a line parallel to a plane
   - e) skew lines

4. Two planes are either parallel or they have a nonempty intersection; there is no other possibility.
   - a) Is this also true of lines? Explain.
   - b) Is this also true of rays? Explain.
   - c) Is this also true of a ray and a plane? Explain.

5. Consider a plane α with a line m in this plane. Also consider a plane β with a line n in this plane. Draw a picture to illustrate each of the following situations.
   - a) Planes α and β intersect. Line m intersects β. Line n does not intersect α.
   - b) Planes α and β intersect. Neither line m nor n intersects the plane of the other.
   - c) Planes α and β intersect and lines m and n intersect.
   - d) Planes α and β intersect. Lines m and n are skew, but each intersects the plane of the other.

16. SMALL ABSTRACT DEDUCTIVE SYSTEMS

It is often instructive to consider small deductive systems, where the objective is to make some proofs in which one's intuition is not so apt to be a distracting influence. Such systems consist of a small set of axioms, referring to nothing in particular. When one does a proof in one of these systems, he has only a very few axioms to consider. Otherwise, the procedure is precisely like that in our previous deductive development of geometry. The following system provides one illustration.

There are two undefined terms, *bee* and *hive*. A hive, however, is understood to be a set of bees. There are three axioms.

Axiom 1 There are at least two bees.

Axiom 2 For any two distinct bees there is one and only one hive containing them.

Axiom 3 For any hive H, there is at least one bee not in H.

EXERCISES

1. Prove that there are at least three bees.

2. Prove that any two distinct hives have at most one bee in common.

3. Prove that there are at least three hives.

4. Since the terms *bee* and *hive* are undefined, we are at liberty to interpret them however we please, provided the interpretation fits the axioms. Let us suppose that a hive is a committee and a bee is any member of the committee. Restate the axioms and the theorems (Exercises 1 through 3), making these substitutions of terms.

5. Suppose that a *bee* is a point and that a *hive* is a line. Restate the axioms and theorems (Exercises 1 through 3) under this interpretation.

Here is another example of a small deductive system. Let us consider a set of points, {*A, B, C, D, E, F, G*}, as the elements of the system. Each of the following subsets will be called a line: {*A, D, B*}, {*B, E, C*}, {*A, F, C*}, {*A, G, E*}, {*B, G, F*}, {*D, G, C*}, and {*D, E, F*}. There are no other lines. (Illus. 2.30.)

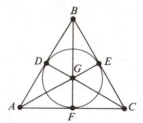

Illus. 2.30

Since *line* is defined (we identified all seven of them) without using betweenness, we can use our lines to define betweenness.

For any points *X, Y, Z* in this system, *X-Y-Z* if and only if the set of these three points is a line.

The reader will note that with this definition, *A-D-B*, *A-B-D*, and *D-A-B* are all true.

EXERCISES

1. Make up your own definition of *segment* in this system.

2. Which of Axioms 2.1 through 2.8 would hold true for this system?

3. Which of Theorems 2.1 through 2.10 would be true for this system?

CHAPTER 3

SEPARATIONS, CURVES, AND SURFACES

1. INTRODUCTION

The next step in this formal development of geometry is to consider how certain of the fundamental figures can be separated, or divided into other figures. Although the axioms we shall use are intuitively simple and perhaps obvious, they provide for a rigorous development, in which the logical shortcomings of Euclid can be overcome. Euclid did not even consider such things as separations, and this was one source of his errors.

2. SEPARATIONS ON A LINE

From the definition of a half-line one can see that, given a point P on a line l, the line l may be described as the union of the point P and the two opposite half-lines, as shown in Illus. 3.1.

Half-line P Half-line

l　　　**Illus. 3.1**

We might also say that the point P separates the line into three sets of points: the two opposite half-lines and a third set consisting of the point P itself. In this kind of separation, the point P is a boundary point. It does not belong to either half-line but produces the separation.

If two points belong to the same half-line, we say they are "on the same side of P." If they are on different half-lines, then they are "on opposite sides of P." Furthermore, it seems intuitively clear that if two points, A and B, are on the same side of P, the segment $\overline{AB}$ is also contained in (is a subset of) the same half-line (or side) and does not contain P. On the other hand, if A and B are on opposite sides of P, then the segment $\overline{AB}$ intersects both half-lines and contains the point P (see Illus. 3.2). Proofs of the above statements are relatively simple.

46

Points A and B on the same side of P

Points A and B on opposite sides of P **Illus. 3.2**

Theorem 3.1 Given a point P on a line l, if two points A and B are on the same side of P, then the segment $\overline{AB}$ is contained in (is a subset of) that side of P and does not contain P.

If A and B are on opposite sides of P, then the segment $\overline{AB}$ intersects both half-lines and contains the point P.

Proof. If A and B are on the same side of P, then by the definition of a half-line and Theorem 2.7, we have $P\text{-}A\text{-}B$ (or $P\text{-}B\text{-}A$). If X is any point in the interior of $\overline{AB}$, then $A\text{-}X\text{-}B$.

By Axiom 2.4, $P\text{-}A\text{-}B \rightarrow B\text{-}A\text{-}P$ and $A\text{-}X\text{-}B \rightarrow B\text{-}X\text{-}A$.

By Axiom 2.5, $B\text{-}A\text{-}P$ and $B\text{-}X\text{-}A \rightarrow B\text{-}X\text{-}P$ and $X\text{-}A\text{-}P$.

By Axiom 2.4, $B\text{-}X\text{-}P$ and $X\text{-}A\text{-}P \rightarrow P\text{-}X\text{-}B$ and $P\text{-}A\text{-}X$.

By the definition of a half-line, either betweenness relation, $P\text{-}X\text{-}B$ or $P\text{-}A\text{-}X$, implies that X is on the same half-line as A and B. Thus $\overline{AB}$ is contained in the same half-line as A and B. Furthermore, $\overline{AB}$ does not contain P. (Why?)

On the other hand, if A and B are on opposite sides of P, then $\overline{AB}$ intersects both half-lines, since A is on one half-line and B is on the other. Segment $\overline{AB}$ contains P and $A\text{-}P\text{-}B$ (or $B\text{-}P\text{-}A$) since by Axiom 2.7 there is exactly one betweenness relation for these three points, and any other relation (such as $A\text{-}B\text{-}P$ or $P\text{-}A\text{-}B$) would imply that A and B are on the same side of P, contrary to hypothesis. Thus, if A and B are on opposite sides of P, $\overline{AB}$ intersects both half-lines and contains the boundary point P.

3. CONVEX SETS

The word *convex* is probably familiar to the reader. If an object is convex, it has no portion of its surface "bent inward." A football is convex, for example, while a spoon is not. Vegetable cans are convex, except that if one is dropped, it may be bent inward and no longer be convex. Figures such as those of Illus. 3.3 are convex. The drawings show a segment contained in each figure. It is characteristic of convex sets that one cannot find a segment with endpoints in the figure unless the entire segment is in the set. The figures shown in Illus. 3.4 are not convex. For each of these figures it is possible to find at least one segment with endpoints in the figure but with part of the segment not in the figure.

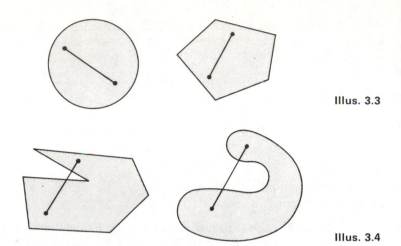

Illus. 3.3

Illus. 3.4

A convex set is now to be defined precisely, on the basis of these ideas.

Definition 3.1 A set S is a *convex set* if and only if, for any two distinct points A and B in S, the segment $\overline{AB}$ is also contained in S.

We have already studied several figures which are convex. Theorem 3.1 shows that a half-line is a convex set, for example. A line is also a convex set, since for any two distinct points A and B on a line, the segment $\overline{AB}$ is also contained in the line. Although it may not seem quite as obvious, a set containing a single point is also a convex set. This last statement may seem more reasonable if one thinks of the property of convexity as: *If* two distinct points A and B are in a set, then $\overline{AB}$ is in the set.

The discussion in the preceding section might now be summarized by saying that any point on a line separates that line into three distinct convex sets: two half-lines and a set containing just the boundary point. The union of the two half-lines is not a convex set, however.

In the sections to follow we will often wish to determine whether or not a set is a convex set. In this regard the following theorem will play an important role.

Theorem 3.2 The intersection of any two convex sets is also a convex set.

Proof. Let us first recall what is meant by the intersection of two sets. If S is a set of points and T is a set of points, then the intersection set $S \cap T$ is the set of all points which are common to S and T. Now suppose that S is a convex set, T is a convex set, and A and B are two points belonging to the intersection, $S \cap T$. Since points A and B belong to $S \cap T$, they belong to S and also to T. But, by hypothesis, S and T are convex sets. Thus $\overline{AB}$ is contained in S and $\overline{AB}$ is also contained in T. By the definition of the intersection of two sets, this means that

$\overline{AB}$ is contained in the intersection $S \cap T$. By Definition 3.1, therefore, $S \cap T$ is a convex set. We therefore conclude that the intersection of any two convex sets is also a convex set.

EXERCISES

1. Which of the geometric figures defined in Chapter 2 are convex sets?
2. Give an example to show that the union of two convex sets is not necessarily a convex set.
3. Explain what you would have to do to show that a set is *not* a convex set.

Rigorous exercise

4. a) Prove that a set containing just one point is a convex set. [*Hint:* Consider Theorem 3.2.]
   b) Prove that the empty set is convex.

4. SEPARATIONS ON A PLANE

From our experience with the separation of a line by a point, it would seem reasonable to expect that a line in a plane separates the plane in an analogous manner. In this case the line would be the boundary and would constitute one of the three sets. The other two sets of points might then be called *half-planes* or *sides*. Such a separation is shown in Illus. 3.5.

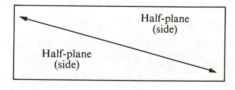

Illus. 3.5

Furthermore, like half-lines, the half-planes appear to be convex sets. That is, if two points, A and B, are in the same half-plane, then the segment $\overline{AB}$ is also in the half-plane and does not intersect the boundary line. On the other hand, if A and B are in opposite half-planes, then $\overline{AB}$ must intersect the boundary line. Both situations are shown in Illus. 3.6.

Although it seems intuitively clear that any line in a plane separates the plane in the manner just described, this does not follow from the present set of axioms. We shall therefore assume this and list it as an axiom.

Axiom 3.1 (Plane Separation Axiom) Given a line l in a plane, the set of all points in the plane not on l consists of two disjoint sets such that (1) each set is a convex set and (2) if A is in one set and B is in the other, then $\overline{AB}$ intersects l.

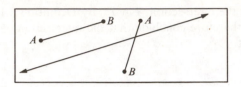

Illus. 3.6

The two sets referred to in the axiom are called *half-planes* or *sides*, and the line *l* is called the *edge* or *boundary* of the half-planes. We will think of two points in the same half-plane as being "on the same side of *l*." Since every half-plane is uniquely determined by a line *l* and a point *A* not on *l*, such a half-plane will be denoted as "*l/A*" (read "the half-plane determined by the line *l* and containing the point *A*"). If we wish to refer to the half-plane opposite *l/A*, we may do so by writing "*l/~A*" (read "the half-plane determined by the line *l* and *not* containing the point *A*"). See Illus. 3.7.

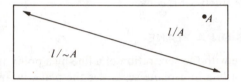

Illus. 3.7

Theorem 3.3 If *P* is a point on a line *l* and *Q* is a point not on this line then the half-line $\overset{\circ}{PQ}$ is contained entirely in one side of the line *l*.*

Outline of proof. Note first that lines *l* and $\overleftrightarrow{PQ}$ lie on a plane determined by *l* and *Q* and that $\overleftrightarrow{PQ}$ intersects *l* at a unique point, which is *P* (Illus. 3.8). Now suppose that *T* is some point on the half-line $\overset{\circ}{PQ}$. Then *T* is on the line $\overleftrightarrow{PQ}$ and lies either in *l/Q* or *l/~Q*. If *T* lies in *l/~Q*, then *Q-P-T*. But this contradicts the assumption that $T \in \overset{\circ}{PQ}$. Thus, every point *T* on $\overset{\circ}{PQ}$ must lie on *l/Q*.

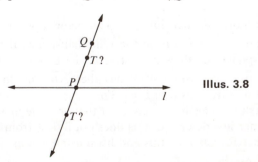

Illus. 3.8

---

* The theorem could also be stated: $P \in l$ and $Q \notin l \rightarrow \overset{\circ}{PQ} \subset l/Q$.

EXERCISES

1. Which kind of separation might be illustrated by
   a) folding a sheet of paper,   b) cutting a piece of string,
   c) the center line on a highway,   d) a performer on a tightrope?
   e) Find some other physical models that illustrate each kind of separation.

2. A plane separates space into three sets in a manner similar to the separations on a line and a plane.
   a) Describe these three sets.
   b) What might we name two of these sets?
   c) If two points, A and B, did not lie on the plane separating space, what possible relationships might exist between the segment $\overline{AB}$ and the boundary plane?

3. Consider a line l on a plane and three points, A, B, and C, on the plane which do not lie on l.
   a) Name the possible locations of these points with respect to the two sides.
   b) Would it be possible to locate the three points so that each is on a side opposite the other two?

4. a) Draw a line l on a plane.
   b) Draw two points, A and B, on one side of l, and a point C on the opposite side.
   c) Draw segments $\overline{AB}$, $\overline{BC}$, and $\overline{AC}$. How many of these segments intersect the line l?
   d) Draw three points R, S, and T on one side of l. Draw segments $\overline{RS}$, $\overline{ST}$, and $\overline{RT}$. How many of these segments intersect l?
   e) From the results of (c) and (d), make a conjecture concerning any three points in the plane which are not on l.

5. Would a ray separate a plane in the manner specified in Axiom 3.1? Explain.

Rigorous exercises

6. Prove that every half-plane contains at least three noncollinear points.

7. Write a complete proof of Theorem 3.3.

Angles

The reader's previous experience with angles may lead him to feel that an angle should be defined to be a rotation, or "an amount of opening," or perhaps a number of degrees. All of these ideas are commonly associated with the word *angle*. A careful development of geometry, however, demands careful and precise definitions; hence none of the more traditional ideas mentioned above are satisfactory. Since an angle is to be a geometric entity, it must be defined in such a way that it is a geometric figure, or a set of points. The following definition, saying that an angle is the union of two rays, is precise and meets the other criteria for a satisfactory definition.

Definition 3.2 An *angle* is a figure consisting of two distinct rays with a common endpoint. If the rays are opposite rays of a line, the angle is called a *straight angle*. The rays are called *sides* and the common endpoint is called the *vertex*.

An angle is made up of rays. The figure of Illus. 3.9(a) is not an angle since it is made up of segments. Nevertheless it *determines* an angle like the one on the right. If A is any point of one ray (except B), B is the endpoint, and C is any point on the other ray (except B), then the angle may be denoted as $\angle ABC$ or $\angle CBA$. (Note that the middle letter always identifies the vertex.) In some instances we might also refer to this angle as simply $\angle B$.

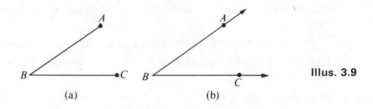

(a)   (b)   **Illus. 3.9**

Would an angle separate a plane into three sets of points? If so, how would we describe the three sets? From Illus. 3.10 it seems clear that an angle* does separate a plane into three sets and that one set of points is the angle itself. We might also observe that the point P belongs to a set which is different from the set containing the points X, Y, and Z.

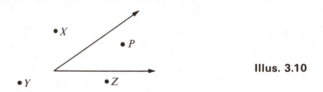

Illus. 3.10

It is not difficult to find names for these sets and we usually refer to the set containing P as the *interior* of the angle, and the set containing X, Y, and Z as the *exterior* of the angle. Our problem now is to formulate a definition for the interior and exterior of an angle, without introducing any new undefined terms, so that if Q is an arbitrary point in the plane (not on the angle), then we can, by applying the definition, determine whether Q is an exterior or an interior point.

———————————

* In the remainder of this section the word "angle" will mean nonstraight angle.

EXERCISES

Exploratory exercises

Refer to Illus. 3.11 in answering Exercises 1 through 6. $\angle ABC$ consists of the rays $\overrightarrow{BA}$ and $\overrightarrow{BC}$. Lines $\overleftrightarrow{BA}$ and $\overleftrightarrow{BC}$ are determined by these rays.

1. Point P is on the same side of $\overleftrightarrow{BC}$ as which point?
2. Point P is on the same side of $\overleftrightarrow{BA}$ as which point?
3. Does every point on $\overset{\circ}{\overrightarrow{BA}}$ lie on the same side of $\overleftrightarrow{BC}$? Why?
4. Does every point on $\overset{\circ}{\overrightarrow{BC}}$ lie on the same side of $\overleftrightarrow{BA}$? Why?
5. Which half-line is in $\overleftrightarrow{BC}/P$?
6. Which half-line is in $\overleftrightarrow{BA}/P$?

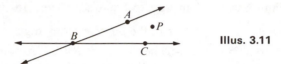

Illus. 3.11

Refer to Illus. 3.12 in answering Exercises 7 through 11.

7. Does the point X lie in $\overleftrightarrow{BC}/A$?
8. Does X lie in $\overleftrightarrow{BA}/C$?
9. Describe the location of Y in relation to $\overset{\circ}{\overrightarrow{BA}}$ and $\overset{\circ}{\overrightarrow{BC}}$.
10. Describe the location of Z in relation to $\overset{\circ}{\overrightarrow{BA}}$ and $\overset{\circ}{\overrightarrow{BC}}$.
11. Describe the location of P in relation to $\overset{\circ}{\overrightarrow{BA}}$ and $\overset{\circ}{\overrightarrow{BC}}$.

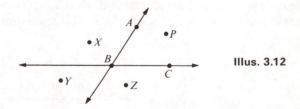

Illus. 3.12

12. Draw two intersecting lines, mark the points A, B, and C as in the drawings for Exercises 1 through 11.
    a) Shade the half-plane containing $\overset{\circ}{\overrightarrow{BA}}$.
    b) Shade the half-plane containing $\overset{\circ}{\overrightarrow{BC}}$.
    c) What is the intersection of these two half-planes?
    d) Is $\angle ABC$ contained in this intersection?

13. Complete the following statement:
    A point Q is in the *interior* of an angle $\angle ABC$ if and only if . . .

14. Assuming an angle separates the plane into three distinct and disjoint sets, if a point is not on the angle and is not in the interior, then it would have to be where?

15. Complete the following statement:
    A point Q is in the *exterior* of an angle if and only if . . .

16. Which of the following statements is true? Explain why the other statement is false.

    a) An angle is the union of two rays with a common endpoint.

    b) An angle is the intersection of two rays with a common endpoint.

17. If R and S are exterior to $\angle ABC$, does it necessarily follow that $\overline{RS}$ is in the exterior of $\angle ABC$? Illustrate your answer with a drawing.

Given below is a definition of the interior and exterior of an angle. Compare this definition with the answers to Exercises 13 and 15 above.

Definition 3.3   A point Q is in the *interior* of an angle $\angle ABC$ if and only if Q is in the intersection of half-planes $\overleftrightarrow{BC}/A \cap \overleftrightarrow{BA}/C$. A point Q is in the *exterior* of an angle $\angle ABC$ if and only if Q is not on the angle and is not in the interior of the angle.

The above definition might also have been stated as follows:*

The *interior* of an angle $\angle ABC$ is the intersection of half-planes $\overleftrightarrow{BC}/A \cap \overleftrightarrow{BA}/C$. The *exterior* of an angle $\angle ABC$ is the set of points which are not on the angle nor in the interior of the angle.

We are now ready to answer the question raised previously as to whether an angle separates a plane. From Axiom 3.1 and Definition 3.3 it can be seen that an angle does separate a plane into three distinct sets: (1) the angle, (2) the interior of the angle, and (3) the exterior of the angle. This separation is shown in Illus. 3.13.

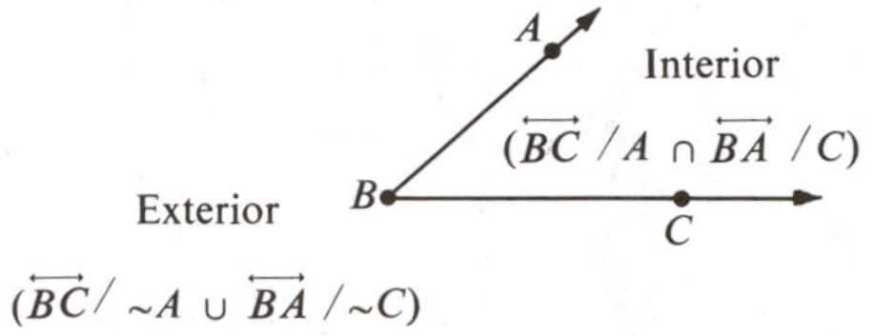

Illus. 3.13

---

* In general, a definition may be stated using either the verb "is" or the connective "if and only if." We usually choose one form over the other in a particular case because it is clearer, simpler, and/or easier to apply. In the case above, the first form of the definition points out somewhat more clearly how it can be applied in determining whether or not a point belongs to the interior or exterior of an angle. The second form of the definition may be preferred because it is more concise and therefore easier to remember.

It would also appear from Illus. 3.13 that the interior of an angle is a convex set and the exterior is not a convex set. Furthermore, if P is a point in the interior of an angle and Q is a point in the exterior of the angle, then the segment $\overline{PQ}$ intersects the angle. All of this is summarized in the following theorem. The proof is left as an exercise.

Theorem 3.4 Given any nonstraight angle, the set of all points in its plane and not on the angle consists of two disjoint sets such that:

a) One of the sets is convex and the other is not.

b) If P is in one set and Q is in the other, then $\overline{PQ}$ meets the angle.

The convex set mentioned in Theorem 3.4 is called the *interior* of the angle, and the other is called the *exterior*.

EXERCISES

1. Describe the separation of a plane by a straight angle. How does this separation differ from the separation produced by a nonstraight angle?

Rigorous exercises

2. Prove Theorem 3.4 (a).

3. Prove Theorem 3.4 (b).

4. Prove that the intersection of the interiors of $\angle ABC$ and $\angle BCA$ is a convex set. (A, B, and C are noncollinear points.)

Triangles

Although a triangle was not formally defined in Chapter 2, we did describe it as the union of three segments, $\overline{AB} \cup \overline{BC} \cup \overline{CA}$, where A, B, and C are distinct noncollinear points. Let us now add this definition to our geometry.

Definition 3.4 For any three distinct noncollinear points A, B, and C, the union of the segments $\overline{AB}$, $\overline{BC}$, and $\overline{CA}(\overline{AB} \cup \overline{BC} \cup \overline{CA})$ is called a *triangle*. The segments are called *sides*, and the points A, B, and C are called *vertices* (sing., *vertex*). Triangle ABC is denoted "$\triangle ABC$."

Since three noncollinear points are contained in one and only one plane, a triangle also determines a unique plane. Does a triangle have angles? A triangle consists of three segments and contains no rays. Hence, a triangle *contains* no angles as subsets. Nevertheless, a triangle "has" angles in the sense that angles are determined by the pairs of sides, the vertex of each angle being the same as a vertex of the triangle [Illus. 3.14(a)]. Moreover, if we consider all three angles of a triangle, we obtain a figure consisting of three lines which intersect in pairs at the vertices of the triangle, as shown in Illus. 3.14(b). This latter figure suggests a way of defining the interior of a triangle. If, for example, we consider the intersection of the half-planes $\overleftrightarrow{BC}/A \cap \overleftrightarrow{AC}/B \cap \overleftrightarrow{BA}/C$, we obtain a set containing all of the points in the "inside" or interior of the triangle. The intersection of these half-planes is shown in Illus. 3.15. The exterior may then be described as the union of the half-planes $\overleftrightarrow{BC}/\sim A \cup \overleftrightarrow{AC}/\sim B \cup \overleftrightarrow{BA}/\sim C$, as shown in Illus. 3.16.

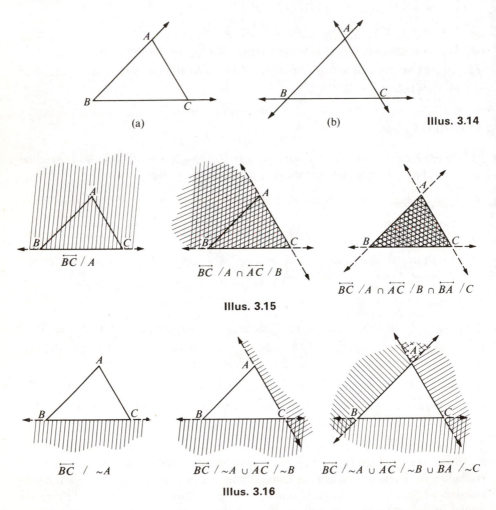

(a)   (b)   **Illus. 3.14**

$\overrightarrow{BC}\ /A$

$\overrightarrow{BC}\ /A \cap \overrightarrow{AC}\ /B$

$\overrightarrow{BC}\ /A \cap \overrightarrow{AC}\ /B \cap \overrightarrow{BA}\ /C$

Illus. 3.15

$\overrightarrow{BC}\ /\sim A$

$\overrightarrow{BC}\ /\sim A \cup \overrightarrow{AC}\ /\sim B$

$\overrightarrow{BC}\ /\sim A \cup \overrightarrow{AC}\ /\sim B \cup \overrightarrow{BA}\ /\sim C$

Illus. 3.16

It will be noted that $\triangle ABC$ does not belong to either the interior or the exterior. Rather, it separates the plane into three distinct sets: the interior, the exterior, and the triangle itself. Thus, if a point in the plane is not on the triangle, it must be in either the interior or the exterior of the triangle, but not in both.

Accordingly, we now give the following definition:

Definition 3.5 The *interior* of a triangle ABC is the intersection of half-planes $\overleftrightarrow{BC}/A \cap \overleftrightarrow{AC}/B \cap \overleftrightarrow{BA}/C$. The *exterior* of a triangle ABC is the set of all points in the plane of the triangle which are not on the triangle and are not in the interior of the triangle.

A triangle separates the plane in which it lies in much the same way that an angle separates a plane. The points not on the triangle consist of two disjoint sets. One of these is the interior of the triangle and the other is the exterior. The next theorem states this fact more precisely. The proof is left as an exercise.

Theorem 3.5 For any triangle:

a) The interior is a convex set.

b) If P is in the interior and Q is in the exterior, then $\overline{PQ}$ meets the triangle.

EXERCISES

1. Rewrite Definition 3.5, using the connective "if and only if."

2. Write an alternative definition of the interior of a triangle which involves the interiors of the angles of the triangle.

3. Could a point be in the interior of an angle of a triangle and also be in the exterior of the triangle?

4. Could an angle of a triangle be a straight angle?

5. If a line in the plane of a triangle $\triangle ABC$ intersects exactly one of the vertices of the triangle, what possibilities are there with regard to the intersection of this line and the sides of the triangle?

6. a) How are the interiors of an angle and a triangle alike?

   b) How are they different? [*Hint:* What geometric figures can be contained in the interior of an angle? The interior of a triangle?]

Rigorous exercises

7. Prove Theorem 3.5.

8. Prove the following: If a line in the plane of a triangle $\triangle ABC$ does not contain a vertex of the triangle but intersects one side of the triangle, then the line also intersects one other side of the triangle.* [*Hint:* See Exercises 3 and 4, page 51.]

9. Prove that the line referred to in Exercise 8 intersects *no more than one* other side of the triangle.

---

* This theorem is usually referred to as *Pasch's Axiom* (strong form), after Morris Pasch, a German mathematician who assumed it as a fundamental axiom in his development of geometry. Although we shall not do it here, it can be shown that Pasch's Axiom is logically equivalent to the Plane Separation Axiom, which means that if we assume either one as an axiom, we can derive the other from it.

5. CURVES

Early in Chapter 2 we defined a *geometric figure* as a nonempty set of points. Since then we have defined many special types of figures such as *segment*, *line*, *angle*, etc., in each case using only our undefined terms and the language of sets. We now wish to consider a special class of geometric figures known as *curves*, in which many of the figures already defined, as well as many others not yet defined, belong. Since a rigorous definition of *curve* is rather complicated and beyond the scope of this book, we will consider the general idea of a curve on an intuitive basis only.

In general, geometric figures can be divided into two classes: sets of points which are curves and sets of points which are not curves. Some examples of curves are shown in Illus. 3.17.

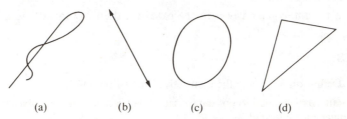

(a)   (b)   (c)   (d)   **Illus. 3.17**

Note that some curves look "straight" while others do not. As a matter of fact, one may be tempted to describe curve (a) as a "curved line" and curve (b) as a "straight line." Neither description has any meaning, however, except as a method of describing *visual* differences in drawings. Furthermore, the use of such descriptions implies that there are different types of lines, and since this is not the case, they are best avoided.

While some curves can be identified by name [for example, curve (b) is a line and curve (d) is a triangle], it would be impossible to identify all of the types of figures that we wish to include in our set of curves. We therefore need to identify at least one characteristic common to all curves, so that figures which we wish to identify as curves will have this characteristic, while all other figures will not.

Let us imagine that we have some special rubber bands that will stretch without limit and will bend in any manner. Furthermore, after stretching and bending, the rubber band always remains in that configuration. In forming one of our "rubber band configurations," we will cut the band no more than once. Some configurations are pictured in Illus. 3.18.

If we were to cut a rubber band and think of stretching it infinitely far in both directions, we would have a real world model of a line. In the configurations above, (a) is cut while (b) is not.

While there is no limit to the different kinds of configurations one could form, they all share a rather simple physical property, which may be described in any one of the following ways: Each configuration is in "one piece," or there are no "breaks" within the configuration; or a rubber band configuration is made up of

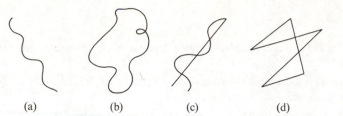

(a)   (b)   (c)   (d)   **Illus. 3.18**

only one piece of rubber band. We do not allow the rubber band to be broken into two or more pieces.

We are now in a position to give an intuitive description of a curve.

A *curve* is a geometric idealization of a rubber band configuration.

Although the curves we have considered so far have been in a plane, it is also possible to have curves in space, that is, curves that do not lie in any one plane. As physical models of such a curve, one might consider a coil spring or a bent coat hanger, in addition to our rubber band configurations.

Closed Curves

Using our rubber band configurations as models, we find that it is easy to differentiate between closed curves and those that are not closed. Very simply, if the rubber band is not cut, it is a model of a closed curve. If the rubber band is cut, it is a model of a curve that is not closed.

Curves (a) and (b) in Illus. 3.17 are not closed, while curves (c) and (d) are closed. You should have no difficulty identifying the closed curves shown in Illus. 3.18.

Simple Closed Curves and Separations

A closed curve that does not "cross itself" is called a simple closed curve. If a closed curve does "cross itself," it is of course a nonsimple closed curve. Again, the term "cross itself" is interpreted in the broad sense, as indicated in Illus. 3.19, where curve (a) is a simple closed curve and curves (b) and (c) are nonsimple closed curves. This again is an intuitive description rather than a definition.

A closed curve (in a plane) that does not cross itself is called a *simple closed curve*.

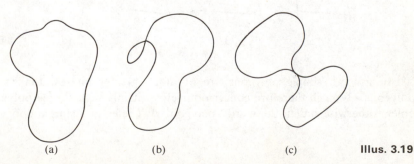

(a)   (b)   (c)   **Illus. 3.19**

EXERCISES

1. Which of the geometric figures defined thus far in our geometry are curves? Which are not?

2. Describe a physical model of a curve on a plane which is (a) not closed, (b) closed and simple, (c) closed and nonsimple.

3. Describe a physical model of a curve that is not on a plane.

4. In what ways does the interior of an angle differ from the interior of a simple closed curve?

5. Could the interior of a simple closed curve be a convex set? Illustrate with drawings.

6. POLYGONS

A polygon is a special type of closed curve. While a rigorous definition is possible, it is preferable to consider an intuitive description of polygon, as we did with curves.

A *polygon* is a closed curve consisting entirely of segments.

Polygons may be classified into two types: *simple* and *nonsimple*. Some examples of each type are shown in Illus. 3.20. Intuitively speaking, a polygon is simple if it does not cross itself. A nonsimple polygon would then be one which does cross itself. While polygon (c) clearly crosses itself, it may appear that (d) does not. Although the word "cross" is not normally used in this manner, we will agree to think of polygon (d) as "crossing itself." One may also use the following intuitive description of simple polygons.

A polygon is called a *simple polygon* if and only if no two sides meet except at a vertex.

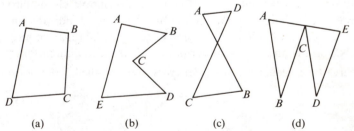

| (a) | (b) | (c) | (d) | **Illus. 3.20** |

In general, simple polygons are of much greater interest than nonsimple polygons. We will therefore concern ourselves mainly with simple polygons, and unless otherwise stated, the word "polygon" will refer to a simple polygon.

Polygons are often classified according to the number of sides, one of the more familiar of these being the triangle, which of course has three sides.* The names of some other kinds of polygons are given below.

Name	Number of Sides
Triangle	3
Quadrilateral	4
Pentagon	5
Hexagon	6
Heptagon	7
Octagon	8
Nonagon	9
Decagon	10
Dodecagon	12

A polygon with four sides is called a quadrilateral. It is possible to classify quadrilaterals according to special properties of the figure. One such property is parallelism. Since the meaning of quadrilateral is intuitively clear, we will make the following a formal definition.

Definition 3.6 A quadrilateral that has *at least one* pair of parallel sides is called a *trapezoid*. A quadrilateral that has two pairs of parallel sides is called a *parallelogram*.

Since a parallelogram has at least one pair of parallel sides, it is actually a special kind of trapezoid. We usually refer to it as a parallelogram, however, to

---

* It is interesting to note that Euclid did not think of triangles as we do today. A polygon consisting of three sides was, according to Euclid, a trilateral (*tri* meaning "three" and *lateral* meaning "side"). Since *quadrus* means "four," a polygon with four sides was of course a quadrilateral. A trilateral was considered to be a special kind of triangle, since it contained three angles. In addition, however, a figure such as the one that is shown below was also considered a triangle, having only three angles, at A, B, and C, the one at D not being considered to belong to the "figure" since what it "contains" is outside the figure. Thus Euclid thought of triangles as consisting of all trilaterals and in addition some quadrilaterals.

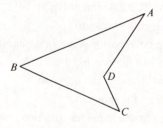

distinguish it from other trapezoids having only one pair of parallel sides. Relations between these kinds of figures are shown in Illus. 3.21(a). Illustration 3.21(b) depicts the fact that the set of all parallelograms is a subset of the set of all trapezoids and that the set of all trapezoids is a subset of the set of all quadrilaterals.

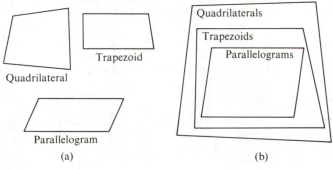

Illus. 3.21

EXERCISES

1. Draw a simple and a nonsimple polygon having

   a) 4 sides,   b) 5 sides,   c) 6 sides,   d) 10 sides.

   e) For each of the polygons in (a) through (d), shade in what appears to be the interior of the polygon.

2. What is the relationship between the number of the vertices and the number of sides of the simple polygons in Exercise 1? Would this relationship hold for a simple polygon with n sides? Explain.

3. Could a triangle be nonsimple? Explain.

4. If a vertex is in the interior of a side of a polygon, what is the least number of sides the polygon can have?

5. Is it possible for a quadrilateral with two pairs of parallel sides to be nonsimple? Explain.

Exploratory exercises

6. a) Draw a simple polygon like the one in Illus. 3.20(a).

   b) Draw two points, A and B, inside the polygon.

   c) Does segment $\overline{AB}$ intersect the polygon?

   d) Would $\overline{AB}$ intersect the polygon for some other choice of the points A and B?

7. a) Draw a simple polygon like the one in Illus. 3.20(b).

   b) Draw two points, A and B, inside the polygon.

   c) Does segment $\overline{AB}$ intersect the polygon?

   d) Would the answer to (c) be the same for any two points inside the polygon?

e) Draw a point C outside the polygon. Does segment $\overline{AC}$ intersect the polygon? Would your answer be the same for any two points, one inside and one outside the polygon?

f) Draw another point D outside the polygon. Does $\overline{CD}$ intersect the polygon? Does the choice of D affect the answer above?

8. a) Could the interior of a simple polygon contain a line? A ray?

   b) Could the exterior of a simple polygon contain a line? A ray?

9. How does the interior of an angle differ from the interior of a triangle?

10. The interior of the polygon in Illus. 3.22 is contained in the half-plane $\overleftrightarrow{EA}/C$.

   a) In what other half-planes is the interior of the polygon contained?

   b) Describe the interior as the intersection of half-planes.

11. The interior of the polygon in Illus. 3.23 is contained in the half-plane $\overleftrightarrow{AB}/D$.

   a) In what other half-planes is the interior of the polygon contained?

   b) Can the interior be described as the intersection of half-planes? Explain.

Illus. 3.22   Illus. 3.23

Simple Polygons and Separations

From the previous exercises one can see that it is fairly easy to identify the regions of a plane which we would wish to call the *interior* and *exterior* of a simple polygon. In fact, one might reasonably conjecture that any simple polygon (of which the triangle is a special case) separates the plane into three distinct sets, one of which is the polygon itself.

Since a triangle is a simple polygon, and we have already ascertained the properties resulting from such a separation, we might be tempted to conclude that all simple polygons separate the plane in the same manner as the triangle. It is not difficult to see, however, that the interiors of some simple polygons are not convex sets (cf. Exercise 7, page 62), and we shall therefore need to identify a property of the interior which is common to all simple polygons. If we consider a polygon like the one shown in Illus. 3.24, it will be observed that, while $\overline{AB}$ intersects the polygon, it is possible to join A and B by a *sequence of connected segments* which do not intersect the polygon. Similarly, any two points C and D in the exterior could be joined in this manner without intersecting the polygon. Generally speaking, a curve that consists entirely of segments is called a *polygonal curve.** It

---

* Observe that the segments do not all have to lie on a plane. Hence it is possible to have a polygonal curve in space as well as in a plane.

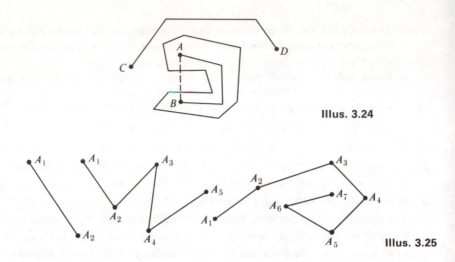

Illus. 3.24

Illus. 3.25

follows that a polygon is simply a closed polygonal curve in a plane. Some polygonal curves are shown in Illus. 3.25. Note that a segment is a polygonal curve.

Using the idea of a polygonal curve, we might say that two points, A and B, are in the same (either exterior or interior) set if and only if there is at least one polygonal curve that joins these points and does not intersect the polygon. The opposite of this situation, two points in different sets, would then be described by the negation of the sentence above: Two points, A and B, are in different sets (one in the interior and one in the exterior) if and only if *every* polygonal curve that joins the points intersects the polygon.

The problem still remains as to how to distinguish between the interior and the exterior of a polygon. While the interiors of some polygons (for example, the triangle) may be defined in terms of the intersection of half-planes, this is not possible for the interior of a polygon of the type shown in Illus. 3.24. One rather simple solution suggested in Exercise 8 of the previous exercises is to merely assume that the interior of a polygon contains no line while the exterior of a polygon contains at least one line.

The preceding discussion may be summarized as follows:

Axiom 3.2† For any simple polygon in a plane, the set of all points in the plane, not on the polygon, consists of two sets such that (1) for each set, if two points,

_____

† Although we shall assume this here, it is actually a special case of the Jordan Curve Theorem, first stated by the French mathematician, Camille Jordan (1838–1922). The Jordan Curve Theorem for all simple closed curves will be discussed in the section on curves. For the proof of the special case above, the reader is referred to B. H. Arnold, *Intuitive Concepts in Elementary Topology*, Prentice-Hall Inc., 1962.

A and *B*, are in the set, there exists at least one polygonal curve in the plane that joins the points and does not intersect the polygon, (2) if *A* is in one set and *B* is in the other, every polygonal curve in the plane that joins these points intersects the polygon.

The set which contains at least one line is called the *exterior* and the set which contains no line is called the *interior*.

EXERCISES

1. Make a table summarizing the properties of a separation of a plane by (a) a line, (b) an angle, (c) a triangle, and (d) any simple polygon. Identify the similarities and differences in these separations.

2. Describe some situations in the physical world which bring to mind the separation of a plane by a simple polygon.

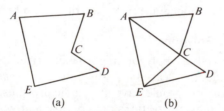

(a)   (b)   **Illus. 3.26**

3. Consider the polygon *ABCDE* and its interior, as shown in Illus. 3.26(a). If we draw segments $\overline{AC}$ and $\overline{EC}$ in the interior, we obtain a figure consisting entirely of the union of triangles and their interiors, as shown in Illus 3.26(b). The polygon is then said to be "triangulated." Triangulate each of the polygons shown in Illus. 3.27. Make a conjecture concerning the relationship between the number of triangles and the number of sides of the polygon. Support your conjecture. Make up other examples if you need to.

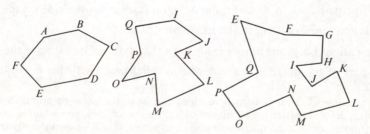

Illus. 3.27

4. Consider the figure in Illus. 3.26(b). Would the union of the interiors of the three triangles be the same as the interior of polygon *ABCDE*? Explain.

5. A simple polygon and a set of parallel rays which intersect the polygon are shown in Illus. 3.28. Some of the endpoints of the rays are in the interior of the polygon and some are in the exterior.

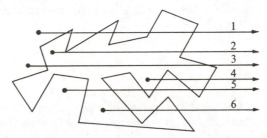

Illus. 3.28

a) How many times does each ray intersect the polygon?

b) Make a conjecture about the relationship between the location of the endpoint of a ray and the number of times the ray intersects the polygon. Draw some other rays if you wish.

c) Which ray does not produce the same pattern as the others? Why?

Convex and Concave Polygons

Intuitively speaking, a *concave* polygon might be described as one in which a side is "caved in." If no sides are "caved in," the polygon is *convex*.* While this is a rather rough description, one would have little difficulty in using it to decide which simple polygons in the previous sections are concave and which are convex. The term "caved in" refers only to the way a drawing "looks," however, and we shall have to translate this idea into more precise language for use in our official vocabulary. To do this, let us consider the polygons (one concave and one not) shown in Illus. 3.29.

It will be noted that *each* side of polygon (a) determines a line which separates the plane into two half-planes and that for each such separation, the remainder of the polygon is contained in only one of these half-planes. For example, side $\overline{AB}$ determines a line $\overleftrightarrow{AB}$ and the remainder of the polygon is contained entirely in the half-plane $\overleftrightarrow{AB}/C$. Moreover, the polygon (except for the side on the boundary line) is contained entirely in the half-plane $\overleftrightarrow{BC}/E$.

In the case of polygon (b) it is easily seen that side $\overline{AB}$ determines the line $\overleftrightarrow{AB}$ and that part of the polygon is contained in $\overleftrightarrow{AB}/E$ and part of it is contained in

---

* The term "convex," as it is used here, does not have the same meaning as the term "convex set," defined earlier. As a matter of fact, a convex polygon is *not* a convex set. While this may seem like an unfortunate choice of terms, it will be seen shortly that every convex polygon is associated with a convex set.

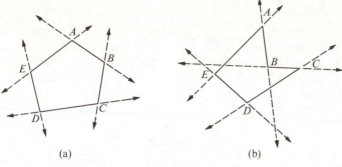

(a)   (b)

Illus. 3.29

the opposite half-plane, $\overleftrightarrow{AB}/C$ (or $\overleftrightarrow{AB}/\sim E$). Thus, there is *at least one* side of polygon (b) which determines two half-planes such that a part of the polygon is contained in each of them.

From the observations above, we note that any simple polygon must possess the characteristics of either (a) or (b), but not of both. We therefore make our definition accordingly.

Definition 3.7 A simple polygon is a *convex* polygon if and only if each side lies on a line determining a half-plane which contains the remainder of the polygon. Any simple polygon which is not convex is a *concave* polygon.

It is now possible to define the interior of a convex polygon.

Definition 3.8 The interior of a convex polygon is the intersection of all half-planes H with the following property: The edge of H contains a side of the polygon and the remainder of the polygon is contained in H.

The following theorem is a direct consequence of the definition above and Theorem 3.2.

Theorem 3.6 The interior of every convex polygon is a convex set.

Another way of distinguishing between a convex and a concave polygon involves the diagonals of a polygon. A diagonal of a simple polygon is defined as follows:

Definition 3.9 A *diagonal* of a polygon is a segment that joins two vertices and is not a side of the polygon.

Some convex and concave polygons are shown in Illus 3.30. The diagonals have been drawn (dashed) for each.

It will be observed that for each of the convex polygons, the diagonals (except

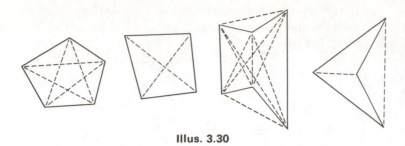

Illus. 3.30

for the endpoints) are all contained in the interior of the polygon. Is this true for the diagonals of the concave polygons?

The Jordan Curve Theorem

From the previous discussion of polygons, one can see that a simple polygon is actually a special type of simple closed curve which consists entirely of segments. As a matter of fact, one might also think of any simple closed curve as being approximated by simple polygons having their vertices on the curve. As the vertices increase in number and become closer together, the polygons become better and better approximations of the curve (see Illus. 3.31). Because of this, we might reasonably expect that any simple closed curve will separate the plane in a manner analogous to the separation produced by any one of the polygons associated with the curve, and Axiom 3.2 might therefore be extended to include all simple closed curves. Such a statement is known as the Jordan Curve Theorem. It has been proved, but the proof is complicated. It may be stated as follows.

Illus. 3.31

Jordan Curve Theorem   For any simple closed curve in a plane, the set of all points in the plane not on the curve consists of two sets such that (1) for each set, if two points, *A* and *B*, are in the set, there exists at least one polygonal curve in the plane that joins the points and does not intersect the curve; (2) if *A* is in one set and B is in the other, every polygonal curve in the plane that joins these points intersects the curve; and (3) one of these sets contains a line (the exterior); the other contains no line (the interior).

EXERCISES

1. Why must every simple polygon be either concave or convex?
2. a) Draw a convex polygon having six sides.
   b) Draw all of the diagonals of the polygon.
   c) How many diagonals are there?
   d) Which part of each diagonal is in the interior? Which part is on the polygon?
3. a) Draw a concave polygon having five sides.
   b) Draw all of the diagonals of the polygon.
   c) How many diagonals are there?
   d) Is any part of a diagonal in the exterior of the polygon?
   e) Are the endpoints of any of the diagonals in the exterior? in the interior? Why?
4. a) Draw polygons with 4, 5, 6, 7, and 8 sides, respectively.
   b) Draw the diagonals of each of the polygons in (a).
   c) Make a conjecture about the relationship between the number of diagonals and the number of vertices of a polygon. Check your conjecture with some other polygons.
   d) Does your conjecture hold for either convex or concave polygons?
5. Earlier it was proved that the interior of a convex polygon is a convex set. What property would a simple closed curve have if its interior is a convex set?

Rigorous exercises

6. Prove that a convex polygon is not a convex set.
7. Prove Theorem 3.6.

7. SURFACES

A plane is an example of the kind of figure we call a surface. There are many other such figures. Their common feature seems to be that they are connected sets in space "one point thick." We do not intend that this serve as a definition of *surface*, but it may help to educate the intuition. Let us look at some examples of surfaces.

Cylinders

Consider a curve on a plane and a line that meets the curve but does not itself lie in the plane, as in Illus. 3.32. Now consider the set of all lines parallel to line m

which contain a point of the curve C. Shown in Illus. 3.33, the union of all these lines is called a *cylindrical surface*. The curve C could be any sort of curve. In particular, it might be a closed curve, a simple closed curve, a nonclosed curve, or a polygon (also shown in Illus. 3.33). Let us now make the following definition.

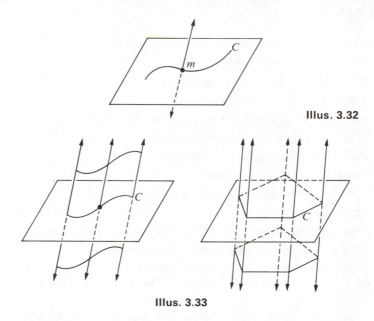

Illus. 3.32

Illus. 3.33

Definition 3.10 If C is any plane curve and m is a line not on the plane but containing a point of C, then the union of m and all lines parallel to m and containing a point of C is called a *cylindrical surface*. The line m is called a *directrix*, and the curve C is called the *generatrix*. Any of the lines parallel to m are called *elements* of the surface.

If we consider a closed cylindrical surface (one for which the generatrix C is a simple closed curve) and a pair of parallel planes, which intersect the surface as in Illus. 3.34,* we obtain a figure called a *cylinder*. The figure in question here consists of that portion of the cylindrical surface between the planes, together with the curves C_1 and C_2 and their interiors. The portion between the planes is called the *lateral surface* of the cylinder and the other regions are called the *bases*. If C_1 and C_2 are polygons, as in Illus. 3.34(b), then a special kind of cylinder is obtained, known as a *prism*.

Definition 3.11 Consider any cylindrical surface for which the generatrix C is a simple closed curve, and two parallel planes, P_1 and P_2, which intersect all elements

---

* One of the curves, C_1 or C_2, may be the generatrix C, but this is not necessary.

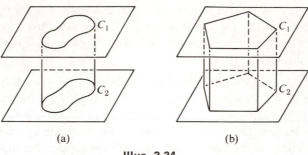

(a)   (b)

Illus. 3.34

of the cylindrical surface, forming curves C_1 in P_1 and C_2 in P_2. The figure consisting of that portion of the cylindrical surface between P_1 and P_2, together with C_1 and C_2 and their interiors, is called a *cylinder*. If C is a polygon, then the cylinder is also called a *prism*. The curves C_1 and C_2 and their respective interiors are called the *bases* and the portion between the bases is called the *lateral surface*. The polygons that constitute the lateral surface of a prism are called its *lateral faces*.

If the bases of a prism are triangles, it is called a *triangular* prism. If they are pentagons, it is called a *pentagonal* prism, and so on.

Cones

Let us now consider a curve C in a plane and a point P off the plane, as in Illus. 3.35. Consider the set of all lines which contain P and also a point of C. The union of these lines is a figure called a *conical surface*. Illustration 3.36 is an example. Note that this surface has two parts, called *nappes*, on opposite sides of the point P. Like cylindrical surfaces, a conical surface is of unlimited extent because its elements are.

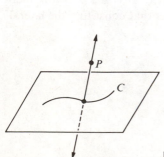

Illus. 3.35

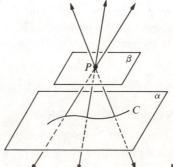

Illus. 3.36

Definition 3.12   If C is a curve in a plane α and P a point not on the plane, then the union of all lines that contain P and a point of C is called a *conical surface*. Any of the lines contained in the surface is called an *element* of the surface. The curve C is called the *generatrix*.

Given a plane β which contains P and is parallel to α, each portion of the conical surface in a half-space determined by β, together with the point P, is called a *nappe* of the surface.

If we consider a conical surface for which the generatrix is a simple closed curve, we can visualize the figure consisting of the simple closed curve and its interior and that portion of the conical surface between the plane α and point P. The point P is also to be included (Illus. 3.37). This kind of figure is called a *cone*. The curve C and its interior constitute the *base* and the rest of the figure is called the *lateral surface*. If the curve C happens to be a polygon, then the cone is also called a *pyramid*.

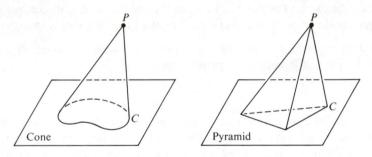

Illus. 3.37

Definition 3.13   Consider any conical surface for which the generatrix C is a simple closed curve. The figure consisting of C and its interior, together with that portion of the conical surface between C and the point P, together with point P, is called a *cone*. The curve C with its interior is called the *base*. The rest of the figure is called the *lateral surface*. The point P is called the *vertex*. If C is a polygon, then the cone is also called a *pyramid*. The polygonal regions that constitute the lateral surface of a pyramid are called its *lateral faces*.

EXERCISES

1. Draw a cylindrical surface where
   a) the generatrix C is a polygonal curve,
   b) C is a simple closed curve but not polygonal.

2. Draw a cylinder which is
   a) a prism,   b) not a prism.
3. Draw a conical surface where
   a) the generatrix C is a nonpolygonal, nonclosed curve,
   b) C is a simple polygon.
4. Draw a cone that is
   a) a pyramid,   b) not a pyramid.
5. Draw
   a) a triangular pyramid,   b) a hexagonal pyramid,
   c) a triangular prism,   d) a pentagonal prism,
   e) a hexagonal prism.
6. Make a cardboard model of a cylinder. To do this, use a piece of stiff cardboard for the plane of the generatrix. Draw a generatrix C on it, and cut a slot the shape of C. Then insert a piece of heavy paper through C.
7. Make a cardboard model of a circular cone. To do this, draw a large circle and cut out a sector OAB as shown (Illus. 3.38). Bring segments $\overline{OA}$ and $\overline{OB}$ together and tape them.

Illus. 3.38

Rigorous exercises

8. Can we prove that the lateral edges of a prism are parallel? Discuss.
9. Can we prove that each of the lateral faces of a prism is bounded by a parallelogram? Discuss.
10. Can we prove that the lateral faces of a pyramid are triangular regions?
11. Can any of the lateral edges of a pyramid be parallel? Discuss.

Polyhedra

Pyramids and prisms consist entirely of polygons and are examples of some particular kinds of closed surfaces that are of considerable interest. Any closed surface that consists entirely of polygonal regions is called a *polyhedron*.

It will be recalled that because of the difficulty in finding an adequate definition of *closed curve*, we agreed to accept the general idea on an intuitive basis. A corresponding difficulty exists in defining a *closed surface*. Since that is beyond the scope of this work, we shall leave this to our intuition and not attempt to give a definition. If, however, we accept the notion of a closed surface, we can then define polyhedron.

Definition 3.14 A *polyhedron* is a closed surface consisting entirely of polygonal regions. The polygonal regions are called *faces*.

Any pyramid or prism will serve as an example of a polyhedron, but there are many others, two of which are shown in Illus. 3.39. The vertices of the polygons are also called *vertices* of the polyhedron. The sides of the polygonal regions are also called *edges* of the polyhedron.

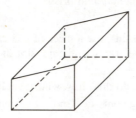 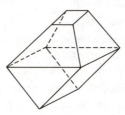

Illus. 3.39

Polyhedra, like polygons, are classified according to the number of boundary elements. In the case of polyhedra, we classify them by the number of faces. The smallest number of faces possible is four, and a polyhedron with four faces is called a *tetrahedron*. Other names are given in the table below.

Name	Number of Faces
Tetrahedron	4
Pentahedron	5
Hexahedron	6
Octahedron	8
Decahedron	10
Dodecahedron	12
Icosahedron	20

It is possible for a polyhedron not to be simple. This is analogous to a non-simple closed curve, but we do not consider the problem of a polyhedron "crossing itself." A *simple* polyhedron, rather, is one of the sort we usually consider, without any holes in it. Such a polyhedron is called a simple one, or it may be called a *simplex*. Illustration 3.40 shows a nonsimple polyhedron, one with a "hole" in it. It is a sort of "polyhedral doughnut." A polyhedron might have several "holes"

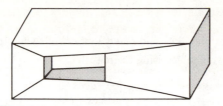

Illus. 3.40

in it. A simplex has the important property that it separates space into three sets: an interior, an exterior, and the polyhedron (boundary) itself.

We shall consider only simple polyhedra here. It is possible for the interior of a polyhedron to be a convex set, but some polyhedra have interiors which are not. A polyhedron might also be called convex when its diagonals are all in its interior (except the vertices). If any part of a diagonal (except its endpoints) is not in the interior, the polyhedron would be called concave. A diagonal, of course, is a segment which joins two vertices and does not lie in any face.

Some polyhedra of special interest are the five *Platonic Solids*, so named because Plato was supposed to have regarded four of them as symbolizing the four "elements": earth, fire, air, and water; the fifth (the dodecahedron) was thought of as the shape which enveloped the entire universe. The five Platonic Solids are in Illus. 3.41. The drawing at the right of each of the polyhedra presents a pattern which can be used to construct a model of each. The polyhedra referred to as the Platonic Solids are *convex regular* polyhedra. A rigorous definition of regular polyhedra (or regular polygons) has not been given. It is sufficient here to think of regular polyhedra as being convex, and consisting only of polygonal regions which have the same size and shape.

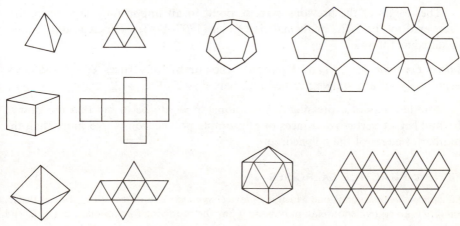

Illus. 3.41

EXERCISES

1. Copy the patterns in Illus. 3.41 and make a model of each of the five Platonic Solids.
2. Complete the following table, using Exercise 1.

	Vertices	Faces	Edges
Tetrahedron	4	4	6
Hexahedron	—	—	—
Octahedron	—	—	—
Dodecahedron	—	—	—
Icosahedron	—	—	—

3. Count the number of vertices, faces, and edges of each of the polyhedra shown in Illus. 3.39. Add your results to the table of Exercise 2.
4. Consider a solid block of wood whose surface is a hexahedron. Suppose one corner is sliced off, forming another face. Find the number of vertices, faces, and edges of this new polyhedron, and add your result to the table of Exercise 2.
5. Consider the block of wood in Exercise 4. If each of the eight corners is sliced off, how many vertices, faces, and edges are there in the new polyhedron? List your result in the table of Exercise 2.
6. Which of the polyhedra in Exercises 1 through 5 are convex polyhedra? Which are regular polyhedra?
7. Study the table you constructed in Exercises 2 through 5. Make a conjecture regarding the relationship between the number of vertices, faces, and edges for each of the polyhedra.

The results of the previous exercise point to an important relationship, first discovered by the Swiss mathematician Euler (1707–1783). This relationship may be stated as follows.

Euler's Formula   If V, E, and F represent the number of vertices, edges, and faces, respectively, of any simple polyhedron, then $V - E + F = 2$.

Another way of expressing this relationship would be to say that the sum of the number of vertices and faces of any simple polyhedron is two more than the number of edges of the polyhedron.

REVIEW PRACTICE EXERCISES

An *equivalence relation* is one which is *reflexive*, *symmetric*, and *transitive*. For example, *equality* is an equivalence relation because it has the following three properties (A, B, and C may refer to sets or numbers):

1. For any A, $A = A$. (reflexive)

2. For any A, B, if $A = B$, then $B = A$. (symmetric)

3. For any A, B, and C, if $A = B$ and $B = C$, then $A = C$. (transitive)

In Exercises 1 through 5, a relation and the set on which it is defined is given. Decide whether the relation is an equivalence relation, and if it is not, tell why.

1. "Is the father of"; males.

2. "Is the same sex as"; people.

3. "Is a multiple of"; natural numbers.

4. "Lives in the same city as"; people.

5. "Has the same birthday as"; people.

6. What is an *order relation*? Give two examples of this type of relation.

JUST FOR FUN

Here is a *dissection* puzzle. Trace the dodecagon and the four segments on another piece of paper. The segments divide the interior into six parts. Cut out the dodecagon and separate the six parts with scissors. Now try to rearrange the parts to form a square.

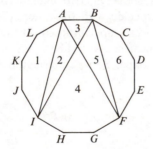

CHAPTER 4

CONGRUENCE

1. INTRODUCTION

The intuitive idea of congruence is quite simple and is the same for any pair of geometric figures. Roughly speaking, two figures are congruent if one of them can be shown to fit or to coincide with the other. One might also describe this by saying that if one figure were superimposed on the other they would coincide. Another way to describe congruence would be to say that two figures are congruent if they are the same size and shape. Some congruent pairs of figures are in Illus. 4.1.

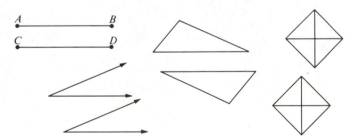

Illus. 4.1

While descriptive words such as "coincide," "fit exactly," and "the same size and shape" may express our intuitive idea of congruence, they cannot very well be used in a satisfactory definition. One might attempt to define congruence by using the idea of superposition. That is, we might say that two figures are congruent if and only if, when one is superimposed on the other, corresponding parts coincide (or fit exactly). In the case of the segment, the endpoints would have to coincide; for triangles, the vertices (and hence the sides) would have to coincide. The difficulty with this approach is that it is based on the assumption that a figure can be moved without changing its internal structure. While this may seem intuitively reasonable, it has long been questioned as a suitable assumption on which the definition of congruence should rest. As a matter of fact, Euclid himself apparently felt a reluctance in using it and did so sparingly in his *Elements*.*

Another approach to the definition of congruence would be to introduce the concept of measurement. In the case of segments, one might say that two segments are congruent if and only if they have the same *length*. Similarly, two angles would be congruent if and only if they have the same angle measure. For triangles, where *both* shape and size are important, one might say they are congruent if and only if their corresponding angles have the same measure and their areas are the same.

While it is possible to treat measurement first and then define congruence in terms of it, our approach will be to develop a theory of congruence first. Congruence will be taken here as undefined.

> Congruence for segments is undefined.
>
> Congruence for angles is undefined.

The basic structure of our system now consists of four undefined concepts: *point*, *between*, *congruence for segments*, and *congruence for angles*. We shall first consider congruence for segments. As we develop the undefined concept of congruence, we shall of course depend on our intuition. In the following exercises the student is expected to utilize his imagination and intuition and to look for some appropriate properties that can be used as axioms.

EXERCISES

Exploratory exercises

1. Suppose we begin with the segment $\overline{AB}$ as shown in Illus. 4.2(a) below. Draw a ray and label the endpoint C. Now open your compass so that one point is on A and the other on B. Then place the point on C and strike an arc on the ray at a point D. Is the segment $\overline{CD}$ congruent to the segment $\overline{AB}$? Use the same procedure to find other segments (with endpoint C) on this ray which are congruent to the segments shown in (b), (c), and (d). Given any segment, could you always find a point X on the ray such that $\overline{CX}$ is congruent to the given segment? Using your compass, repeat the procedure for finding a point D on the ray such that $\overline{CD}$ is congruent to $\overline{AB}$. Does it appear as though you got the same point as the first time? How many points do there appear to be on the ray such that the segment determined by C and that point is congruent to $\overline{AB}$?

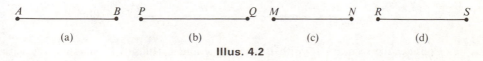

| A | B | P | Q | M | N | R | S |

| (a) | (b) | (c) | (d) |

Illus. 4.2

---

* Compare with *The Thirteen Books of Euclid's Elements*, translation with introduction and commentary by Sir Thomas L. Heath, Dover Publications, New York, 1956, pp. 225–231.

2. In each part of Illus. 4.3 decide which segments are congruent. If you need to, use your compass to help you decide.

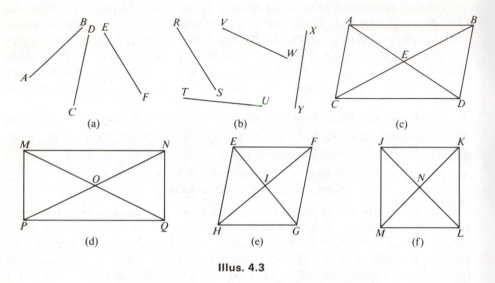

Illus. 4.3

3. When we found the point D on the ray in Exercise 1, we said that $\overline{CD}$ was congruent to $\overline{AB}$. Could we also have said that $\overline{AB}$ was congruent to $\overline{CD}$? Or suppose, for Exercise 2(a), you said that $\overline{AB}$ is congruent to $\overline{EF}$, and someone else said that $\overline{EF}$ is congruent to $\overline{AB}$. Would both statements be true?

4. Would it seem sensible to say that any segment is congruent to itself? Why or why not?

5. Draw a segment $\overline{AB}$. Draw a ray with endpoint C and use your compass to find a point D on the ray such that $\overline{AB}$ is congruent to $\overline{CD}$. Draw another ray with endpoint E and use your compass to find a point F on this ray such that $\overline{CD}$ is congruent to $\overline{EF}$. Now use your compass to find a point G on the second ray (with endpoint E) such that $\overline{AB}$ is congruent to $\overline{EG}$. What can you say about the points F and G? If one segment is congruent to a second and the second is congruent to a third, what relationship would you expect between the first and the third segments?

Illus. 4.4

6. Consider the segment $\overline{AC}$ with B between A and C (Illus. 4.4). Draw a ray with endpoint D. Use your compass to find a point E on the ray such that $\overline{AB}$ is congruent to $\overline{DE}$. Next put the point of the compass on E and find a point F on the ray such that $\overline{BC}$ is congruent to $\overline{EF}$. Now put the point of the compass on D and find a

point G on the ray such that $\overline{AC}$ is congruent to $\overline{DG}$. What is the relationship between the points F and G? What is the relationship between the segments $\overline{AC}$ and $\overline{DF}$?

7. Consider the same segments as in Exercise 6. Draw a ray with endpoint D. Use your compass to find a point E on the ray such that $\overline{AB}$ is congruent to $\overline{DE}$, as before. Next find a point F such that $\overline{AC}$ is congruent to $\overline{DF}$. Is $\overline{BC}$ congruent to $\overline{EF}$? Draw a new ray, with endpoint G. Find a point I on this ray such that $\overline{GI}$ is congruent to $\overline{AC}$. Next find a point H on $\overline{GI}$ such that $\overline{HI}$ is congruent to $\overline{BC}$. Is $\overline{GH}$ congruent to $\overline{AB}$?

2. CONGRUENCE FOR SEGMENTS

We shall begin formally by listing the assumptions which describe the basic properties of congruence for segments. The symbol "$\cong$" will mean "is congruent to" and we will write "$\overline{AB} \cong \overline{CD}$" to say that "$\overline{AB}$ is congruent to $\overline{CD}$."

Axiom 4.1 For any segments $\overline{AB}$, $\overline{CD}$, and $\overline{EF}$
  a) $\overline{AB} \cong \overline{AB}$;
  b) if $\overline{AB} \cong \overline{CD}$, then $\overline{CD} \cong \overline{AB}$;
  c) if $\overline{AB} \cong \overline{CD}$ and $\overline{CD} \cong \overline{EF}$, then $\overline{AB} \cong \overline{EF}$.

The properties described in Axiom 4.1 are called the *reflexive, symmetric,* and *transitive* properties, respectively. Any relation that has these three properties is called an *equivalence relation*. Thus Axiom 4.1 could be stated: Congruence for segments is an equivalence relation. Equality is another example of an equivalence relation.

An important property of an equivalence relation is that it partitions or divides a set into subsets (called equivalence classes) such that *every* element in the set belongs to one and only one subset. In the present example, if we consider the set of all segments in our geometry, the congruence relation for segments divides or partitions this set into subsets, each subset consisting of segments which are all congruent to one another.

Axiom 4.2 Given a segment $\overline{AB}$ and a ray $\overrightarrow{CD}$, there is exactly one (a unique) point E on $\overrightarrow{CD}$ such that $\overline{AB} \cong \overline{CE}$. (Illus. 4.5.)

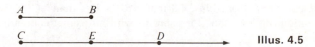

Illus. 4.5

Axiom 4.2 is sometimes referred to as the "compass" axiom.

The next two axioms are often known as the "addition" and "subtraction" axioms for segments. These names are metaphoric, of course, since we have not defined addition or subtraction for segments.

Axiom 4.3 (Segment "Addition") If A-B-C, A'-B'-C', $\overline{AB} \cong \overline{A'B'}$, and $\overline{BC} \cong \overline{B'C'}$, then $\overline{AC} \cong \overline{A'C'}$. (See Illus. 4.6.)

Illus. 4.6

Axiom 4.4 (Segment "Subtraction") If A-B-C, A'-B'-C', $\overline{AB} \cong \overline{A'B'}$, and $\overline{AC} \cong \overline{A'C'}$, then $\overline{BC} \cong \overline{B'C'}$. (See Illus. 4.6.)

EXERCISES

1. Axiom 4.1 states that congruence for segments is an equivalence relation. Consider the relation "is a subset of." Is this an equivalence relation? Explain.

2. Consider the following statement:

   Given a segment $\overline{AB}$ and a point C on a line l, there are exactly two points E and E' on l such that $\overline{AB} \cong \overline{CE}$ and $\overline{AB} \cong \overline{CE'}$.

   How does this statement compare with Axiom 4.2? Could we use this statement in place of Axiom 4.2? See if you can improve the wording of this statement.

Illus. 4.7

3. With reference to Illus. 4.7:
   a) If $\overline{AB} \cong \overline{CD}$, what can you conclude? Why?
   b) If $\overline{AC} \cong \overline{BD}$, what can you conclude? Why?

Illus. 4.8

Exploratory exercise

4. One can "see" from Illus. 4.8 that $\overline{AB}$ "is less than" $\overline{CD}$. See if you can make a definition of "less than" for segments without using the notion of measurement.

Order for Segments

As an immediate application of the congruence relation for segments, we now define the order relations of *less than* and *greater than* for segments. Consider segments $\overline{AB}$ and $\overline{CD}$ in Illus. 4.9. It is clear from the drawing that $\overline{AB}$ is not congruent to $\overline{CD}$; in fact it is not difficult to see that $\overline{AB}$ is "less" than $\overline{CD}$. How does one arrive at such a conclusion? It may be surprising that the conclusion is based on the undefined concepts of *congruence* and *betweenness*.

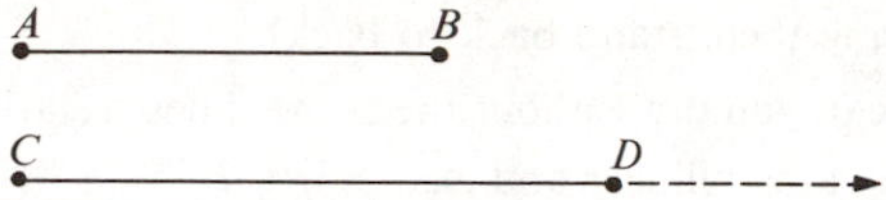

Illus. 4.9

To conclude that the segment $\overline{AB}$ is less than the segment $\overline{CD}$, one might reason that if he "placed" $\overline{AB}$ next to $\overline{CD}$, "lining up" the points A and C, a part of segment CD would be left over. To rephrase this more precisely we would say, "By Axiom 4.2 there is a unique point E on $\overrightarrow{CD}$ such that $\overline{AB} \cong \overline{CE}$, and in this case E is between C and D." On the other hand, we might also say, "By Axiom 4.2 there is a unique point F on $\overrightarrow{AB}$ such that $\overline{CD} \cong \overline{AF}$, and in this case B is between A and F." From this we would conclude that $\overline{CD}$ is greater than $\overline{AB}$. Note that neither conclusion is based on any sort of measurement process, but each rests entirely on the concepts of congruence and betweenness.

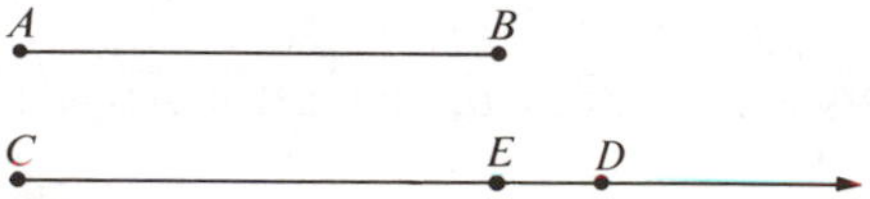

Illus. 4.10

Definition 4.1

a) A segment $\overline{AB}$ is *less than* a segment $\overline{CD}$ if and only if the point E on $\overrightarrow{CD}$ for which $\overline{AB} \cong \overline{CE}$ is between C and D. (Illus. 4.10.)
b) $\overline{AB}$ is *greater than* $\overline{CD}$ if and only if D is between C and E. (Illus. 4.11.)
c) To say that "$\overline{AB}$ is less than $\overline{CD}$" we may write "$\overline{AB} \prec \overline{CD}$." To say that "$\overline{AB}$ is greater than $\overline{CD}$" we may write "$\overline{AB} \succ \overline{CD}$."

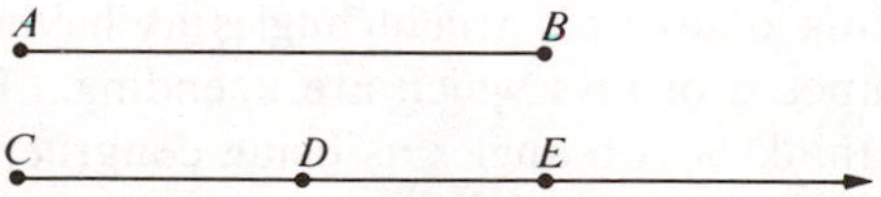

Illus. 4.11

The following theorem is almost an immediate consequence of Definition 4.1. The proof is left as an exercise.

Theorem 4.1 For any segments $\overline{AB}$ and $\overline{CD}$, exactly one of the following is true:

$$\overline{AB} \cong \overline{CD}, \qquad \overline{AB} \prec \overline{CD}, \qquad \text{or} \qquad \overline{AB} \succ \overline{CD}.$$

EXERCISES

👓👓 1. Give three examples from everyday life where the concepts of congruence and betweenness are used to make a comparison (for example, comparing the heights of two people by having them stand back to back).

2. What conclusions can you draw about the betweenness relations among three points, A, B, and C, if they are collinear and $\overline{AB} \succ \overline{AC}$?

3. In Axioms 4.3 and 4.4, replace "A-B-C" with "A, B, and C are any three points." Draw figures to illustrate any conclusions you might make which differ from those given in the axioms.

👓👓 *Rigorous exercises*

4. Prove that for any segments $\overline{AB}$, $\overline{CD}$, and $\overline{EF}$, if $\overline{AB} \cong \overline{CD}$ and $\overline{EF} \cong \overline{CD}$, then $\overline{AB} \cong \overline{EF}$.

5. Prove that if A-B-C-D, A'-B'-C'-D', $\overline{AB} \cong \overline{A'B'}$, $\overline{BC} \cong \overline{B'C'}$, and $\overline{CD} \cong \overline{C'D'}$, then $\overline{AD} \cong \overline{A'D'}$.

6. Prove Theorem 4.1. [*Hint:* For each of the three possibilities show that if this one holds, then the other two cannot.]

7. Prove that for any segments $\overline{AB}$, $\overline{CD}$, and $\overline{EF}$, if $\overline{AB} \prec \overline{CD}$ and $\overline{CD} \cong \overline{EF}$, then $\overline{AB} \prec \overline{EF}$.

8. Prove that for any segments $\overline{AB}$, $\overline{CD}$, and $\overline{EF}$, if $\overline{AB} \prec \overline{CD}$ and $\overline{CD} \prec \overline{EF}$, then $\overline{AB} \prec \overline{EF}$.

9. Is $\prec$ for segments an equivalence relation?

3. CONGRUENCE FOR ANGLES

We observed at the beginning of the chapter that the intuitive notion of congruence is the same for all geometric figures. That is, any figure is congruent to another if they can be shown to coincide in some way. In particular we thought of congruence for segments in terms of being "the same length."

Turning now to congruence for angles, we might observe first that while the general notion of "coincidence" or of being "the same size" is the same, it would be meaningless to think of two congruent angles as having "the same length," since an angle is composed of rays which are unending. Rather, it would seem clearer intuitively to think of two angles as being congruent whenever they have "the same opening." This suggests that congruence for segments and congruence for angles are actually two different relations, and this is indeed the case. Since the general idea is the same, however, we will use the word "congruence" to name both relations. And since it will be clear from the notation whether we are referring to segments or angles, we will also continue to use the symbol "$\cong$" to mean "is congruent to," and write "$\angle A \cong \angle B$" to say that $\angle A$ is congruent to $\angle B$.

As in the case of segments, we shall first explore the intuitive idea of congruence for angles as an initial step in formulating the basic properties we wish for our new relation.

EXERCISES

👁 *Exploratory exercises*

1. Decide which angles are congruent in Exercises 2(c), (d), (e), and (f) (p. 80). Can you use your compass to help you decide?

2. Consider the angle $\angle ABC$ as shown in Illus. 4.12(a) below, where $\overline{AB} \cong \overline{BC}$. Draw a ray and label the endpoint E. Find a point F on the ray such that $\overline{AB} \cong \overline{BC} \cong \overline{EF}$. Place the point of your compass on E and draw an arc through F as shown in Illus. 4.12(b). Next place one point of your compass on C and the other on A. Using this compass opening, place one point on F and draw an arc so that it intersects the arc passing through F, as shown in Illus. 4.12(c). Label the point of intersection D and draw $\overrightarrow{ED}$. Is $\angle DEF \cong \angle ABC$? Follow this procedure for the angles shown in Illus. 4.13. Given any angle and an arbitrary ray, do you think you could always find a second ray with the same endpoint such that the angle would be congruent to the given angle?

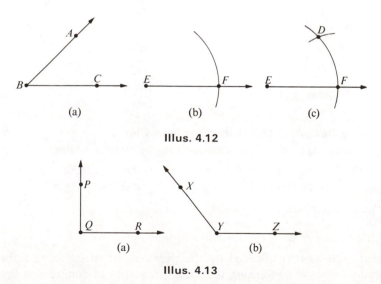

Illus. 4.12

Illus. 4.13

3. Consider $\angle ABC$ and ray $\overrightarrow{PQ}$ as shown in Illus. 4.14. Line $\overleftrightarrow{PQ}$ separates the plane into two half-planes, h_1 and h_2. How many half-lines $\overrightarrow{PX}$ are there in h_1 such that $\angle XPQ \cong \angle ABC$? How many are there in h_2?

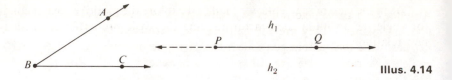

Illus. 4.14

4. a) Could an angle be congruent to itself?

   b) Would every angle be congruent to itself? If so, what kind of relation is angle congruence?

5. If $\angle ABC \cong \angle DEF$ and $\angle DEF \cong \angle GHI$, how are $\angle ABC$ and $\angle GHI$ related?

6. Consider a straight angle $\angle ABC$ and a ray $\overrightarrow{DE}$. How many rays are there with endpoint D such that the angle determined by $\overrightarrow{DE}$ and this ray is congruent to $\angle ABC$?

7. Consider $\angle ABC$ and $\angle CBD$ as shown in Illus. 4.15.

   a) Draw a ray $\overrightarrow{QR}$. Draw a ray $\overrightarrow{QP}$ such that $\angle PQR \cong \angle CBD$.

   b) Draw a half-line $\overset{\circ}{\overrightarrow{QT}}$ in $\overleftrightarrow{QP}/\sim R$ such that $\angle TQP \cong \angle ABC$.

   c) Draw a half-line $\overset{\circ}{\overrightarrow{QX}}$ in $\overleftrightarrow{QR}/T$ such that $\angle XQR \cong \angle ABD$.

   d) What observations can you make?

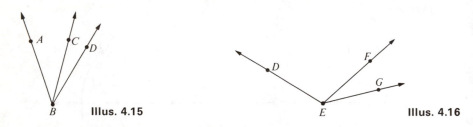

Illus. 4.15   **Illus. 4.16**

8. Consider $\angle DEF$ and $\angle FEG$ as shown in Illus. 4.16.

   a) Draw a ray $\overrightarrow{MN}$. Draw a ray $\overrightarrow{MP}$ such that $\angle NMP \cong \angle FEG$.

   b) Draw a half-line $\overset{\circ}{\overrightarrow{MQ}}$ in $\overleftrightarrow{MP}/N$ such that $\angle QMP \cong \angle DEG$.

   c) Draw a half-line $\overset{\circ}{\overrightarrow{MR}}$ in $\overleftrightarrow{MN}/\sim P$ such that $\angle RMN \cong \angle DEF$.

   d) What observations can you make?

Assumptions about Congruence for Angles

Experience such as that afforded by the previous set of exercises leads to the assumptions we make concerning the basic properties of congruence for angles. We now list them formally.

Axiom 4.5 For any angles $\angle A$, $\angle B$, and $\angle C$

a) $\angle A \cong \angle A$;

b) if $\angle A \cong \angle B$, then $\angle B \cong \angle A$;

c) if $\angle A \cong \angle B$ and $\angle B \cong \angle C$, then $\angle A \cong \angle C$.

Any line on a plane determines two half-planes. A ray on a plane also determines two half-planes. They are the half-planes determined by the line on which the ray lies. The next axiom guarantees the existence of congruent angles. Given an angle and a ray, we can, on either half-plane determined by that ray, find a half-line that determines an angle congruent to the given one.

Axiom 4.6

a) For any nonstraight angle $\angle ABC$ and ray $\overrightarrow{B'C'}$ on the edge of a half-plane h_1, there is a unique ray $\overrightarrow{B'A'}$, with A' in h_1 such that $\angle ABC \cong \angle A'B'C'$. (Illus. 4.17.)

b) For any angles $\angle A$ and $\angle B$, if $\angle A$ is a straight angle, then $\angle A \cong \angle B$ if and only if $\angle B$ is a straight angle.

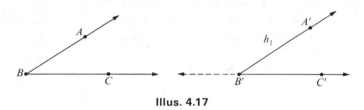

Illus. 4.17

The latter part of this axiom might also be stated: All straight angles are congruent and no nonstraight angle is congruent to a straight angle.

The next two axioms are often called the "addition" and "subtraction" axioms for angles. These names are metaphoric of course, since we have not defined addition or subtraction for angles.

Axiom 4.7 (Angle "Addition") If D is interior to $\angle ABC$ (nonstraight), D' is interior to $\angle A'B'C'$ (nonstraight), $\angle ABD \cong \angle A'B'D'$ and $\angle DBC \cong \angle D'B'C'$, then $\angle ABC \cong \angle A'B'C'$. (Illus. 4.18.)

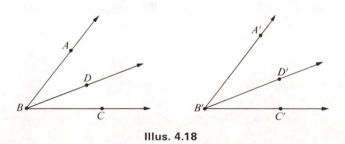

Illus. 4.18

Axiom 4.8 (Angle "Subtraction")

a) If D is interior to $\angle ABC$ (nonstraight), D' is interior to $\angle A'B'C'$ (nonstraight), $\angle ABD \cong \angle A'B'D'$, and $\angle ABC \cong \angle A'B'C'$, then $\angle DBC \cong \angle D'B'C'$. (Illus. 4.18.)

b) For any straight angles, $\angle ABC$ and $\angle A'B'C'$, if D is any point not on $\angle ABC$, D' is any point not on $\angle A'B'C'$, and $\angle ABD \cong \angle A'B'D'$, then $\angle DBC \cong \angle D'B'C'$.

EXERCISES

1. a) Explain in your own words the meaning of each part of Axiom 4.5.
   b) What other way might Axiom 4.5 be stated?

2. What would be wrong with Axiom 4.6(a) if the restriction that $\angle ABC$ is not straight were omitted?

3. a) With regard to Axiom 4.6(a), if h_2 is the half-plane opposite h_1, is there a unique ray $\overrightarrow{B'A''}$ with A'' in h_2 such that $\angle ABC \cong \angle A''B'C'$?
   b) Does the axiom as stated allow for both possibilities?

4. According to Axiom 4.6(b), what can you conclude if $\angle P \cong \angle Q$ and $\angle P$ is not a straight angle?

5. Consider a point X in the plane of an angle $\angle ABC$.
   a) $\angle ABC$ separates the plane into three sets. Name them.
   b) Can the point X be a member of more than one of these sets?
   c) If X is in the interior of $\angle ABC$, what can you conclude about $\overset{\circ\longrightarrow}{BX}$?
   d) If X is in the exterior of $\angle ABC$, what can you conclude about $\overset{\circ\longrightarrow}{BX}$?

Rigorous exercises

6. Prove that if

$$\angle E \cong \angle F \quad \text{and} \quad \angle G \cong \angle F, \quad \text{then} \quad \angle E \cong \angle G.$$

7. In Illus. 4.19, if $\angle ABC \cong \angle DBE$, what other angles are congruent? How do you know?

8. In Illus. 4.19, if $\angle FGI \cong \angle JGH$, what other angles are congruent? How do you know?

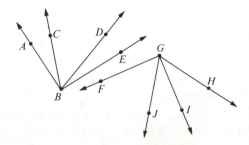

Illus. 4.19

⑨ *Exploratory exercise*

9. One can "see" from Illus. 4.20 that ∠*ABC* is "less than" ∠*DEF*. See if you can make a definition of "less than" for angles without using the notion of measurement.

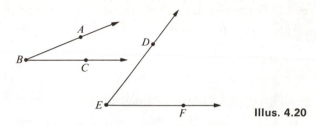

Illus. 4.20

Order for Angles

As with segments, it is possible to define the order relations, *less than* and *greater than*, for angles. To do this, one uses the concept of the interior of an angle.

Definition 4.2 Suppose ∠*ABC* and ∠*DEF* are any nonstraight angles, and consider ∠*GEF*, where $\overleftrightarrow{EG}$ is in the half-plane $\overleftrightarrow{EF}/D$ and ∠*GEF* ≅ ∠*ABC*. Then

a) ∠*ABC* is *less than* ∠*DEF* if and only if $\overrightarrow{EG}$ is in the interior of ∠*DEF*. (Illus. 4.21.)

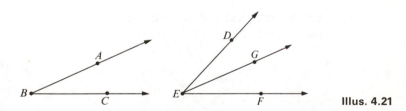

Illus. 4.21

b) ∠*ABC* is *greater than* ∠*DEF* if and only if $\overrightarrow{EG}$ is in the exterior of ∠*DEF*. (Illus. 4.22.)

c) If ∠*ABC* is not straight and ∠*DEF* is straight, then ∠*ABC* is less than ∠*DEF*. As for *less than* and *greater than* with segments, we will denote "∠*A* less than ∠*B*" as "∠*A* < ∠*B*" and "∠*A* greater than ∠*B*" as "∠*A* > ∠*B*."

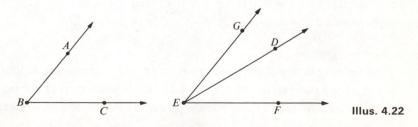

Illus. 4.22

As for segments, given any two angles, one must be less than the other or else they are congruent. This is stated in the following theorem. The proof is left as an exercise.

Theorem 4.2 For any angles $\angle ABC$ and $\angle A'B'C'$, exactly one of the following is true: $\angle ABC \cong \angle A'B'C'$, $\angle ABC < \angle A'B'C'$, or $\angle ABC > \angle A'B'C'$.

EXERCISES

1. Is $<$ for angles an equivalence relation? Why?

2. In Illus. 4.23, if $\angle ABC$ and $\angle FGH$ are straight angles and $\angle DBC \cong \angle FGE$, what other angles are congruent? How do you know?

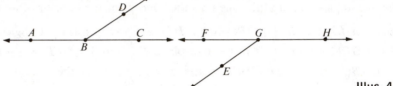

Illus. 4.23

3. In Illus. 4.24, if $\angle ONP$ and $\angle RSU$ are straight angles and $\angle MNP \cong \angle RST$, what other angles are congruent? How do you know?

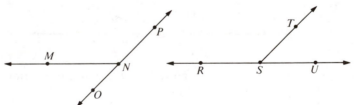

Illus. 4.24

Rigorous exercises

4. Prove Theorem 4.2.

5. Prove that if $\angle A < \angle B$ and $\angle B \cong \angle C$, then $\angle A < \angle C$.

6. Prove that if $\angle A < \angle B$ and $\angle B < \angle C$, then $\angle A < \angle C$.

Exploratory exercises

7. Sometimes two angles share a common ray or side. For example, in Illus. 4.25(a) the ray $\overrightarrow{BC}$ is a common side of $\angle 1$ and $\angle 2$. Decide which pairs of numbered angles in (b), (c), and (d) have a common side. If there is a common side, name it.

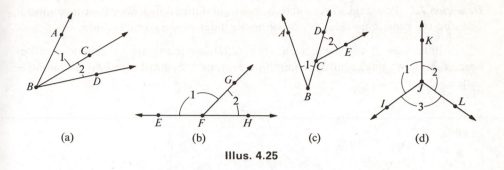

(a)   (b)   (c)   (d)

Illus. 4.25

8. a) Each of the common sides in Exercise 7 determines two half-planes. In each case, how are the remaining sides situated with respect to the common side?

   b) Consider ∠*ABD* and ∠*CBD* in Illus. 4.26. Name the common side. How are the remaining sides situated with respect to the common side?

            Illus. 4.26   **Illus. 4.27**

9. Consider two distinct lines $\overleftrightarrow{AC}$ and $\overleftrightarrow{DE}$ which intersect at a point *B* (Illus. 4.27).

   a) How many rays are there with endpoint *B*?

   b) Name the nonstraight angles determined by the intersecting lines.

   c) Name a pair of nonstraight angles which do not have a common side. How many such pairs are there?

   d) How are the angles in each of the pairs in (c) related? Can you support your conjecture with reasons?

More about Angles

At the beginning of this section we introduced an undefined congruence relation for angles and accepted a set of assumptions which describe the basic properties of this relation. Order for angles was then defined in terms of these assumptions and other previously accepted concepts. We now turn our attention to further relationships among angles.

In the previous exercises we saw that two angles may share a common side and in some cases the remaining sides are on opposite sides (or in opposite half-planes, except for the vertex) of the common side. Two such angles are said to be *adjacent*.

Definition 4.3 Two angles are said to be *adjacent* if and only if they have a common side and the remaining sides are on opposite sides of the common side.

In Illus. 4.28 $\angle ABC$ is adjacent to $\angle CBD$ in each figure. It should also be noted that if two angles have a common side, then they must also have a common vertex.

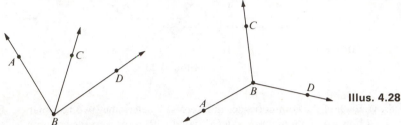

Illus. 4.28

If two distinct lines intersect, they determine four nonstraight angles. Any two of these angles that are not adjacent are called *vertical angles*. In Illus. 4.29 $\angle ABC$ and $\angle DBE$ are vertical angles; $\angle ABD$ and $\angle CBE$ are also vertical angles.

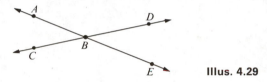

Illus. 4.29

Definition 4.4 Two angles, $\angle ABC$ and $\angle DBE$, are *vertical angles* if and only if C-B-D and A-B-E, or A-B-D and C-B-E.

The next theorem is almost an immediate consequence of the axioms and definitions given in this section.

Theorem 4.3 Vertical angles are congruent.

Proof. Consider the vertical angles $\angle ABE$ and $\angle CBD$ in Illus. 4.30 below. By definition of vertical angles, A-B-D and E-B-C, and hence $\angle ABD$ and $\angle EBC$ are straight angles. Since $\angle ABC \cong \angle ABC$ by Axiom 4.5, it follows from Axiom 4.8 that $\angle ABE \cong \angle CBD$. Thus vertical angles are congruent.

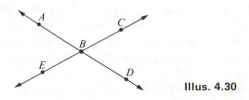

Illus. 4.30

If two angles are adjacent and the sides opposite the common side are opposite rays of a line, the angles are said to be *supplementary*. Further, any angle which is congruent to one of these angles is supplementary to the other.   More precisely we define supplementary angles as follows.

Illus. 4.31

Definition 4.5   ∠*A* is *supplementary* to ∠*BCD* if and only if ∠*A* ≅ ∠*DCE*, where *E* is a point such that *B-C-E*. (See Illus. 4.31.)

It should also be noted that ∠*BCD* is supplementary to ∠*A* (*supplementary* is a symmetric relation).   In summary, any angle which is congruent to ∠*BCD* is supplementary to any angle which is congruent to ∠*DCE*.

The next theorem follows easily from Definition 4.5 and the axioms of this section.   The proof is left as an exercise.

Theorem 4.4   If two angles are congruent, then their supplements are also congruent.

Using the concept of congruence and Definition 4.5, we can now state what we mean by a *right angle*.

Definition 4.6   A *right angle* is an angle that is congruent to every angle that is supplementary to it.

Definition 4.7   Two lines are *perpendicular* to each other if and only if they intersect and at least one angle determined by the intersection is a right angle.   (See Exercise 7 ff.)

We shall often wish to speak of perpendicular segments, rays, etc.   In Illus. 4.32 the figures of each pair are perpendicular.

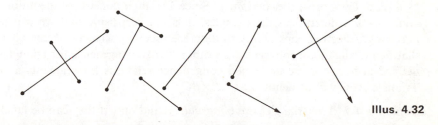

Illus. 4.32

After studying the preceding drawings, the reader may be able to give a definition like the following.   Completion of the definition is left as an exercise.

Definition 4.8   Any two figures of the following kinds, segments, rays, half-lines, and lines, are *perpendicular* if and only if . . .

EXERCISES

1. In Illus. 4.33, $\angle ADB \cong \angle EDF$. What other angles are congruent? How do we know from Definition 4.4 that $\angle ADB$ and $\angle EDF$ are not vertical angles?

2. Name two supplementary angles in Exercise 1 that are not adjacent angles.

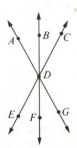

Illus. 4.33

3. Complete Definition 4.8.

4. According to Definition 4.6, if $\angle A$ is supplementary and congruent to $\angle B$, then $\angle A$ is a right angle. Is $\angle B$ also a right angle? Support your answer with reasons.

Rigorous exercises

5. Prove Theorem 4.4.

6. Prove that any right angle is supplementary to itself.

7. Prove that if two distinct lines intersect such that one of the angles determined by the intersection is a right angle, then the other three are also right angles.

4. CONGRUENCE FOR TRIANGLES

Thus far we have considered two kinds of congruence relations, one for segments and one for angles. Both are undefined relations and both have been described by a set of basic properties (axioms). Since a triangle consists of the union of three segments and determines three angles, it is not surprising that we can use these two congruence relations jointly to describe precisely what we mean when we say that two triangles are congruent. In other words, congruence for triangles will be defined in terms of the undefined congruence relations for segments and angles. Triangle ABC will be denoted as " $\triangle ABC$."

Definition 4.9 Two triangles are congruent if and only if they can be labeled ABC and $A'B'C'$ such that $\overline{AB} \cong \overline{A'B'}$, $\overline{BC} \cong \overline{B'C'}$, $\overline{CA} \cong \overline{C'A'}$, $\angle A \cong \angle A'$, $\angle B \cong \angle B'$, $\angle C \cong \angle C'$. We write " $\triangle ABC \cong \triangle A'B'C'$" to say that $\triangle ABC$ is congruent to $\triangle A'B'C'$.

Usually the notation is chosen so that one can tell which pairs of angles and which pairs of sides are congruent. For example, if $\triangle RST \cong \triangle XYZ$, we know that $\angle R \cong \angle X$, $\angle S \cong \angle Y$, $\angle T \cong \angle Z$, $\overline{RS} \cong \overline{XY}$, $\overline{ST} \cong \overline{YZ}$, and $\overline{TR} \cong \overline{ZX}$. It should be noted that Definition 4.9 does not specify that the two triangles are different. Thus, it is possible that points A', B', and C' are the same points (in some order) as A, B, and C. Having noted this, it is easy to prove the following theorem.

Theorem 4.5 Congruence for triangles is an equivalence relation.

Occasionally it is helpful to use marks in a drawing to indicate which pairs of segments or angles are congruent to each other. For example, in Illus. 4.34, the single dash indicates that $\overline{RS} \cong \overline{XY}$, the double dash, that $\overline{ST} \cong \overline{YZ}$, etc. For angles we will use "arcs" in the same way; for example, the single "arc" indicates that $\angle R \cong \angle X$, etc.

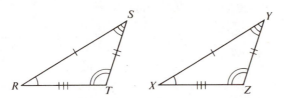

Illus. 4.34

EXERCISES

1. Decide which triangles are congruent in Exercise 2, Illus. 4.3, parts (c), (d), (e), and (f), page 80.
2. How many congruent pairs of sides are there in a pair of congruent triangles? How many congruent pairs of angles are there in a pair of congruent triangles?
3. There is at least one pair of congruent triangles in each of the following figures. Identify this congruence, using the proper notation. For example, in Illus. 4.35(a) $\triangle ABD \cong \triangle FEC$.

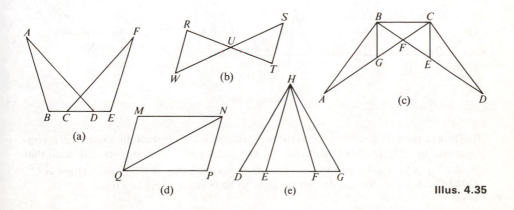

Illus. 4.35

👀 *Rigorous exercise*

    4. Prove Theorem 4.5.

👀 *Exploratory exercises*

    5. In the pairs of triangles in Illus. 4.36 some pairs of segments or angles are identified as congruent. Using a compass, decide which of the remaining pairs are congruent.

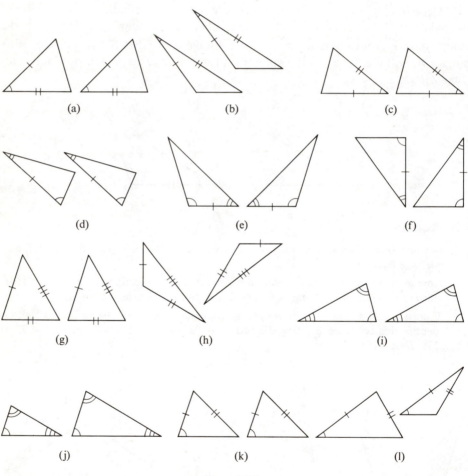

Illus. 4.36

    6. Draw a triangle and label the vertices A, B, and C. Construct an angle that is congruent to $\angle A$, and label the vertex A'. On one ray of $\angle A'$ find a point B' such that $\overline{A'B'} \cong \overline{AB}$. Find a point C' on the other ray such that $\overline{A'C'} \cong \overline{AC}$. Draw $\overline{B'C'}$. What other congruences do there appear to be between these two triangles?

The Side-Angle-Side Axiom

The reader may have noted in the preceding exercise that whenever certain combinations of congruent pairs occurred, the remaining pairs were also congruent. For example, when two sides and the included angle of one triangle were congruent, respectively, to two sides and the included angle of the other, the remaining pairs were also congruent. On the other hand, if the three pairs of angles were congruent, the remaining pairs (the three pairs of segments) were not necessarily congruent.

These observations, although few, do provide us with a conjecture about a possible relationship between congruence for segments and congruence for angles. While further experimentation might help to strengthen our convictions with regard to such a conjecture, it would be impossible for us to prove it. We shall therefore assume one of these relationships and proceed from there.

Axiom 4.9 (SAS) Given $\triangle ABC$ and $\triangle A'B'C'$, if $\overline{AB} \cong \overline{A'B'}$, $\overline{AC} \cong \overline{A'C'}$, and $\angle A \cong \angle A'$, then $\triangle ABC \cong \triangle A'B'C'$.

This assumption says that if two sides and the "included" angle (that is, the angle determined by the two sides in question) of one triangle are congruent to the corresponding two sides and the included angle of another triangle, then the remaining pairs of angles and segments are also congruent. This of course is, by Definition 4.9, the same as concluding that the two triangles are congruent. We shall refer to this as the Side-Angle-Side Axiom, or simply SAS.

A Basic Congruence Theorem

In the previous exercises the reader may have noted other combinations of congruent pairs that seemed to provide sufficient information for concluding that the triangles were congruent. For example, whenever two angles and the included side (that is, the side that lies on a ray of each of the angles in question) of one triangle were congruent to two angles and the included side of another, the remaining pairs were congruent and hence the triangles were congruent. We can now show that if we have the SAS axiom, then the relationship just described must also hold.

Theorem 4.6 (ASA) If, for $\triangle ABC$ and $\triangle A'B'C'$, $\angle A \cong \angle A'$, $\angle C \cong C'$, and $\overline{AC} \cong \overline{A'C'}$, then $\triangle ABC \cong \triangle A'B'C'$.

Outline of proof. If we can show that $\overline{AB} \cong \overline{A'B'}$, then by the SAS axiom (4.9), we can conclude that $\triangle ABC \cong \triangle A'B'C'$. Our approach to the proof will be as follows. We know there is a unique point D' on the ray $\overrightarrow{A'B'}$ such that $\overline{AB} \cong \overline{A'D'}$. We will show that D' and B' are really the same point (that is, $D' = B'$) and thus $\overline{AB} \cong \overline{A'B'}$. Then, by the SAS assumption, $\triangle ABC \cong \triangle A'B'C'$. In the proof below, supporting reasons have been omitted and are left as an exercise for the student.

Suppose D' is the unique point on the ray $\overrightarrow{A'B'}$ such that $\overline{AB} \cong \overline{A'D'}$ (see Illus. 4.37).   Since $\angle A \cong \angle A'$ and $\overline{AC} \cong \overline{A'C'}$, then $\triangle ABC \cong \triangle A'D'C'$.   Thus $\angle C \cong \angle D'C'A'$.   But, since $\angle B'C'A' \cong \angle C$, then $\angle B'C'A' \cong \angle D'C'A'$, and therefore $\overrightarrow{C'D'} = \overrightarrow{C'B'}$.   Hence $D' = B'$ and $\overline{AB} \cong \overline{A'B'}$.   It therefore follows that $\triangle ABC \cong \triangle A'B'C'$.

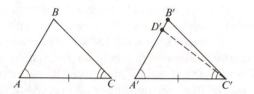

Illus. 4.37

It should be noted that D' could have been represented on either side of B' on the ray $\overrightarrow{A'B'}$ in Illus. 4.37.   The important thing to keep in mind is that D' represents that *unique* point on the ray $\overrightarrow{A'B'}$ such that $\overline{AB} \cong \overline{A'D'}$, and although we suspect that D' will turn out to be the same point as B', we do not know where it is initially; we know merely that it does exist and is unique.

EXERCISES

Rigorous exercises

1. Complete the proof of Theorem 4.6 by supplying the supporting reasons for each assertion.

2. Prove that if $\overleftrightarrow{CD}$ and $\overleftrightarrow{AE}$ intersect at B such that A-B-E, C-B-D, $\overline{AB} \cong \overline{BE}$, and $\overline{CB} \cong \overline{BD}$, then $\triangle ABC \cong \triangle EBD$ (Illus. 4.38).

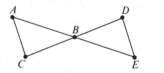

Illus. 4.38

3. Prove that if $\angle ADC \cong \angle CBA$ and $\angle ABD \cong \angle CDB$, then $\triangle DAB \cong \triangle BCD$ (Illus. 4.39).

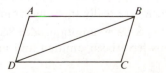

Illus. 4.39

4. Prove that if $\overline{EF} \cong \overline{EH}$, $\overline{FG} \cong \overline{GH}$, $\angle DFG \cong \angle CHG$, F-G-H, E-F-D, and E-H-C, then $\angle EGF$ and $\angle EGH$ are right angles (Illus. 4.40).

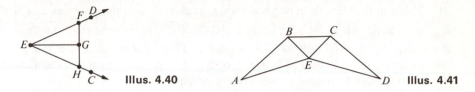

Illus. 4.40   Illus. 4.41

5. Prove that if $\overline{BE} \cong \overline{CE}$, $\overline{AB} \cong \overline{CD}$, and $\angle ABE \cong \angle DCE$, then $\angle CEA \cong \angle BED$ (Illus. 4.41).
6. Prove that if $\overline{AB} \cong \overline{CD}$, $\angle A \cong \angle D$, and $\angle ACF \cong \angle DBE$, then $\triangle AFC \cong \triangle DEB$ (Illus. 4.42).

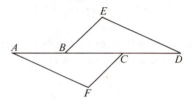

Illus. 4.42

7. When the SAS axiom was added to our list of formal assumptions, we actually said more than was necessary. It would have been possible, for example, to have listed the following axiom instead, in which we conclude only that the two remaining pairs of angles are congruent.

Given $\triangle ABC$ and $\triangle A'B'C'$, if $\overline{AB} \cong \overline{A'B'}$, $\overline{AC} \cong \overline{A'C'}$, and $\angle A \cong \angle A'$, then $\angle B \cong \angle B'$ and $\angle C \cong \angle C'$.

Axiom 4.9 would then be called the SAS theorem, since it can be proved, using the axiom above. Prove the "SAS theorem" (Axiom 4.9), using the axiom above.

Another Basic Congruence Theorem

The reader may have noted from the exercises on pages 95 and 96 that in each case, if the three pairs of corresponding sides were congruent, the triangles were congruent. This relationship is referred to as the "Side-Side-Side" Theorem (SSS).

Before considering the proof of the SSS theorem, we will need to prove a theorem about a special kind of triangle known as an *isosceles* triangle. We begin with the following definition.

Definition 4.10 *An isosceles* triangle is a triangle with at least two congruent sides. Given any two congruent sides, the remaining side is called the *base*, and the angles opposite the congruent sides are called the *base angles*.

Theorem 4.7 The base angles of an isosceles triangle are congruent.

Outline of proof. Consider the isosceles triangle shown in Illus. 4.43, where $\overline{AB} \cong \overline{AC}$. Think of the triangle in two ways: one, as $\triangle ABC$, and the other, as $\triangle A'B'C'$. By hypothesis, $\overline{AB} \cong \overline{AC}$, and since $\overline{AC} \cong \overline{A'B'}$, it follows that $\overline{AB} \cong \overline{A'B'}$. Similarly, $\overline{AC} \cong \overline{A'C'}$. Also $\angle A \cong \angle A'$. Thus $\triangle ABC \cong \triangle A'B'C'$, and it follows that $\angle B \cong \angle B'$ and hence $\angle B \cong \angle C$. We have now proved that if a triangle is isosceles, its base angles are congruent.

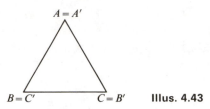

Illus. 4.43

We shall now prove the SSS theorem. There are three cases to be considered. The first case is proved below, and the proofs of the second and the third case are left as an exercise.

Theorem 4.8 (SSS) If, for $\triangle ABC$ and $\triangle A'B'C'$, $\overline{AB} \cong \overline{A'B'}$, $\overline{AC} \cong \overline{A'C'}$, and $\overline{BC} \cong \overline{B'C'}$, then $\triangle ABC \cong \triangle A'B'C'$.

Proof. CASE 1 (see Illus. 4.44). Let us begin by outlining the general procedure of the proof. Consider $\triangle ABC$ and $\triangle A'B'C'$ as shown in Illus. 4.44(a) and (b). By hypothesis, $\overline{AB} \cong \overline{A'B'}$, $\overline{AC} \cong \overline{A'C'}$, and $\overline{BC} \cong \overline{B'C'}$. If we can show that one pair of corresponding angles are congruent, then by the SAS Axiom we can conclude that the triangles are congruent. We consider $\triangle ABC$ and $\triangle ADC$ in Illus. 4.44(c), where $\triangle ADC \cong \triangle A'B'C'$. We then proceed to show that $\angle ABC \cong \angle ADC$, which, by the SAS axiom, would mean that $\triangle ABC \cong \triangle ADC$. And since $\triangle ADC \cong \triangle A'B'C'$, we therefore conclude that $\triangle ABC \cong \triangle A'B'C'$.

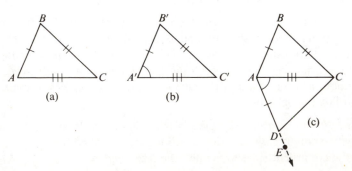

(a) (b) (c)

Illus. 4.44

By Axiom 4.6 there exists a unique ray $\overrightarrow{AE}$ on $\overleftrightarrow{AC}/\sim B$ such that $\angle EAC \cong \angle A'$.
From Axiom 4.2 ("compass" axiom) we know there exists a unique point D on
$\overrightarrow{AE}$ such that $\overline{AD} \cong \overline{A'B'}$. Since, by hypothesis, $\overline{AC} \cong \overline{A'C'}$, it follows from the
SAS axiom that $\triangle ADC \cong \triangle A'B'C'$.

Now consider the segment $\overline{BD}$, $\triangle ABD$, and $\triangle CBD$, as shown in Illus. 4.45.
Since $\triangle ADC \cong \triangle A'B'C'$, $\overline{AD} \cong \overline{A'B'}$, and by hypothesis $\overline{A'B'} \cong \overline{AB}$. It therefore follows from Axiom 4.1(c) that $\overline{AD} \cong \overline{AB}$. Similarly, $\overline{DC} \cong \overline{BC}$. We now
know (Definition 4.10) that $\triangle ABD$ and $\triangle CBD$ are isosceles triangles. From
Theorem 4.7 it follows that $\angle ABD \cong \angle ADB$ and $\angle CBD \cong \angle CDB$, which
allows us to conclude (Axiom 4.7) that $\angle ABC \cong \angle ADC$. By SAS we conclude
that $\triangle ABC \cong \triangle ADC$. But since we already know that $\triangle ADC \cong \triangle A'B'C'$, it
follows from Theorem 4.5 that $\triangle ABC \cong \triangle A'B'C'$.

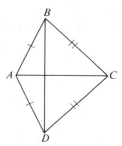

Illus. 4.45

CASE 2. In the first case, $\overline{BD}$ intersects $\overline{AC}$ at an interior point. It is possible,
of course, that $\overline{BD}$ may not intersect $\overline{AC}$ at all, as shown in Illus. 4.46. The proof
is almost identical to the one for Case 1.

CASE 3. In the third case, $\overline{BD}$ intersects $\overline{AC}$ at A, as shown in Illus. 4.47.

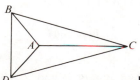   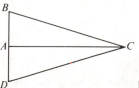

Illus. 4.46   **Illus. 4.47**

We now have three sets of conditions, SAS, ASA, and SSS, which are sufficient
for concluding that two triangles are congruent. More precisely, we have shown
that *if* SAS is a sufficient condition for concluding that two triangles are congruent,
then ASA and SSS are also sufficient conditions for concluding that two triangles
are congruent. The former student of high school geometry may recall using these
conditions to construct a triangle congruent to a given triangle.

The reader may have noted that there are six possible combinations of congruent conditions associated with any two triangles, the other three being AAA, SSA, and AAS. Exercise 5, page 96, provided examples to show that AAA and SSA are *not* sufficient to conclude that two triangles are congruent. The remaining combination, AAS, while not so intuitively obvious as the first three, is also a sufficient condition for concluding that two triangles are congruent. Although we shall state this theorem now, the proof must be postponed until later since it depends on some concepts not yet developed.

Theorem 4.9 (AAS) If, for $\triangle ABC$ and $\triangle A'B'C'$, $\angle BAC \cong \angle B'A'C'$, $\angle ABC \cong \angle A'B'C'$, and $\overline{BC} \cong \overline{B'C'}$, then $\triangle ABC \cong \triangle A'B'C'$.

EXERCISES

1. State the converses of Theorems 4.6 and 4.8. Which ones are true?

2. Draw two triangles that are not congruent but whose corresponding angles are congruent.

3. Which of the following congruence conditions are sufficient for concluding that two triangles are congruent?

   a) SAA   b) SSA   c) AAA

   d) ASA   e) SSS   f) AAS

Rigorous exercises

4. Prove that if $\overline{AB} \cong \overline{AC}$ and $\overline{DB} \cong \overline{DC}$, then $\triangle ABD \cong \triangle ACD$ (Illus. 4.48).

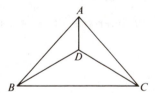

Illus. 4.48

5. Prove that if $\overline{AB} \cong \overline{CD}$ and $\overline{AD} \cong \overline{BC}$, then $\angle DAB \cong \angle BCD$ and $\angle ADC \cong \angle ABC$ (Illus. 4.49).

Illus. 4.49

6. Prove that if in $\triangle TRV$ and $\triangle TSU$, $\overline{RS} \cong \overline{UV}$, $\overline{ST} \cong \overline{TU}$, $\overline{SX} \cong \overline{XU}$, $T\text{-}X\text{-}W$, $R\text{-}W\text{-}V$, and $S\text{-}X\text{-}U$, then $\angle TWV$ is a right angle (Illus. 4.50).

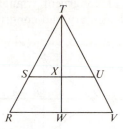

Illus. 4.50

7. Prove that if $\overline{BE} \cong \overline{BD}$, $\angle A \cong \angle C$, and $\angle AED \cong \angle CDE$, then $\triangle ABE \cong \triangle CBD$ (Illus. 4.51).

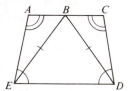

Illus. 4.51

8. Prove that if, in $\triangle ABC$, $\overline{AB} \cong \overline{BC} \cong \overline{CA}$, then $\angle A \cong \angle B \cong \angle C$.

9. Complete the proof of Theorem 4.7.

10. State the converse of Theorem 4.7 and prove it.

11. Prove Case II of Theorem 4.8.

12. Prove Case III of Theorem 4.8.

5. MORE ABOUT RIGHT ANGLES AND PERPENDICULARS

Since right angles have been defined, it may seem reasonable to conclude that they exist. From an intuitive standpoint, it may not seem sensible to define something that doesn't exist. The fact is, however, that a definition in no way establishes the existence of the thing defined. For example, segments exist in our geometry because of the axioms we formulated in Chapter 2 and not because of the definition of *segment* itself.

We will now show that right angles do exist in our geometry. It will be noted that the proof of their existence depends in part on the SAS axiom.

Theorem 4.10 Right angles exist.

Outline of proof. Consider a line l, a point A on l, and a point P not on l. Choose an arbitrary point O on l, determining $\overrightarrow{OP}$ and $\angle POA$. There exists a point Q

on the opposite side of *l* such that $\angle QOA \cong \angle POA$ and $\overline{OP} \cong \overline{OQ}$ (Illus. 4.52). *PQ* intersects *l* at a point *B*, and *P-B-Q*. If $B \neq O$, then we have $\triangle PBO \cong \triangle QBO$. By the definitions of congruent triangles and supplementary angles, $\angle PBO$ and $\angle QBO$ are supplementary and congruent, and thus are right angles.

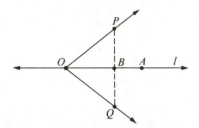

Illus. 4.52

Since *P* was an arbitrary point, there are two other possibilities, as shown in Illus. 4.53. The proof of each of these cases is left as an exercise.

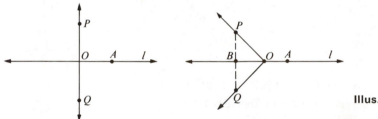

Illus. 4.53

It will be recalled that we assumed all straight angles are congruent and that no nonstraight angle is congruent to a straight angle. It is apparently true, analogously, that all right angles are congruent and no nonright angle is congruent to a right angle. We shall outline a proof of that fact now.*

Theorem 4.11 All right angles are congruent; no nonright angle is congruent to a right angle.

Outline of proof. Consider two right angles $\angle A$ and $\angle B$, with adjacent supplements $\angle A'$ and $\angle B'$, respectively [Illus. 4.54(a)]. By Theorem 4.2, $\angle A$ is either less than, greater than, or congruent to $\angle B$. Let us assume that $\angle A \prec \angle B$. Then, if $\angle B'' \cong \angle B$ and $\angle B'''$ is an adjacent supplement of $\angle B''$, we have the situation shown in Illus. 4.54(b). Now, $\angle B' \cong \angle B'''$ and also, $\angle B \cong \angle B'''$. Thus $\angle A \prec \angle B'''$. (Why?) But $\angle B''' \prec \angle A'$. (Why?) Therefore $\angle B'' \prec \angle A$ and $\angle A \prec \angle A$. (Why?) Thus $\angle A \prec \angle B$ is not possible. By a similar argument,

---

* In Euclid's development this was an axiom. Moreover, he also assumed that no nonright angle is congruent to a right angle but never stated that fact, either as an axiom or theorem.

$\angle A > \angle B$ is not possible. The only remaining possibility is that $\angle A \cong \angle B$, which was to be shown.

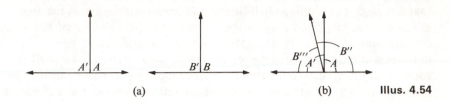

(a)   (b)   **Illus. 4.54**

Suppose $\angle TAR$ is congruent to a right angle, $\angle BPQ$ (Illus. 4.55). Then it is also congruent to $\angle CPQ$. Since congruent angles have congruent supplements, $\angle TAR \cong \angle TAS$, and therefore $\angle TAR$ is a right angle. It follows (the contrapositive) that if an angle is not a right angle, then it is not congruent to a right angle.

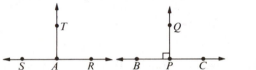

Illus. 4.55

EXERCISES

Rigorous exercises

1. Prove the two remaining cases of Theorem 4.10.

2. Write a complete proof of Theorem 4.11.

Exploratory exercises

3. a) Draw a segment $\overline{AB}$ and, with a compass and straightedge, bisect it to find its midpoint P.

   b) Can this be done for any segment?

   c) Does a segment have more than one midpoint?

4. a) Draw an acute angle, and construct its bisector.

   b) Draw an obtuse angle and construct its bisector.

   c) Draw a right angle and construct its bisector.

   d) Does any nonstraight angle have a bisector?

   e) Does an angle have more than one bisector?

6. BISECTORS OF SEGMENTS AND ANGLES

It seems fairly obvious, intuitively, that every segment has a midpoint. In fact it seems clear that a segment has one and *only one* midpoint. Similarly, it appears that any angle has a unique half-line in its interior which bisects the angle. These are important facts, not only in themselves, but also to some of the succeeding development. Thus it is important that we consider them here. The logical development consists of a series of theorems, some of which we shall prove and some of which we shall not. Proofs of all of them are possible and may be found in other books,* but some of the proofs are rather involved and will not be given here.

The first of these theorems says that if we have a segment joining two points on the sides of an angle, as in Illus. 4.56, then any half-line from the vertex that is in the interior of the angle must intersect that segment. This theorem is sometimes known as the "crossbar theorem." We state it here without proof.

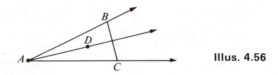

Illus. 4.56

Theorem 4.12 ("Crossbar Theorem") If a point D is in the interior of $\angle BAC$, then $\overset{\circ}{\overrightarrow{AD}}$ intersects $\overline{BC}$.

It is now possible to prove an important theorem about perpendiculars, which says in effect that from a point on a line there is a unique perpendicular to that line. This assumes, of course, that we are considering only geometry on a plane. The theorem is stated more precisely as follows. The proof is left as an exercise.

Theorem 4.13 Given a line l on a plane α and a point P on l, there is one and only one line m on α which contains P and is perpendicular to l.

The next theorem concerns exterior angles of a triangle. We shall outline a proof of it, but first we shall need a definition of *exterior* angle.

Definition 4.11 An angle that is adjacent and supplementary to an angle of a triangle is called an *exterior angle* of the triangle.

The two angles of the triangle not adjacent to the exterior angle are called *opposite interior angles*.

Theorem 4.14 (Exterior Angle Theorem) Any exterior angle of a triangle is greater than either of the opposite interior angles.

---

* Compare with M. L. Keedy *et al.*, *Exploring Geometry*, New York: Holt, Rinehart, & Winston, 1967.

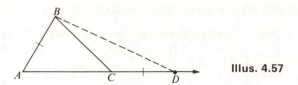

Illus. 4.57

Outline of proof. Consider any triangle △*ABC*, and an exterior angle ∠*BCD*, where *A-C-D* (Illus. 4.57). We shall show that ∠*BCD* ≇ ∠*ABC* and ∠*BCD* ⊀ ∠*ABC* and hence that ∠*BCD* ≻ ∠*ABC*. On $\overrightarrow{AC}$ there is a point *D* such that *A-C-D* and $\overline{CD}$ ≅ $\overline{AB}$. Now suppose that ∠*BCD* ≅ ∠*ABC*. Then, △*ABC* ≅ △*DCB* and ∠*CBD* ≅ ∠*ACB*. But this means that ∠*ABD* is a straight angle, whence *A*, *B*, and *D* are collinear. This is impossible, however, for it would mean that two distinct lines ($\overleftrightarrow{AB}$ and $\overleftrightarrow{AC}$) intersect in the two points, *A* and *D*. Thus ∠*BCD* ≇ ∠*ABC*.

If ∠*BCD* were less than ∠*ABC*, there would be a half-line $\overset{\circ}{\overrightarrow{BE}}$ in the interior of ∠*ABC* such that ∠*EBC* ≅ ∠*BCD*. Now $\overset{\circ}{\overrightarrow{BE}}$ meets $\overline{AC}$ at some interior point *F*, by Theorem 4.12. Thus we have △*FBC* with an interior angle congruent to an exterior angle, which we have just shown is impossible.

Since an exterior angle is not congruent to or less than the opposite interior angle, it must be greater. In other words, ∠*BCD* ≻ ∠*ABC* (Illus. 4.58).

A similar proof can be given for the other opposite interior angle and also for the other exterior angle at *C*.

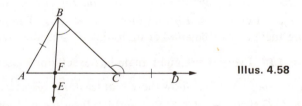

Illus. 4.58

Now that we have this theorem, we find it fairly easy to prove Theorem 4.9. The proof will be left to the reader. We are also in a position to prove that every segment has a unique midpoint and that every angle has a unique bisector. We shall first make the necessary definitions and then outline the proofs of the theorems.

Definition 4.12 A point *C* of a segment $\overline{AB}$ such that $\overline{AC}$ ≅ $\overline{CB}$ is called a *midpoint* (or *bisector*) of $\overline{AB}$.

Definition 4.13 A half-line $\overrightarrow{BD}$ interior to an angle ∠*ABC* such that ∠*ABD* ≅ ∠*DBC* is called a *bisector* of ∠*ABC*.

Theorem 4.15 Every segment has one and only one midpoint.

Outline of proof. We first show there is at least one midpoint. Consider a segment $\overline{AB}$. There exist points C and D on opposite sides of $\overleftrightarrow{AB}$ such that $\angle CAB \cong \angle DBA$ and $\overline{AC} \cong \overline{BD}$ (Illus. 4.59). Now, $\overline{CD}$ meets $\overline{AB}$ at an interior point E. For suppose $\overline{CD}$ met $\overline{AB}$ at some exterior point, call it E', such that $A\text{-}B\text{-}E'$. Then exterior angle $\angle ABD$ would be less than the opposite interior angle $\angle BE'D$ of $\triangle BE'D$, which would be a contradiction of Theorem 4.14. Hence E is an interior point. By Theorem 4.9 (AAS), $\triangle ACE \cong \triangle BDE$; thus $\overline{AE} \cong \overline{EB}$, and E is a midpoint of $\overline{AB}$.

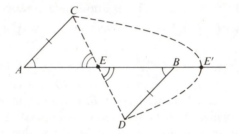

                                                                            Illus. 4.59

We now show that E is the only midpoint of $\overline{AB}$. Suppose there is another point E' such that $\overline{AE'} \cong \overline{E'B}$. Then either $A\text{-}E\text{-}E'\text{-}B$ or $A\text{-}E'\text{-}E\text{-}B$. Suppose that $A\text{-}E\text{-}E'\text{-}B$. Then by definition of *less than for segments*, $\overline{AE} < \overline{AE'}$ and also $\overline{AE} < \overline{E'B}$. But $\overline{E'B} < \overline{EB}$ and also $\overline{E'B} < \overline{AE}$. This is a contradiction and therefore there cannot be another midpoint E', and the midpoint E is unique.

Theorem 4.16 Every nonstraight angle has exactly one bisector.

Outline of proof. We first show there is at least one bisector. Consider nonstraight angle $\angle ABC$ such that $\overline{AB} \cong \overline{BC}$, and let D be the midpoint of $\overline{AC}$ (Illus. 4.60). The point D is in the interior of $\angle ABC$ and $\triangle ABD \cong \triangle CBD$. Thus $\angle ABD \cong \angle CBD$ and $\overrightarrow{BD}$ is a bisector of $\angle ABC$.

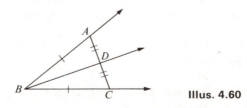

                                                            Illus. 4.60

We now show that $\overrightarrow{BD}$ is the only bisector of $\angle ABC$. Suppose there is another bisector $\overrightarrow{BE}$. Then E is in the interior of $\angle ABC$ and $\overrightarrow{BE}$ intersects $\overline{AC}$ at a point D'. It then follows that

$$\triangle ABD' \cong \triangle CBD' \qquad \text{and} \qquad \overline{AD'} \cong \overline{D'C}.$$

Thus D' is the midpoint of $\overline{AC}$, and $D' = D$, which means that ray $\overrightarrow{BE}$ and ray $\overrightarrow{BD}$ are the same.

EXERCISES

1. The midpoint or bisector of a segment is a point. Define a *perpendicular bisector* of a segment.
2. What kind of a triangle is $\triangle ABC$ if the angular bisector of each angle of the triangle is perpendicular to the side opposite the angle?
3. What kind of triangle would we have if the perpendicular bisector of every side bisects the opposite angle of the triangle?
4. What kind of triangle would we have if the perpendicular bisectors of two sides intersected at the midpoint of the third side of the triangle?
5. What kind of triangle would we have if the perpendicular bisectors of the sides all met in a point in the exterior of the triangle?
6. Will the angle bisectors always meet in a point interior to the triangle? Explain.

Rigorous exercises

7. Write a complete proof of Theorem 4.13.
8. Write a complete proof of Theorem 4.14.
9. Write a complete proof of Theorem 4.15.
10. Write a complete proof of Theorem 4.16.

7. FURTHER CLASSIFICATION OF POLYGONS

In Chapter 3 simple polygons were classified according to the number of sides: a triangle having three sides; a quadrilateral, four sides; etc. Using the concept of parallelism, special kinds of quadrilaterals were identified, such as trapezoids and parallelograms. Now that we have two new concepts, congruence for segments and congruence for angles, it is possible to extend our classification of polygons.

Angles and Triangles

Earlier in this chapter we defined a right angle as one that is congruent to every angle that is supplementary to it. We now define two other kinds of angles.

Definition 4.14 An angle that is less than a right angle is called an *acute* angle. An angle that is greater than a right angle but less than a straight angle is called an *obtuse* angle.

Triangles can be classified according to their angles.

Definition 4.15 An *acute* triangle is a triangle whose angles are all acute.
A *right* triangle is a triangle with one right angle. The side opposite the right angle is called the *hypotenuse*.
The other two sides are called the *legs* of the triangle.
An *obtuse* triangle is a triangle with one obtuse angle.

Triangles can also be classified according to their sides.

Definition 4.16 A *scalene* triangle is a triangle with no pairs of congruent sides.
An *isosceles* triangle is a triangle with two or more congruent sides.
An *equilateral* triangle is a triangle whose three sides are all congruent.

Quadrilaterals

By definition a quadrilateral is a polygon with four sides. If at least one pair of sides is parallel, it is called a *trapezoid*; and if both pairs of sides are parallel, it is called a *parallelogram*. Using congruence for segments and angles, we can now identify special kinds of parallelograms.

Definition 4.17 A parallelogram whose sides are all congruent is called a *rhombus* (plural, *rhombi*). A parallelogram containing at least one right angle is called a *rectangle*.
A rhombus with at least one right angle is called a *square*.

EXERCISES

👀 1. Construct six triangles, one of each kind in Definitions 4.15 and 4.16.

   2. A scalene triangle may be an acute triangle. Could a right triangle also be a scalene triangle? Construct triangles that illustrate the various combinations between Definitions 4.15 and 4.16.

   3. Draw a diagram like the one in Illus. 3.21(b) (p. 62), illustrating the relationship between all kinds of quadrilaterals.

   4. Do we have any guarantee that rhombi exist? That rectangles exist? Discuss.

👀 *Rigorous exercises*

   5. Prove that a triangle is equilateral if and only if it is equiangular. [*Hint:* See Exercises 8 and 10, p. 103.]

   6. Prove that the diagonals of a rhombus are perpendicular bisectors of each other. [*Hint:* Consider Theorem 4.13.]

7. Prove that
   a) no triangle can have more than one right angle,
   b) no triangle can have more than one obtuse angle.

8. A GENERAL CONCEPT OF CONGRUENCE

At the beginning of the chapter we introduced two undefined congruence relations
—one for segments and another for angles. Later these two undefined relations
were used to define congruence for triangles. Let us now consider how we might
extend the general concept of congruence to other types of geometric figures.

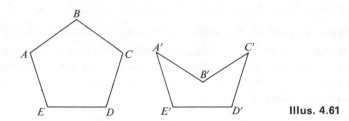

Illus. 4.61

Suppose we begin by considering polygons. Since two triangles are congruent
if and only if all three sides are pairwise congruent, one might wish to say that any
two polygons are congruent if and only if all sides are pairwise congruent. While
this may appear at first to be reasonable, it can be seen from Illus. 4.61 that the
idea would not be satisfactory. From the drawings one can see that in addition
to having the sides pairwise congruent, we would also want the angles associated
with the polygons to be pairwise congruent. Thus it is rather easy to describe
our intuitive idea of congruence for simple polygons. One merely notes that all
sides of one polygon are pairwise congruent to a corresponding side of the other,
and all angles are likewise pairwise congruent.

Now let us consider figures in general. Since an arbitrary curve may contain no
segments, we cannot speak of congruent sides. Similarly, an arbitrary figure may
have no angles associated with it, and thus we cannot speak of congruent angles.
What does it mean, then, to say that two figures are congruent? Suppose we
approach the problem as follows. Consider a figure C, as shown in Illus. 4.62(a),
and "construct" another figure which we feel would be called congruent to it.
Choose an arbitrary point P on or near the figure C and consider segments determined by P and other points on C, for example, Q, R, S, and T. Now choose
arbitrary points P' and Q' such that $\overline{PQ} \cong \overline{P'Q'}$ and construct points R', S',
and T' such that $\overline{PR} \cong \overline{P'R'}$, $\overline{PS} \cong \overline{P'S'}$, $\overline{PT} \cong \overline{P'T'}$, etc., and $\angle RPQ \cong \angle R'P'Q'$,
$\angle SPQ \cong \angle S'P'Q'$, etc. If this process could be continued for every point on
figure C, one would obtain a figure C' like that in Illus. 4.63. The points P, P'
and Q, Q', etc., are said to be *corresponding* points, and in general, for any point

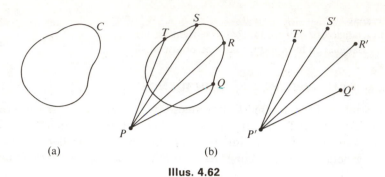

(a)   (b)

Illus. 4.62

X on curve C, there is a unique combination of a segment $\overline{PX}$ and an angle $\angle XPQ$ (relative to the two half-planes determined by $\overleftrightarrow{PQ}$) which determines a unique point X' on figure C'. Similarly, any point X' on figure C' determines a unique point X on C in the same manner. There is thus a one-to-one correspondence between the points of C and C'.

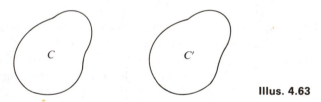

Illus. 4.63

EXERCISES

1. Why must we distinguish half-planes to obtain a one-to-one correspondence between the points of two congruent figures? Support your reasons with drawings.

2. Illustrate with drawings the one-to-one correspondence described above for two congruent rhombi. Two congruent concave pentagons.

3. Consider two congruent curves C and C', with points P, Q, and R on curve C and corresponding points P', Q', and R' on curve C'. What can be said about the relationship between

   a) $\overline{PQ}$ and $\overline{P'Q'}$ b) $\overline{PP'}$ and $\overline{QQ'}$
   c) $\overline{PR}$ and $\overline{P'R'}$ d) $\overline{PQ}$ and $\overline{P'R'}$
   e) $\overline{QR}$ and $\overline{Q'R'}$ f) $\angle PQR$ and $\angle P'Q'R'$
   g) $\angle RPQ$ and $\angle Q'P'R'$?

If two figures are related like those of the preceding section, we shall wish to call them congruent. Before making a general definition of congruence, however, we will need to develop our ideas a little further. Consider any pair of figures related to each other, like those of Illus. 4.63. If we choose any two points, V and W, on figure C, as in Illus. 4.64, and find the corresponding points V' and W' on figure C', they determine segments $\overline{VW}$ and $\overline{V'W'}$. By the SAS Axiom, triangles PVW and $P'V'W'$ are congruent, and $\overline{VW} \cong \overline{V'W'}$. In other words, the segment determined by any two points of one figure is congruent to the segment determined by the corresponding points of the other. This seems to be the principal feature of figures related in this way. It should also be noted that the correspondence between figures C and C' is not altered by the fact that they are oriented differently because of a rotation.

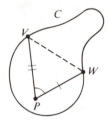
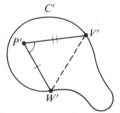

Illus. 4.64

In a similar manner we would expect to be able to "flip" one of the figures and still demonstrate a one-to-one correspondence between the points of C and C'. This is shown in Illus. 4.65.

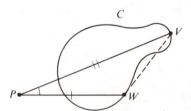
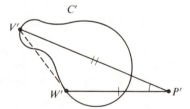

Illus. 4.65

We are now ready to make a useful and general definition of congruence.

Definition 4.18 Two figures are said to be *congruent* if and only if there exists a one-to-one correspondence between their points such that the segment determined by any two points of one figure is congruent to the segment determined by the corresponding points of the other figure.*

It should be noted that there may be many one-to-one correspondences between the points of two sets. If we are to apply this definition to show that two figures,

---

* It may also be said that the two figures are *isometric*, and that such a one-to-one correspondence is *distance preserving* (cf. Chapter 9).

such as circles, are congruent, we must show that there exists a particular one-to-one correspondence, and it must be one in which a segment joining any pair of points of one figure is congruent to the segment joining the corresponding pair of points of the other figure.

EXERCISES

1. Draw a point P' on a sheet of paper and construct a triangle congruent to the one below (Illus. 4.66), using the method described at the beginning of this section. Which pairs of corresponding points are of most interest? Why?

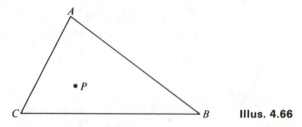

Illus. 4.66

2. When we proved that the base angles of an isosceles triangle were congruent, we viewed the triangle in two ways, as $\triangle ABC$ and $\triangle A'B'C'$, and then showed that $\triangle ABC \cong \triangle A'B'C'$ (Illus. 4.67). Describe the one-to-one correspondence between the points of these triangles which could be established, using P and P', where $\overline{PB} \cong \overline{P'B'}$ and $\angle PBA \cong \angle P'B'A$.

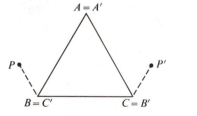

Illus. 4.67

Rigorous exercises

3. Prove that every figure is congruent to itself.

4. Prove that congruence as defined in Definition 4.18 is an equivalence relation.

5. Show that if two triangles are "congruent" according to Definition 4.18, then they are congruent according to our earlier definition of triangle congruence, and conversely.

6. Show that if two convex pentagons have corresponding sides and corresponding angles congruent, then they are congruent.

9. CIRCLES AND SPHERES

Circles are among the most interesting of geometric figures, as are their analogs in space, spheres. Let us begin with a definition of a circle.

Definition 4.19 Given a point O in a plane α and a segment $\overline{AB}$, the set of all points X on α such that $\overline{OX} \cong \overline{AB}$ is called a *circle*. Point O is called the *center* of the circle and any segment $\overline{OX}$, where X is on the circle, is called a *radius* of the circle.

It should be noted that the set of points constituting a circle does not contain the center. A circle is a simple closed curve, and no point of its interior is a point of the circle. Similarly, the radii of a circle are not subsets of it. In fact, for any radius, one endpoint lies on the circle and no other point of it is on the circle. We often say that a circle is *determined* by a point (the center) and a segment (determining the length of the radii).

We now make several definitions, giving names of important kinds of figures associated with circles.

Definition 4.20 A segment having both endpoints on a circle is called a *chord* of the circle.

A chord that contains the center of a circle is called a *diameter* of the circle.

Definition 4.21 Circles in the same plane having the same center are called *concentric* circles.

Definition 4.22 A line in the plane of a circle that contains one and only one point of the circle is said to be a *tangent* to the circle. The point of intersection is called the *point of tangency*.

From the general definition of congruence given in the previous section, it would seem a reasonable conjecture that two circles are congruent if they have congruent radii.

Theorem 4.17 Two circles are congruent if they have congruent radii.

While it is not difficult to prove this theorem, the proof of the converse requires some additional theorems not yet proved.

EXERCISES

1. a) Draw a circle with radius $\overline{AB}$.
   b) Draw a chord of the circle that is not a diameter.
   c) Draw a chord of the circle that is a diameter.
   d) Draw a circle that is congruent to the circle in (a) above.

e) Draw two other circles that are concentric to the circle in (d) above.

2. a) Draw a circle with radius $\overline{OA}$.

   b) Choose two points B and C on the circle such that $\overline{BC}$ is not a diameter. Are B, C, and O collinear? If B, C, and O are not collinear, what kind of triangle would $\triangle BOC$ be? Why could $\triangle BOC$ not be a scalene triangle?

3. Use the concepts of *less than* and *greater than for segments* to

   a) define the interior of a circle,

   b) define the exterior of a circle.

4. Does the interior of a circle appear to be a convex set? Does the exterior? Explain.

5. Which of the following statements do you think are true? Illustrate your answer.

   a) The intersection of a circle with a line may be empty.

   b) Two circles may intersect in exactly one point.

   c) Two circles may intersect in exactly two points.

   d) Two circles may intersect in exactly three points.

   e) If a line intersects a circle at one point, then it intersects the circle at two points.

👓 *Rigorous exercises*

6. Prove that if $\overline{AB}$ is a diameter and O is the center, then A-O-B.

7. Prove that if two distinct circles are concentric, then they are not congruent. Prove that the converse is false.

8. Prove that the center of a circle bisects every diameter of the circle.

9. Consider Illus. 4.68 below. Prove that if $\overline{OC} \perp \overline{AB}$, then $\overline{AC} \cong \overline{CB}$.

10. Prove the converse of Exercise 9.

11. Prove Theorem 4.17.

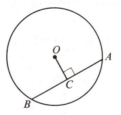

Illus. 4.68

Spheres

If we extend our concept of a circle to space, we obtain a sphere, defined as follows.

Definition 4.23 Given a point O and a segment $\overline{AB}$, the set of all points X in space such that $\overline{OX} \cong \overline{AB}$ is called a *sphere*. The point O is called the *center* of the sphere and for any point X on the sphere, $\overline{OX}$ is called a *radius* of the sphere.

The essential difference between this definition and the one for a circle is that here we consider points in space, whereas in the case of the circle we consider only those on a plane. The sphere is a familiar surface and may be represented approximately by the surface of a basketball. It is well to note again that the interior (for example, the inside of a ball) is not a part of the sphere, nor is the center. Definitions 4.20 through 4.22 are essentially the same for a sphere, except that the word "circle" is to be replaced by the word "sphere." Definition 4.22 needs slightly more modification in that the line need only intersect the sphere at exactly one point. A line cannot be in the "plane" of a sphere. A plane which intersects a sphere at exactly one point is also said to be tangent to the sphere.

Theorem 4.18 If a plane contains the center of a sphere whose radii are congruent to $\overline{AB}$, then the intersection of the plane and the sphere is a circle having the same center and radii also congruent to $\overline{AB}$.

Outline of proof. We need only to recall the definitions of a circle and a sphere. The intersection is illustrated in Illus. 4.69. The set of all points X in the intersection (being on both the sphere and the plane) is such that $\overline{OX}$ is congruent to the radius of the sphere, and since X is in the plane and O is also in the plane, $\overline{OX}$ is in the plane. (Why?) Thus all points X lying in the intersection also lie on a circle in the plane with center O and radius $\overline{OX}$ (which is the radius of the sphere).

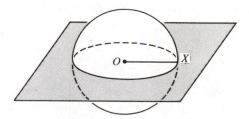

Illus. 4.69

The establishment of the previous theorem makes it sensible to give the following definition.

Definition 4.24 The intersection of a sphere and a plane that contains the center of the sphere is called a *great circle*.

Great circles are so named because they are the circles of greatest radius that lie on a sphere. For example, the equator and all of the meridians (longitude circles) one sees drawn on a globe are representations of great circles. On the other hand, parallels (latitude circles), with the exception of the equator, are not great circles (see Illus. 4.70).

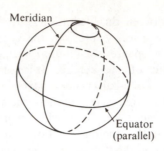

Meridian

Equator
(parallel)   **Illus. 4.70**

A parallel (other than the equator) might be described as the intersection of the sphere and a plane which does not contain the center of the sphere and is parallel to the plane passing through the equator. While this description is correct, it does not tell us what type of geometric figure a parallel is. As any student of geography "knows" from observation of the world globe in his classroom, parallels are circles. They are called "small circles" of the sphere. From the theorems to follow, we shall see that he is indeed correct.

We now list the following theorems, the first of which is a congruence theorem for right triangles. The reader may note that it deals with the SSA condition where the angles concerned are right angles.

Theorem 4.19 If, for two right triangles, the hypotenuse and one leg of one triangle are congruent to the hypotenuse and the corresponding leg of the other triangle, then the triangles are congruent.

Outline of proof. Consider $\triangle ABC$ and $\triangle DEF$ with congruent parts as indicated in Illus. 4.71.

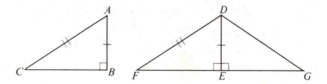

Illus. 4.71

There exists a point G on ray $\overrightarrow{FE}$ such that $F\text{-}E\text{-}G$ and $\overline{EG} \cong \overline{BC}$. Since $\angle DEG$ is a right angle, $\triangle ABC \cong \triangle DEG$. (Why?) Thus $\overline{DG} \cong \overline{AC} \cong \overline{DF}$, $\angle G \cong \angle F \cong \angle C$, and $\triangle ABC \cong \triangle DEF$. The details are left as an exercise.

Theorem 4.20 For any line l and any point P not on l there is one and only one line m that contains P and a point of l and is perpendicular to l.

Outline of proof. Choose any point Q on l and consider $\overrightarrow{QP}$ (Illus. 4.72). If $\overrightarrow{QP} \perp l$, then there exists a line, $\overleftrightarrow{PQ}$, perpendicular to l. If $\overrightarrow{QP}$ is not perpendicular to l, then $\angle AQP$ or $\angle BQP$ is acute. Let us suppose it is the latter. Then there

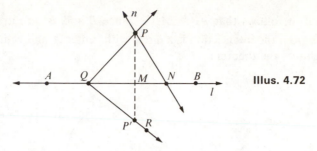

Illus. 4.72

exists $\overset{o}{QR}$ on $l/\sim P$ such that $\angle BQP \cong \angle BQR$. On $\overset{o}{QR}$ there is a point P' such that $\overline{QP} \cong \overline{QP'}$. Moreover, $\overline{PP'}$ meets l at some point M. Triangles QMP and QMP' are congruent; hence $\angle QMP \cong \angle QMP'$, and both angles are right angles. Therefore $\overleftrightarrow{PP'} \perp l$.

Now suppose there were another line, n, containing P and perpendicular to l. It would meet l at a point N different from M. This would violate the exterior angle theorem, and thus there is a unique perpendicular to l containing P.

Before proceeding to the next theorem we will need to consider what it means to say that a line is perpendicular to a plane. This is shown in Illus. 4.73. The definition is as follows:

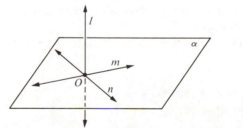

Illus. 4.73

Definition 4.25 A line is *perpendicular* to a plane α at a point O if and only if the line intersects the plane at O and is perpendicular to every line in the plane that lies on the point O.

While it is not crucial to the present development, it might be noted that it is possible to prove that if line l is perpendicular to at least two of the lines mentioned in the definition above, then line l is perpendicular to the rest of them.

Theorem 4.21 If a plane α intersects a sphere s in more than one point, then the intersection of the plane and the sphere is a circle.

Outline of proof. In Illus. 4.74, O is the center of the sphere, A is in plane α, $\overline{OA}$ is perpendicular to plane α, and B and B' are arbitrary points in the intersection of the plane and the sphere. Since $\overline{OB}$ and $\overline{OB'}$ are congruent, and $\overline{OA}$ is congruent to itself, it follows that $\overline{AB} \cong \overline{AB'}$. Since B and B' are arbitrary points in the intersection, the intersection is a circle with center A and radius congruent to $\overline{AB}$. This proves the theorem.

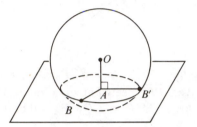

Illus. 4.74

EXERCISES

1. Define the interior and exterior of a sphere.
2. Make a conjecture about the possible intersections of two spheres. A line and a sphere.
3. Does the interior of a sphere appear to be a convex set? Does the exterior? Explain.
4. Make a conjecture about the relationship between a diameter of a sphere and a diameter of a great circle of the sphere.
5. Do all great circles of a sphere intersect? Support your answer.

Rigorous exercises

6. Complete the proof of Theorem 4.18.
7. Complete the proof of Theorem 4.19.
8. Complete the proof of Theorem 4.20.
9. Complete the proof of Theorem 4.21.

10. SURFACES: AN EXTENSION

In Chapter 3 we considered various kinds of surfaces such as cylinders, prisms, cones, pyramids, and polyhedra. Since the concept of congruence had not been introduced at that time, we were unable to consider some special kinds of cylindrical and conical surfaces that involve the idea of perpendicular lines and planes. In this section we will extend the ideas presented in Chapter 3.

Cylinders

If a directrix and the elements of a cylindrical surface, cylinder, or prism are all perpendicular to the plane containing the curve C, we say they are *right* cylindrical surfaces, *right* cylinders, or *right* prisms. In the case of a cylinder, if the simple closed curve C is a circle, the surface is called a *circular* cylinder. If the elements are perpendicular to the plane *and* the curve C in the plane is a circle, we obtain a *right circular* cylinder as shown in Illus. 4.75(a). The "tin" can is probably the best known model of a right circular cylinder.

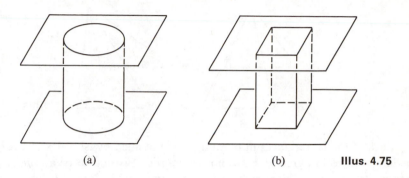

(a)   (b)   **Illus. 4.75**

If the curve C in the plane is a rectangle, the prism is called a *parallelepiped*. If, in addition, the elements are perpendicular to the plane, it is called a *rectangular parallelepiped*, as illustrated in Illus. 4.75(b). Most boxes, cartons, etc., are models of rectangular parallelepipeds.

Cones

A general description of a cone was given in Chapter 3. If the curve C in the plane is a circle, the cone is called a *circular cone*. In addition, if the segment from the point (vertex) P to the center of the circle C is perpendicular to the plane, we call it a *right circular cone*. This type of cone is in Illus. 4.76(a). Some ice cream cones are models of right circular cones.

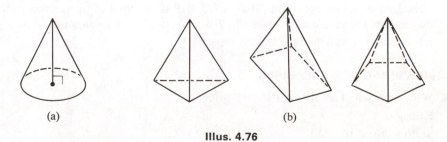

(a)   (b)

Illus. 4.76

You will recall that if the curve C in the plane is a polygon, the cone is also called a *pyramid*. Pyramids may be classified according to the kind of polygon C

in the plane.   *Triangular*, *rectangular*, and *pentagonal pyramids* are shown in Illus. 4.76(b).

Some pyramids have a regular polygon as the boundary of the base. An equilateral triangle is a regular polygon and so is a square. The pyramids in Illus. 4.77 each have a regular polygon (square) as the boundary of the base.

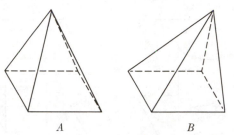

Illus. 4.77

A B

Each of the lateral faces in pyramid A is bounded by isosceles triangles, while the lateral faces in pyramid B are not. Pyramid A is said to be a *regular pyramid* because its base is bounded by a regular polygon and its lateral faces are bounded by isosceles triangles.

EXERCISES

1. Draw a right cylindrical surface where C is
   a) a polygonal curve,   b) a simple closed curve, not polygonal.
2. Draw a right cylinder which is
   a) circular,   b) a triangular prism.
3. Draw a parallelepiped which is
   a) rectangular,   b) not rectangular.
4. Draw a regular pyramid with
   a) a pentagonal region for a base,   b) a triangular region for a base.
5. Consider the regular pyramid in Illus. 4.77. If P is the point of intersection of the diagonals of the square in the base and T is the vertex, make a conjecture about the relationship between the base and the segment $\overline{PT}$.

REVIEW PRACTICE EXERCISES

1. Write exponential notation for the following.
   Example. $7 \times 7 \times 7 = 7^3$.
   a) $10 \times 10 \times 10 \times 10$ b) 6×6
   c) $2 \times 2 \times 2 \times 2 \times 2$ d) $2 \times 4 \times 8$
2. Write exponential notation for the following.
   Example. $\dfrac{1}{2} \times \dfrac{1}{2} \times \dfrac{1}{2} = \dfrac{1}{2^3} = 2^{-3}$.

a) $\frac{1}{10} \times \frac{1}{10} \times \frac{1}{10} \times \frac{1}{10}$ 　　　　 b) $\frac{1}{4} \times \frac{1}{4} \times \frac{1}{4} \times \frac{1}{4} \times \frac{1}{4} \times \frac{1}{4}$

c) $\frac{1}{2} \times \frac{1}{2} \times \frac{1}{5} \times \frac{1}{5}$ 　　　　 d) $\frac{1}{2} \times \frac{1}{4} \times \frac{1}{8}$

3. Write expanded numerals for the following.

Example. $725.48 = 7 \times 10^2 + 2 \times 10 + 5 + 4 \times 10^{-1} + 8 \times 10^{-2}$.

a) 423 　　　　 b) 1609

c) 14,506.32 　　　　 d) 1,453,072.803

4. In the decimal numeration system the place values are powers of ten, as illustrated by the expanded numerals in Exercise 3. In the binary-numeration system the place values are powers of two. The binary numeral 11011_2 thus represents

$$1 \times 2^4 + 1 \times 2^3 + 0 \times 2^2 + 1 \times 2 + 1 \times 1, \quad \text{or} \quad 27_{10}.$$

We say that $11011_2 = 27_{10}$, because "11011_2" and "27_{10}" are names for the same number. For each of the binary numerals below, write a decimal numeral that names the same number.

a) 1011_2 　　　　 b) 1001_2 　　　　 c) 1111_2

d) 101011_2 　　　　 e) 100011_2 　　　　 f) 1000000_2

5. In the decimal numeration system the place values to the right of the decimal point are $1/10$, $1/10^2$, $1/10^3$, and so on. In the binary-numeration system they are $1/2$, $1/2^2$ $(1/4)$, $1/2^3$ $(1/8)$, and so on. The binary numeral $.111_2$ thus represents

$$1 \times \tfrac{1}{2} + 1 \times \tfrac{1}{4} + 1 \times \tfrac{1}{8}, \quad \text{or} \quad \tfrac{7}{8} \, (.875_{10}).$$

We say that $.111_2 = .875_{10}$, because "$.111_2$" and "$.875_{10}$" are names for the same number. For each of the binary numerals below, write a base-ten numeral (fractional or decimal) that names the same number.

a) $.1_2$ 　　　　 b) $.101_2$ 　　　　 c) $.011_2$

d) $.1001_2$ 　　　　 e) $.0001_2$ 　　　　 f) $.01011_2$

6. A carpenter finds the length of a board to be $3\frac{13}{32}$ ft. Express the length of this board with a binary numeral.

11. MAPS

The word "map" is derived from the Latin *mappa* meaning napkin or cloth (on which maps were sometimes painted). A *map* may be described as a drawing or representation of all or part of the surface of the earth. Although more nearly a rotational ellipsoid, the earth is usually thought of as a sphere, and the problem of symbolizing this surface on a map is therefore one of projecting the points of a sphere onto a part of a plane. Since spherical surfaces cannot be "flattened" without tearing or stretching, some distortion is inevitable, and a perfect projection is therefore impossible.

In a previous section a *great circle* was defined as the intersection of a sphere and a plane that contains the center of the sphere. If a plane intersects a sphere in more than one point and does not contain the center of the sphere, the intersection is called a *"small circle."* Parallels, except for the equator, are small circles whose intersecting planes are parallel to the plane of the equator. The equator and all of the meridians are great circles. Parallels, as the name implies, do not intersect; all meridians intersect at the north and south poles. Furthermore, all parallels intersect all of the meridians at right angles. These relationships are shown in Illus. 4.78.

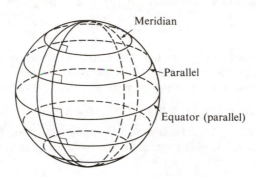

Illus. 4.78

A coordinate system such as that provided by parallels and meridians is essential since it enables us to describe the location of features on the surface of the earth. As a matter of fact, one of the basic problems of map making is the determination of the relationship between the meridians and parallels on the earth and the segments which represent them on a map. There are several methods of representation, called projections, which might be tried, some of the more common being the *cylindrical*, *conical*, and *azimuthal* projections.

Cylindrical Projections

Consider a sphere enveloped by a right circular cylinder as shown in Illus. 4.79(a). If the surface of the sphere is projected onto the cylinder and the cylinder is then cut and spread out, we have a *cylindrical projection*. Such a projection originates from a point or points of perspectivity. For example, one might think of the center of the sphere as such a point, and a cross section of the sphere and cylinder shown in Illus. 4.79(b) illustrates how such a projection works. It is easy to see that the entire earth's surface could not be projected on a cylinder of finite length.

Another possibility would be to consider the axis of the earth as the origin of perspectivity. The cross section in Illus. 4.79(c) shows such a projection. Here the spherical surface is projected on a cylinder of finite length, and while there is still distortion, it is not so severe as with the first method. Although the meridians converge to the poles on the sphere, their projections would be parallel for each method.

A variation of these methods is the *Mercator Projection*, named after the Flemish geographer, Gerardus Mercator, 1512–1594. In this projection* all

---

* Actually, a map projection may be thought of as any orderly network of meridians and parallels and need not be a perspective projection. Mercator merely began with the idea of a cylindrical projection and then manipulated the parallels to produce the map characteristics he desired.

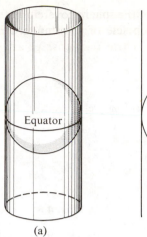

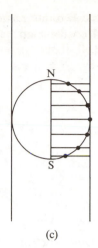

(a)   (b)   (c)

Illus. 4.79

meridians are represented by parallel vertical segments and all parallels by parallel horizontal segments. Parallels and meridians intersect at right angles. One of the principal features of the Mercator projection is that it is *conformal*, which means that an angle between two lines on the earth has the same measure as the angle between the corresponding lines on a map. Since the meridians on the map do not converge as they do on the sphere, the proper relation between angles is maintained by increasing the spacing of the parallels toward the poles. This, of course, also results in a noticeable distortion of size toward the poles. A Mercator projection map of the world is shown in Illus. 4.80.

Conical Projection

In this type of projection a portion of a sphere is projected onto a portion of a cone, as shown in Illus. 4.81. The circle of tangency (where the cone touches the

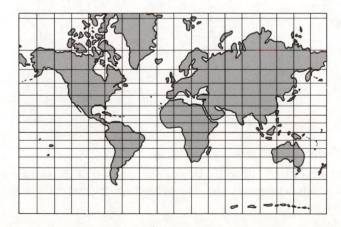

Illus. 4.80

sphere) is called the *standard parallel*. Not all of the sphere covered by the cone is produced on the map. Rather, a strip on each side of the standard parallel is projected.

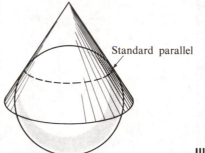

Standard parallel

Illus. 4.81

The *Lambert Conformal Projection* is a conic projection in which two standard parallels are used, the projection then being limited between these two parallels [see Illus. 4.82(a)]. The meridians are represented by lines which converge toward a point outside of the map (that is, the projected point of the vertex of the cone). The parallels appear as arcs which are parts of concentric circles, whose center is the point of convergence of the meridians. This type of projection is particularly useful for maps of countries or continents in the temperate zones because of the relatively small amount of distortion. A Lambert conformal projection of the United States is shown in Illus. 4.82(b).

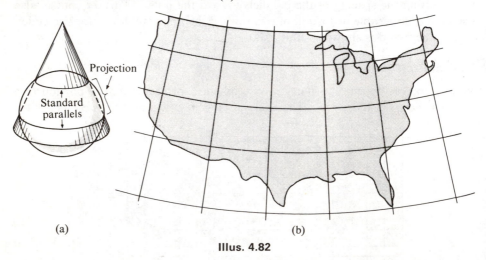

Projection

Standard parallels

(a)   (b)

Illus. 4.82

Azimuthal Projection

An azimuthal projection is obtained by projecting a part of the surface of the earth onto a plane which is tangent to it. The projection originates from some selected point of perspectivity, as shown in Illus. 4.83. The plane may be tangent at the poles, the equator, or any other point on the sphere.

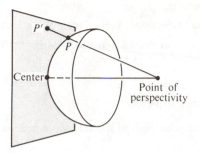

Illus. 4.83

There are several variations of this type of projection. If the point of perspectivity is the center of the sphere, we obtain a *gnomonic* projection. While there are extreme distortions of size and shape, it has the advantage that all great circles appear as segments (that is, are "straight"), thus making it useful for plotting transoceanic sailing and flying routes.

If the point of perspectivity is infinitely far with respect to the two surfaces, we obtain an *orthographic* projection. Although there is considerable distortion at the sides, everything is seen in correct proportion, since the map actually gives the appearance of a three-dimensional globe. Orthographic projections of a hemisphere from a polar and oblique view are shown in Illus. 4.84.

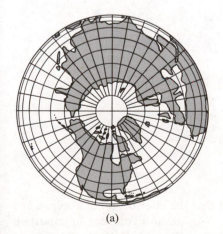

(a)   (b)

Illus. 4.84

EXERCISES

1. Study a Mercator projection map of the world and make a conjecture about the relative areas of Greenland, Alaska, and Mexico. Then look up the actual area of each. Discuss your results.

2. What is an equal area projection? How does it differ from a conformal projection?

3. Why would there be almost no difference between a Lambert conformal projection and a Mercator projection for a state map?

REVIEW PRACTICE EXERCISES

1. Write a decimal numeral for each of the following rational numbers.

   Example. $\frac{18}{12} = 1.5$.

   a) $\frac{19}{4}$ b) $\frac{3}{40}$ c) $\frac{825}{66}$ d) $\frac{585}{27}$ e) $\frac{4}{3}$ f) $\frac{5}{9}$

2. Write a fractional numeral for each of the following rational numbers.

   Example. $3.2 = \frac{16}{5}$.

   a) 12.6   b) .325   c) 4.125   d) $.166\bar{6}\ldots$ e) 17.3   f) $.11\bar{1}\ldots$

3. The decimal numeral for a rational number either terminates, as in 12.6, or it repeats as in $.11\bar{1}\ldots$ When a repeating decimal numeral is given, a fractional numeral can be found as follows (see Appendix for another example):

   Example 1. Find a fractional numeral for $.33\bar{3}\ldots$

$$
\begin{aligned}
n &= .333\ldots \\
10n &= 3.333\ldots \\
\hline
9n &= 3 \qquad \text{(subtracting } n \text{ from } 10n) \\
n &= \tfrac{1}{3}
\end{aligned}
$$

   Example 2. Find a fractional numeral for $.833\bar{3}\ldots$

$$
\begin{aligned}
n &= .8333\ldots \\
10n &= 8.333\ldots \\
100n &= 83.333\ldots \\
\hline
90n &= 75 \qquad \text{(subtracting } 10n \text{ from } 100n) \\
n &= \tfrac{5}{6}
\end{aligned}
$$

   Write a fractional numeral for each of these numbers.

   a) $.77\bar{7}\ldots$ b) $.99\bar{9}\ldots$ c) $.499\bar{9}\ldots$ d) $.45\overline{45}\ldots$

12. CONIC SECTIONS

When we discussed spheres earlier, we proved that the intersection of a sphere and a plane (not tangent to the sphere) is a circle. We will now consider an important set of curves, known as the conic sections, determined by various intersections of a plane with a right circular conical surface. Recall that a right circular conical surface consists of both nappes and is of unlimited extent, as shown in Illus. 4.85.*

Let us now consider the intersection of a cone with: (a) a plane that intersects both nappes of the cone (not on axis); (b) a plane that is parallel to a line (element) of the cone; and (c) a plane that intersects only one nappe, is not parallel to a line of the cone, and is not perpendicular to the axis of the cone.

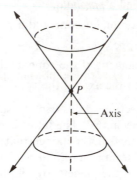

Illus. 4.85

These intersections are shown in Illus. 4.86. The intersection described in (a) is called a *hyperbola*; it consists of two branches and is not closed. The intersection described in (b) is called a *parabola*. It has only one branch and is, like the hyperbola, not closed. The intersection described in (c) is called an *ellipse*. Like the parabola, it is contained in only one nappe but is a closed curve.

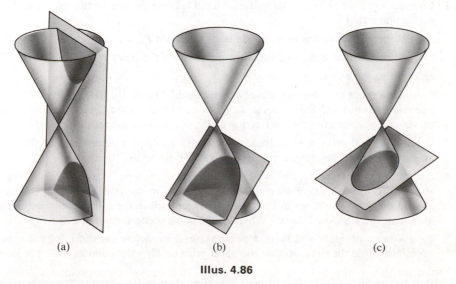

(a)   (b)   (c)

Illus. 4.86

* In the discussion to follow we will refer to the right circular conical surface as simply a cone. Both uses of the word "cone" are common.

We now give the following definitions.*

Definition 4.26

a) The intersection of a right circular cone and a plane not on the axis of the cone that intersects both nappes is called a *hyperbola*.

b) The intersection of a right circular cone and a plane that is parallel to an element of the cone is called a *parabola*.

c) The intersection of a right circular cone and a plane that:

    i) intersects only one nappe,

    ii) is not parallel to an element of the cone, and

    iii) is not perpendicular to the axis of the cone is called an *ellipse*.

Knowledge of the conic sections dates back to such Greek mathematicians as Euclid, Apollonius, and Menaechmus. The latter is reputed to have been the first to discover the curves (about 350 B.C.). Since that time they have been studied by many mathematicians. The famous English mathematician Newton, for example, proved that the paths of the planets are elliptical. The cables of a suspension bridge (with uniform weight distribution) hang in the shape of a parabola, and the path of an unresisted projectile is also a parabola. The properties of these curves are best studied using analytical methods.

EXERCISES

1. Consider the right circular cone in Illus. 4.85. Describe the intersection of this surface with a plane that

    a) is perpendicular to the axis and contains the point *P*,

    b) is perpendicular to the axis and does not contain the point *P*,

    c) contains the axis.

2. Consider a plane that begins parallel to (and distinct from) the axis (of the surface in Exercise 1 above) and then is rotated clockwise. Describe the type and order of the curves one obtains from the intersection of the plane and the conical surface.

3. Given a conical surface, as in Illus. 4.85, would all ellipses determined by different intersecting planes be congruent? Parabolas? Hyperbolas? Circles?

4. Use a thick piece of cardboard as backing for a piece of paper. Place two thumbtacks about 4 in. apart, make a loop of string about 10 in. around, and place the loop around the tacks as shown in Illus. 4.87. Put a pencil in the loop, and, keeping the loop taut, draw a closed curve. Which of the conic sections have you drawn?

5. Take a piece of paper and mark a point about 2 in. below the center of the upper edge. Now fold the paper so that the upper edge touches the point as shown in Illus.

---

* Although the three curves defined here are referred to as the conic sections, there are other curves which are obtained from the intersection of a plane with a cone (see Exercise 1, p. 118).

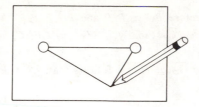

Illus. 4.87

4.88. Refold at different angles, each time so that the edge touches the point. What curve is outlined by the creases in the paper?

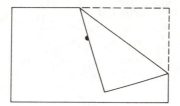

Illus. 4.88

6. Draw a circle about 3 in. in diameter on a piece of paper. Mark a point about 1 in. above the circle as shown in Illus. 4.89. Now fold the paper so that the point touches the circle. Refold many times, each time so that the point touches the circle. The creases in the paper outline two curves. What are they?

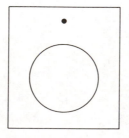

Illus. 4.89

A GEOMETRIC FALLACY

Here is a rather famous argument that all triangles are isosceles. What is wrong with the argument?

Consider $\triangle ABC$ with $\overleftrightarrow{BE}$ the perpendicular bisector of $\overline{AC}$ and the angular bisector of $\angle B$ as in Illus. 4.90.

If the angular bisector does not intersect $\overleftrightarrow{DE}$, they are parallel, in which case the angle bisector is perpendicular to $\overline{AC}$, and $\triangle ABC$ is isosceles.

If the angle bisector intersects $\overleftrightarrow{DE}$ at a point F, then consider segments $\overline{FA}$ and $\overline{FC}$, and perpendiculars $\overline{FG}$ and $\overline{FH}$ as shown in Illus. 4.90. Now $\triangle BFG \cong \triangle BFH$, by AAS

(Theorem 4.9), and therefore $\overline{BG} \cong \overline{BH}$, and $\overline{FG} \cong \overline{FH}$. Also, $\triangle AFD \cong \triangle CFD$, by SAS, and therefore $\overline{FA} \cong \overline{FC}$. Finally, $\triangle FGA \cong \triangle FHC$, by Theorem 4.19, and therefore $AG \cong \overline{CH}$.

Since $\overline{BG} \cong \overline{BH}$ and $\overline{GA} \cong \overline{HC}$, it follows that $\overline{AB} \cong \overline{BC}$, and therefore $\triangle ABC$ is isosceles.

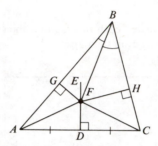

Illus. 4.90

CHAPTER 5

MEASURES

In the previous chapters we have considered properties of geometric figures that are not concerned with size. Definitions were made of "less than" and "greater than" for segments and angles, but never have we asked questions such as "just how large is a segment?" or "how *much* greater is one angle than another?" Now these kinds of questions will be considered and, as we shall see, numbers thus come into geometry. The following development, although based on familiar notions such as the use of a ruler, is largely theoretical. We shall leave a discussion of practical difficulties in measuring actual objects to a later chapter. A review of numeration will be needed before we tackle the problem of measures itself.

1. NUMERATION

Decimal numeration can be explained by means of expanded notation as follows.* The symbol "352.416" means

$$3 \times 10^2 + 5 \times 10^1 + 2 \times 10^0 + 4 \times 10^{-1} + 1 \times 10^{-2} + 6 \times 10^{-3}.$$

The first digit to the right of the decimal point represents a number of tenths, the second one, a number of hundredths, the third one, a number of thousandths, and so on. To the left of the decimal point the place values are one, ten, one hundred, one thousand, and so on.

Numerals of the Hindu-Arabic type, but with nondecimal bases, can be explained in a similar way. The base-seven numeral "352.416_7" means

$$3 \times 7^2 + 5 \times 7^1 + 2 \times 7^0 + 4 \times 7^{-1} + 1 \times 7^{-2} + 6 \times 7^{-3}.$$

The first digit to the right of the septimal point represents a number of sevenths, the second, a number of 49ths, and so on. In the sequel we shall be particularly

---

* See Appendix for another review of numeration.

interested in binary, or base-two, numeration. In base-two symbolism, the only digit symbols are "0" and "1." The meaning of binary numerals can be explained as above for other bases. For example, "1011.101_2" means

$$1 \times 2^3 + 0 \times 2^2 + 1 \times 2^1 + 1 \times 2^0 + 1 \times 2^{-1} + 0 \times 2^{-2} + 1 \times 2^{-3}.$$

The place values to the right of the binary point are one-half, one-fourth, one-eighth, one-sixteenth, and so on. To the left of the point, they are one, two, four, eight, sixteen, and so on.

We assume that the reader is already familiar with the system of rational numbers.* The system of rational numbers is a field, and any of the numbers in the system can be named by a fractional symbol having integers for numerator and denominator. (The latter cannot, of course, be zero.) In fact, one may show that a number is rational by finding such a name for that number.

Example 1. Show that $3\frac{1}{6}$ is a rational number.

$$3\tfrac{1}{6} = 3 + \tfrac{1}{6} = 3 \cdot \tfrac{6}{6} + \tfrac{1}{6} = \tfrac{18}{6} + \tfrac{1}{6} = \tfrac{19}{6}.$$

Since we have named the number $3\frac{1}{6}$ by a fractional symbol having integers for numerator and denominator, we know the number is rational.

Example 2. Show that 101.1101_2 is a rational number.

$$101.1101_2 = 1 \times 2^2 + 0 \times 2^1 + 1 \times 2^0 + 1 \cdot \tfrac{1}{2} + 1 \cdot \tfrac{1}{4} + 0 \cdot \tfrac{1}{8} + 1 \cdot \tfrac{1}{16}$$

$$= 5 + \tfrac{13}{16} = \tfrac{93}{16}.$$

We have shown that the number in question can be named "$\frac{93}{16}$." It is therefore a rational number.

All rational numbers are namable with decimal (or any base) numerals. In some cases, the numerals are endless; for example, $\frac{1}{3} = 0.333\ldots$ In other cases, the numerals terminate; for example, $\frac{1}{4} = 0.25$. When the decimal numeral for a rational number does not terminate, however, it repeats. The same is true for a numeral of any base.

Any endless and nonrepeating numeral (of any base) represents a *real* number that is not rational. In other words, an endless and nonrepeating numeral represents an *irrational* number. (The system of real numbers, so-called, is a field that is an extension of the system of rational numbers.) It follows that no irrational number can be named by a fractional numeral with integers for numerator and denominator, because if a number can be named that way, it is rational. Similarly, irrational numbers cannot be named by a terminating or repeating numeral of any base, because if a number can be named that way, it is rational. Thus the decimal (or other base) numeral for an irrational number is endless and non-repeating.

---

* See Appendix for a review of the system of rational numbers.

When an endless, repeating numeral is given, it is a fairly simple task to find a fractional numeral for the number. The procedure is the same regardless of the base of the repeating numeral.

Example 1

$$\begin{aligned} \text{If} \qquad n &= .666666\ldots, \\ \text{then} \qquad 10n &= 6.66666\ldots \qquad \text{(Multiplying by ten)} \\ \text{and} \qquad 9n &= 6. \qquad\qquad\quad \text{(Subtracting)} \\ \text{Thus} \qquad n &= \tfrac{2}{3}. \end{aligned}$$

Example 2

$$\begin{aligned} \text{If} \qquad n &= .454545\ldots, \\ \text{then} \qquad 100n &= 45.454545\ldots \\ \text{and} \qquad 99n &= 45. \\ \text{Thus} \qquad n &= \tfrac{5}{11}. \end{aligned}$$

Example 3

$$\begin{aligned} \text{If} \qquad n &= .100100100\ldots_2, \\ \text{then} \qquad 8n &= 100.100100\ldots_2 \\ \text{and} \qquad 7n &= 100_2, \text{ or } 4. \\ \text{Thus} \qquad n &= \tfrac{4}{7}. \end{aligned}$$

(Multiplying by 2 moves the binary point one place. Instead of "$8n$" we could write "$1000_2 n$" if we wish to stay with base-two notation throughout)

EXERCISES

1. Write expanded numerals for each of the following numbers.
   a) 25314.178   b) 31042.78031   c) 3154.1054_7 d) 1100110.1011_2

2. Show that each of the following numbers is rational by naming it with a fractional symbol with integers for numerator and denominator.
   a) $31\tfrac{1}{5}$ b) 4.732   c) 101.11_2 d) 10001.101101_2

3. Show that each of the following numbers is rational by naming it with a fractional symbol with integers for numerator and denominator.
   a) $0.888\ldots$ b) $0.11111\ldots$
   c) $0.75757575\ldots$ d) $1011.1100110011001100\ldots_2$

2. MEASURES OF SEGMENTS

To answer the question "how long is a segment?" we shall need something with which to compare it. We shall choose, arbitrarily, some segment as a basis of comparison, assigning to it the number 1. Since it is assigned the number 1, it will be called a *unit*. Suppose the segment $\overline{AB}$ in Illus. 5.1 is chosen to be a unit, and we then wish to find how long the larger one, $\overline{CD}$, is. By Axiom 4.2 (the compass axiom) we know that on ray $\overrightarrow{CD}$ there is a unique point Q_1 such that $\overline{AB} \cong \overline{CQ_1}$. If Q_1 were the same as D, we would stop. In our example, however, C-Q_1-D, and on $\overrightarrow{Q_1 D}$ there is a unique point Q_2 such that $\overline{AB} \cong \overline{Q_1 Q_2}$. More points, Q_3, Q_4, and so on, can be found in this manner. Suppose that $Q_4 = D$ (Q_4 and D are the same point). Segment $\overline{CD}$ is then "divided" into four segments, each congruent to $\overline{AB}$. We would then say that $\overline{CD}$ is four times as long as $\overline{AB}$ and assign to it the number 4. Numbers that are thus assigned to segments are called *measures*. The larger segment here is said to have "measure 4." We also say that its *length* is 4, or that the distance from C to D is 4.*

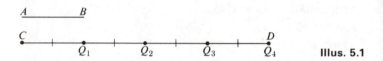

Illus. 5.1

Subunits

In the preceding example, the point Q_4 was also an endpoint of the segment $\overline{CD}$ being measured. Most often none of the points Q_1, Q_2, etc., will coincide with an endpoint of a segment being measured. Illustration 5.2 shows, for example, this more usual situation. Here $\overline{AB}$ is again the unit segment and Q_3 lies between C and D, while Q_4 lies beyond D (such that C-D-Q_4). In this kind of situation we use segments smaller than a unit, called *subunits*. If we bisect $\overline{AB}$ (which we know is possible by Theorem 4.15), we obtain a segment $\overline{AP}$ which we shall call a *half-unit*, assigning it the measure $\frac{1}{2}$. Now on $\overrightarrow{Q_3 Q_4}$ there is a unique point S such that $\overline{Q_3 S} \cong \overline{AP}$ and, if (as in this case) $S = D$, we assign a nonintegral rational number as the measure of $\overline{CD}$. In this example, the measure of $\overline{CD}$ is $3\frac{1}{2}$.

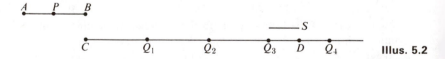

Illus. 5.2

If the point S should not be the same as D, then we can bisect a half-unit to obtain a quarter-unit and proceed much as before. By continued bisecting, we may obtain many subunits, eighth-units, sixteenth-units, and thirty-second-units,

---

* The reader may be asking "4 what—inches, feet, or yards?" In this theoretical development we shall use only one unit segment, hence a name for it, such as "inch" is not necessary. Later, when several units of different size are considered, we shall need to use unit names.

for example. To make decimal arithmetic easier, we might wish to use tenth-units, hundredth-units, and so on. At this stage in our formal development, however, we have no theorem analogous to Theorem 4.15, assuring us that tenth-units exist, so we shall consider only binary subunits for the moment.

The assigning of measures, as described in the preceding examples, is purely theoretical and applies only to the abstract geometric entities that we have created in our formal development. The process conforms, however, to what we might do in measuring a physical object such as a stick or a string. In such cases we might mark off unit distances, using a compass or a rod. If a unit rod were used, it would be placed in successive end-to-end positions along the object being measured. The ruler was invented to provide a shorter procedure. A ruler is simply a stick with unit segments marked on it. It is laid along the object being measured and the measure is read from the instrument.

Although there are some fine points to be considered, it should seem intuitively reasonable that in general we now seem to have a way of assigning measures to segments. The measure of a segment will be denoted in two ways. The measure of a segment $\overline{PQ}$ (which is a number) may be named

$$(1)\ \ m(\overline{PQ}) \quad \text{or} \quad (2)\ \ PQ.$$

EXERCISES

In Exercises 1 through 5, use the segment $\overline{AB}$ in Illus. 5.3 as a unit.

1. a) With a compass, construct a segment with measure 6.

   b) Bisect the unit segment to obtain a half-unit. Then bisect a half-unit to obtain a quarter-unit.

   c) Construct a segment $\overline{CD}$ such that $m(\overline{CD}) = 4\frac{3}{4}$.

   d) Draw a segment with measure less than $3\frac{1}{2}$ but greater than $3\frac{1}{4}$.

A _____ B **Illus. 5.3**

2. Using a compass, find the measures of the segments in Illus. 5.4. Write binary numerals for your answers.

   a) _____

   b) _____

   c) _____

   d) _____   **Illus. 5.4**

3. Make a ruler from a piece of paper or cardboard and mark half-, quarter-, and eighth-units on it.

4. Use the ruler of Exercise 3 to measure the segments in Illus. 5.5. Write binary numerals for your answers.

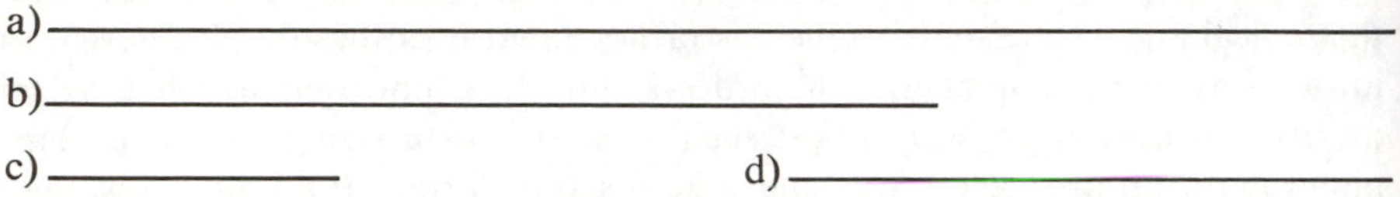

a) ————————————————————————————————

b) ——————————————————————————

c) ——————————   d) ————————————————

Illus. 5.5

5. Use a compass to find the measure of this segment. Use subunits as small as necessary.

————————————————————————————————————

6. Consider and discuss the following statement: Theoretically (if a compass point could be made small enough), one could find an exact measure of any segment, no matter what unit is used, by taking small enough subunits.

Some Fine Points to be Considered

Thus far our ideas fit in with the usual way measurements are obtained. While it may appear that no problems remain, there are several situations that merit further attention.

Suppose, for example, that we choose a very short unit segment to measure a very long segment, as indicated in Illus. 5.6. We begin by locating points Q_1, Q_2, Q_3, and so on, as before, and although there will be many such points, it seems reasonable that eventually we will reach the point D, or a point beyond it. While this may indeed seem reasonable or "intuitively obvious," it has been demonstrated that we cannot prove, from previous axioms, that we shall always reach the end of a segment in the measuring process. Mathematicians have in fact proved this by constructing examples of geometric systems in which all of the previous axioms hold but in which it is not always possible to reach the end of a segment, as we do in our measuring process. Thus we shall need to make this an axiom. It bears the name of Archimedes, who was the first to recognize the need for it.

Illus. 5.6

Axiom 5.1 (Archimedes' Axiom) If $\overline{AB}$ and $\overline{CD}$ are any segments, and on $\overrightarrow{CD}$ we choose points Q_1, Q_2, Q_3, and so on such that $\overline{CQ_1}$, $\overline{Q_1Q_2}$, $\overline{Q_2Q_3}$, and so on are all congruent to $\overline{AB}$, as in Illus. 5.6, then eventually we shall find points Q_n and Q_{n+1} for which $D = Q_n$ or D is between Q_n and Q_{n+1}.

Now that this problem has been solved by introducing Axiom 5.1, it might seem that for any unit and any segment to be measured we can always find a measure. That measure would be the sum of a whole number and a number less than 1. To find the whole number we place the $n + 1$ points as described in Axiom 5.1 and count to determine n. This is the whole number part of the measure. To find the number less than 1, we begin to use subunits. Since we know that every segment can be bisected, we use half-units, quarter-units, eighth-units, and so on, just as we do in using an ordinary ruler. Illustration 5.7 shows, for example, a situation in which n is 5, and we also have $\frac{1}{2} + \frac{0}{4} + \frac{1}{8} + \frac{0}{16} + \frac{1}{32}$. Thus the part of the measure less than 1 is $\frac{21}{32}$. The measure is $5\frac{21}{32}$. If we use binary numeration, we may express this measure as 101.10101_2. In the following discussion, we shall use binary numeration in this manner.

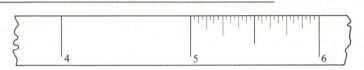

Illus. 5.7

The next important question is the following: As we use smaller and smaller subunits, will we eventually "come out even"? In other words, is there always a subunit small enough that its endpoint will coincide with the endpoint of the segment being measured? The answer to this question is not needed for the present development and will be considered in the next chapter. For now we may simply consider two cases and decide how measures will be assigned in each case. If we do "come out even," then we assign a number as a measure in the manner of the above example. If, on the other hand, we do not "come out even," we shall have to do something else. Let us first list the possible cases, keeping in mind that we always obtain a sequence of 0's and 1's in the binary numeral, as in the preceding example.

CASE 1.   The numeral ends.

CASE 2.   The numeral does not end.

 Subcase a)   The unending numeral is a repeating one.

 Subcase b)   The unending numeral does not repeat.

The problem is how to assign measures, and in case 1 we have solved the problem. Let us now consider the first subcase of 2. If we obtain an unending and repeating binary numeral, then that numeral represents a rational number. We assign that rational number to the segment as its measure. If the binary numeral neither ends nor repeats [subcase (b)], then it represents an irrational number and we assign that irrational number as the measure of the segment.

If the measure of a segment is a rational number, then we can name it with a fractional symbol, a decimal symbol, or a symbol of some other kind, including a binary numeral. If the measure is irrational, on the other hand, we generally have no convenient way of naming it. In any case, the problem of how to assign measures to segments is solved. We now have a theoretical way of assigning a unique real number to any segment.

EXERCISES

1. Suppose that in measuring a segment, as described above, we find that n is 31 and that when one half-unit and one eighth-unit have been applied, the process ends (the end of the eighth-unit is the same as the end of the segment). Find the measure and show that it is a rational number by naming it with a fractional symbol with natural number numerator and denominator.

2. Suppose that, in measuring a segment, $n = 87$ and that subunits are applied as follows: 1 eighth, 1 sixty-fourth, 1 five hundred twelfth, and so on, in an endless repeating pattern. Find the measure of the segment, naming it with a repeating binary numeral. Show that the measure is a rational number by naming it also with a fractional symbol with natural number numerator and denominator.

3. Construct a binary numeral that does not end or repeat. This represents an irrational number.

   a) Can this number be named by a fractional numeral with natural number numerator and denominator? Why?

   b) Can this number be named with a terminating numeral of the Hindu-Arabic type for some base b? With a nonterminating but repeating numeral? Why?

4. We have not proved that we can divide a unit into ten congruent parts. Let us suppose that this can be done and we wish to use decimal subunits, tenth-units, hundredth-units, and so on. In this case it will be more convenient to use decimal numeration rather than binary. In measuring a segment, we find $n = 48$ and subunits are applied as follows: 3 tenths, 2 hundredths, 3 thousandths, 2 ten-thousandths, and so on, in an endless, repeating pattern. Find the measure of the segment, naming it with a decimal numeral. Show that the measure is rational by naming it also with a fractional symbol with natural number numerator and denominator.

5. When we use half-, quarter-, eighth-units, and so on, it is convenient to use binary numeration. When we use tenth-, hundredth-, thousandth-units, and so on, it is convenient to use decimal numeration.

   a) For what kinds of subunits would base-three numeration be most convenient?

   b) For what kinds of subunits would base-b numeration be most convenient?

6. If the measure of a segment is $0.101010\ldots_2$, we could name the measure also with a fractional numeral or with a base-b numeral of any base. For which numbers b would we obtain a terminating numeral? What kinds of subunits would correspond naturally to those numeration bases?

7. If the measure of a segment is $4.333\ldots_{10}$, for what numeration bases would we obtain a terminating numeral? What kinds of subunits would correspond naturally to those numeration bases?

8. In the preceding text we have investigated, theoretically, the assigning of measures to segments, considering several cases. We have not shown that all cases are possible. If they are not, we may have wasted some time. Is there any logical mistake in considering cases in this manner?

3. MEASURES OF UNIONS OF SEGMENTS

Now that we have a way of assigning a measure to any segment, we may easily consider unions of a finite number of segments and the assigning of measures to such figures. If a figure consists of several disjoint segments, whether or not they are on a line, as in Illus. 5.8, we add the measures of the individual segments and agree to assign this sum as the measure of the figure. In Illus. 5.8(a) the measure of the figure is 7 and in Illus. 5.8(b) the measure of the figure is 11.

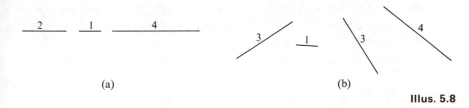

(a)   (b)

Illus. 5.8

It is important, in the preceding example, that the segments concerned be disjoint. If a figure consists of the union of nondisjoint segments, then the measure may not be the sum of the measures of the parts. For example, $m(\overline{AD})$ in Illus. 5.9 is not $m(\overline{AB}) + m(\overline{CD})$, even though $\overline{AD} = \overline{AB} \cup \overline{CD}$.

Illus. 5.9

EXERCISES

In Exercises 1 through 4 use this segment as a unit. ────────

1. Draw five different figures, each consisting of disjoint segments, and each having a measure of 10.

2. Draw a figure consisting of two collinear segments with measures 3 and 5, respectively, such that the figure has measure

   a) 8,   b) 5,   c) 6,   d) 7.

3. a) Draw a segment $\overline{AD}$ of measure 8.

   b) On the segment, mark points B and C such that A-B-C, $AB = 2$, and $BC = 3$.

   c) What is the measure of $\overline{AC}$? Of $\overline{BD}$? Of $\overline{CD}$?

   d) What is $AC + BD$?

e) What is $\overline{AC} \cup \overline{BD}$? What is $\overline{AC} \cap \overline{BD}$?

f) Find $AC + BD - BC$.

4. a) Draw a segment $\overline{AB}$ of measure 6, and on it mark point C such that $AC = 4$.

   b) What is $\overline{AC} \cup \overline{CB}$?

   c) What is the measure of $\overline{AC}$? Of $\overline{CB}$?

   d) What is $m(\overline{AC} \cup \overline{CB})$?

   e) This indicates that the measure of a point [here, $m(\overline{AC} \cap \overline{CB})$] should be what?

5. a) Sizes of nails are designated by assigning numbers (measures) to them. For example, there are "4-penny nails," "8-penny-nails," and so on. Is an 8-penny nail twice as long as a 4-penny nail?

   b) In devising measures of segments we have the (somewhat special) situation as follows: If a segment $\overline{AB}$ is n "times as long" as $\overline{CD}$ (meaning that n segments congruent to $\overline{CD}$ placed end-to-end will match $\overline{AB}$), then $m(\overline{AB})$ is a number n times as large as $m(\overline{CD})$. Why? Discuss the advantages of this kind of measure over the kind described in part (a).

   c) Find some other kinds of measures that are similar to that of part (a).

👁👁 *Exploratory exercises*

6. If two different segments, $\overline{AB}$ and $\overline{CD}$, have the same measure,

   a) are they necessarily congruent ($\overline{AB} \cong \overline{CD}$),

   b) would it be correct to say that $\overline{AB} = \overline{CD}$? $AB = CD$?

7. If two segments are congruent, do they necessarily have the same measure?

8. Consider the following and discuss its possible truth. Give a plausibility argument for your position. For any segments $\overline{AB}$ and $\overline{CD}$, if $\overline{AB} \prec \overline{CD}$, then $m(\overline{AB}) < m(\overline{CD})$.

9. Discuss the possible truth of the converse of the statement in Exercise 8.

4. SOME IMPORTANT PROPERTIES OF MEASURES

Now that we have developed a theory of assigning measures to segments and have had some experience with it, we are in a position to ask, and perhaps answer, questions such as those raised in some of the preceding exercises. Intuitively it seems clear, for example, that two segments are congruent if they have the same measure, and conversely. Also, if one segment is less than another, then its measure must be a smaller number, and conversely. These things are true and can be proved. We shall list them as theorems here because of their fundamental importance but omit the proofs.

Theorem 5.1 For any segments $\overline{AB}$ and $\overline{CD}$, $\overline{AB} \cong \overline{CD}$ if and only if $m(\overline{AB}) = m(\overline{CD})$.

Theorem 5.2 For any segments $\overline{AB}$ and $\overline{CD}$, $\overline{AB} \prec \overline{CD}$ if and only if $m(\overline{AB}) < m(\overline{CD})$.

We can also assign measures to certain figures other than segments. In the previous section we agreed to assign measures to figures which are unions of a finite number of disjoint segments in a certain way, by adding the individual measures. Now let us consider unions of segments that are not disjoint. Segments $\overline{AB}$ and $\overline{CD}$, as shown in Illus. 5.10, for example, have the nonempty intersection $\overline{CB}$ and their union is the segment $\overline{AD}$. If we should add the measures of $\overline{AB}$ and $\overline{CD}$ in this case, we would not obtain the measure of their union. The number obtained would be too large. If, however, we should subtract the measure of $\overline{CB}$, we would then obtain the measure of $\overline{AB} \cup \overline{CD}$, or $\overline{AD}$. Thus

$$m(\overline{AB} \cup \overline{CD}) = m(\overline{AB}) + m(\overline{CD}) - m(\overline{AB} \cap \overline{CD}).$$

The measure of the union of the two segments is the sum of their measures minus the measure of their intersection. This is true of any segments oriented as in Illus. 5.10. We shall list this as a theorem, although we shall not give a proof of it.

$A \qquad C \qquad\qquad B \qquad\qquad D$ **Illus. 5.10**

Theorem 5.3 For any collinear segments $\overline{AB}$ and $\overline{CD}$ for which $\overline{AB} \cup \overline{CD}$ is a segment, $m(\overline{AB} \cup \overline{CD}) = m(\overline{AB}) + m(\overline{CD}) - m(\overline{AB} \cap \overline{CD})$.

Let us now consider the union of two segments, $\overline{AB}$ and $\overline{BC}$, as in Illus. 5.11. In this case the union of the segments is a segment, as in the preceding example; that is, $\overline{AB} \cup \overline{BC} = \overline{AC}$. The intersection of the segments, $\overline{AB} \cap \overline{BC}$, however, is not a segment but consists of the single point B. Is it true in this case that $m(\overline{AB} \cup \overline{BC}) = m(\overline{AB}) + m(\overline{BC}) - m(\overline{AB} \cap \overline{BC})$? We have not considered assigning measures to single points, hence we cannot answer this question. If we assign measures to single points, then how shall we do it? This situation provides us with the motive for and the means of defining a measure for a point. If we wish to preserve the pattern of Theorem 5.3, we must assign the number zero as the measure of any single point. Let us agree to do this. In addition, if $\overline{AB} \cap \overline{CD} = \emptyset$, then $m(\overline{AB} \cap \overline{CD}) = 0$.

$A \qquad\quad B \qquad\qquad\qquad\qquad\qquad C$ **Illus. 5.11**

We shall now state concisely a way of assigning a measure to the union of any two segments, collinear or noncollinear.

For any segments $\overline{AB}$ and $\overline{CD}$, $m(\overline{AB} \cup \overline{CD}) = m(\overline{AB}) + m(\overline{CD}) - m(\overline{AB} \cap \overline{CD})$.

To find the measure of the union of three segments, we first find the measure of the union of two of them and then find the measure of that union with the third segment (Illus. 5.12).

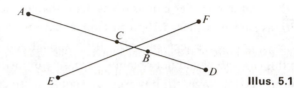

Illus. 5.12

Example. If $m(\overline{AB}) = 4$, $m(\overline{CD}) = 3$, $m(\overline{CB}) = 1$, and $m(\overline{EF}) = 5$, what is the measure of $\overline{AB} \cup \overline{CD} \cup \overline{EF}$?

We first find the measure of the union of two of the segments, say $\overline{AB} \cup \overline{CD}$, or $\overline{AD}$. This is $m(\overline{AB}) + m(\overline{CD}) - m(\overline{BC})$, which is $4 + 3 - 1$, or 6. Now consider $\overline{AD} \cup \overline{EF}$. Its measure is $m(\overline{AD}) + m(\overline{EF}) - m(\overline{AD} \cap \overline{EF})$. This is $6 + 5 - m(\overline{AD} \cap \overline{EF})$. Since $\overline{AD} \cap \overline{EF}$ is a single point, its measure is 0. Hence the measure of the figure is 11.

Any figure consisting of the union of a finite number of segments now has a measure which can be assigned as in the preceding example.* If a figure is the union of four segments, we would find the measure of the union of three of them as we have just done, and then find the measure of the union of that figure with the remaining segment.

EXERCISES

1. Find the measure of the figure in Illus. 5.13, using the method of the preceding example, by considering the figure to be:
   a) $\overline{AB} \cup \overline{BC} \cup \overline{BD}$,   b) $\overline{AC} \cup \overline{BD}$,   c) $\overline{AB} \cup \overline{BC} \cup \overline{CD}$.

Illus. 5.13

---

* Although no proof is given here, the measure so assigned is unique.

2. Find the measure of the figure in Illus. 5.14 by considering it to be:

a) $\overline{AC} \cup \overline{CE} \cup \overline{FG} \cup \overline{CN}$,   b) $\overline{AC} \cup \overline{CD} \cup \overline{DE} \cup \overline{GD} \cup \overline{BF} \cup \overline{CN}$.

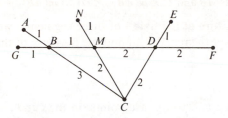

Illus. 5.14

3. a) Find the measure of the polygonal curve in Illus. 5.15.

b) Describe a way to find the length of any polygonal curve.

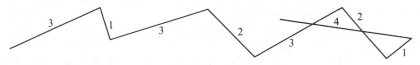

Illus. 5.15

Exploratory exercise

4. A nonpolygonal curve approximated by a polygonal curve is shown in Illus. 5.16.

a) What number would seem reasonable as an approximation of the measure of the curve?

Illus. 5.16

b) How could you obtain a better approximation to the length of the nonpolygonal curve?

5. MEASURES OF ONE-DIMENSIONAL FIGURES

Any union of a finite number of segments is a one-dimensional figure, and can be assigned a measure. There are other one-dimensional figures, such as the curve in Illus. 5.17. Measures can be assigned to nonpolygonal curves, although the manner of doing it is not always simple. In many cases, the length can be approximated by taking the length of a polygonal curve, as shown.

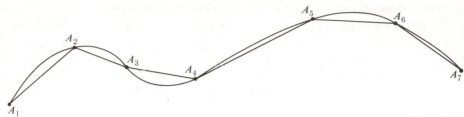

Illus. 5.17

The points A_1, A_2, and so on, are points on the curve, and these determine a polygonal curve. Ordinarily, the more points chosen and the closer together they are, the better will be the approximation of the length of the curve. We may often say that the length of the polygonal curve "approaches" that of the other curve as the number of vertices increases indefinitely and the maximum distance between adjacent vertices decreases toward zero.

Not every curve can be assigned a length. But if each curve of a family of curves can be assigned a length, then the union of a finite number of them can also. Any such figure is known as a *one-dimensional figure*. The one-dimensional measure, or length, of a polygon is known as its *perimeter*, and the one-dimensional measure of a circle is called its *circumference*.

EXERCISES

1. Draw a polygonal curve to a fairly large scale, and find its length in centimeters.

2. Draw a polygonal curve which has 8 sides and a length of 12 cm.

3. Draw a polygon to a fairly large scale and find its perimeter, both in centimeters and in inches.

4. a) Trace the curve in Illus. 5.18 on a piece of paper. Mark the endpoints and then four more points approximately equally spaced. Connect the points to form a polygonal curve. Approximate the length of the curve by measuring the polygonal curve.

Illus. 5.18

   b) Draw five more points on the curve, between the ones you have already marked. Draw a new polygonal curve, and again approximate the measure by measuring the polygonal curve. Compare your two approximations.

5. Draw a one-dimensional figure which is composed entirely of segments but which is not a polygonal curve. Find its measure.

Exploratory exercise

6. a) Using a compass, draw a circle with a 10-cm diameter. Mark four points on the circle, about equally spaced. Approximate the circumference by finding the perimeter of the polygon.

   b) Mark four more points between the others, draw another polygon with 8 sides, and again approximate the circumference by measuring the polygon.

   c) Repeat part (b), introducing 8 more points.

   d) Divide your answer to part (c) by the length of a diameter of the circle.

Circles

The circumference of a circle may be approximated as in Exercise 6 above by in-scribing a polygon and finding its perimeter. The circumference is found to be a number about three times the length of a diameter. We use the Greek letter π to denote the ratio of circumference to diameter.

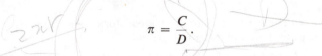

$$\pi = \frac{C}{D}.$$

This definition is an agreement to use the symbol "π" in this fashion.

An important fact, which we shall not attempt to establish at this point (cf. Chapter 6), is that π is the same number for all circles. If one circle, for example, has a diameter twice as long as that of another, then its circumference is also twice that of the first. Thus "π" is not a variable but, rather, a name of some particular number.

The number π has been approximated to a great many decimal places. It is approximately 3.1415926535. It is also approximated sometimes as $\frac{22}{7}$. This number π has been shown not to be a rational number, which means that the decimal numeral for the number does not terminate or repeat. The proof of this fact is quite difficult.

The definition of "π" above is given by an equation from which we may obtain the following:

$$C = \pi \cdot D, \qquad C = 2 \cdot \pi \cdot r.$$

The second equation follows from the first by the fact that a diameter of a circle is twice as long as a radius. These two equations are often regarded as formulas for finding the circumference of a circle.

EXERCISES

1. Find the circumference of a circle having a 4-in. diameter,

   a) using 3.14 as an approximation for π,

   b) using $\frac{22}{7}$ as an approximation for π.

2. Find the decimal numeral for $\frac{22}{7}$. As an approximation to π, it is correct to how many decimal places?

3. a) Draw two concentric circles having diameters 10 cm and 20 cm long, respectively. Draw six diameters of the larger circle.

   b) Connect the ends of these diameters to form a polygon inscribed in the larger circle. Similarly, connect the points where these diameters meet the smaller circle to form a polygon inscribed in the smaller circle.

   c) By measuring, find the perimeters of the two polygons and compare them.

4. How many turns of the handle are required to raise the old oaken bucket from the bottom of the well if the drum on which the rope is wound has a 2-ft diameter and the well is 30 ft deep?

5. a) Two circles have radii of 10 in. and 100 in., respectively. If the length of each radius is increased by 1 in., how does this affect the circumference of each?

   b) If the circumference of a circle is $2 \cdot \pi \cdot r$ in., what will it be if we add x in. to the radius?

6. Imagine a band stretched tight around the equator of the earth. If a 10-ft strip is added to the band, how will this affect the radius?

7. If the radius of a wheel is 14 in., how many revolutions does it make in 1 mi?

8. How many revolutions will gear A make for every revolution of gear B (Illus. 5.19)?

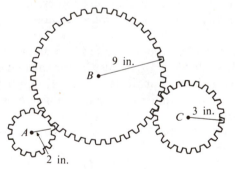

9 in.

B

C 3 in.

A

2 in.

Illus. 5.19

9. Here is a well-known argument that $\pi = 2$. Illustration 5.20 resembles the yin-yang symbol of the Orient. Suppose that diameter $AB = 2$. Then the large semicircle has measure π. The two semicircles (from A to C and from C to B) each have measure $\pi/2$. Thus their total length is π. The sum of the measures of the four next smallest semicircles is also π. In a similar manner, if we continue to construct sets of semicircles (8, 16, 32, etc.) the total length of each set is π. Now if we continue these constructions, the "curve" (set of semicircles) will always have length π; however, as the radii of the semicircles gets smaller, the "curve" gets closer and closer to the diameter $\overline{AB}$. Thus $\pi = 2$. What is wrong with this argument?

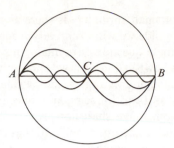

6. ANGLE MEASURE

To answer the question "How large is an angle?" we shall need something with which to compare it. When measures of segments were devised, we chose a segment, arbitrarily, as a *unit* and compared other segments with it. To devise angle measures we shall proceed in an analogous manner. We shall choose, arbitrarily, some angle to serve as a unit. Its measure will of course be 1. Suppose the smaller angle in Illus. 5.21 is the unit angle and we then wish to find the measure of the larger angle. By Axiom 4.6, there is, on the side of $\overleftrightarrow{QR}$ containing P, a unique ray $\overrightarrow{QT_1}$ for which $\angle B \cong \angle RQT_1$. Also, on the side of $\overrightarrow{QT_1}$ containing P (on the side opposite R) there is a unique ray $\overrightarrow{QT_2}$ for which $\angle B \cong \angle T_1QT_2$. This process can be continued until finally a ray $\overrightarrow{QT_n}$ is found which coincides with $\overrightarrow{QP}$ or which lies in the exterior of $\angle PQR$. If $\overrightarrow{QT_n} = \overrightarrow{QP}$, then the number n will be assigned as the measure of the large angle.

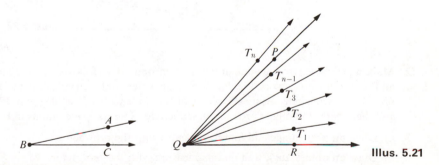

If $\overset{\circ}{\overrightarrow{QT_n}}$ is in the exterior of the angle, then the measure of $\angle PQR$ will be a number between $n - 1$ and n. In this event we begin to consider subunits, as we did for measures of segments. Since every angle can be bisected, we know that half-units, quarter-units, eighth-units, and so on, can certainly be used. Other kinds of subunits are, however, more commonly used.

As in developing measures for segments, certain questions arise in connection with angle measures. If subunits are necessary, does one eventually "come out even" by taking a subunit small enough? Is there any guarantee that by placing enough unit angles side by side we will eventually reach the other side of the large angle? The answers to these questions are analogous to the corresponding ones for segments. No matter how small a unit we choose, we can reach the other side of any angle by placing enough unit angles side by side. This does not have to be an assumption for angles, as it was for segments, since it follows from Archimedes' Axiom and the other axioms. We shall not prove it here, however. When subunits are used, we may not "come out even." In that event, the endless pattern of subunits (and the corresponding numeral) may repeat or it may not. If repetition occurs, the angle being measured will be assigned a rational number for its measure. If not, then an irrational number will be assigned.

EXERCISES

1. Use the angle in Illus. 5.22 as a unit.

    a) Copy it and cut it out with scissors. Draw a ray to serve as one side of an angle. Then using the unit angle, construct an angle with measure 6.

    b) Bisect the unit angle to obtain a half-unit, and bisect a half-unit to obtain a quarter-unit.

    c) Construct an angle $\angle B$ for which $m(\angle B) = 3\frac{3}{4}$.

    d) Draw an angle whose measure is between $4\frac{1}{2}$ and $4\frac{3}{4}$.

Illus. 5.22

2. Make a protractor from a piece of paper or cardboard, as follows: Draw a large circle and cut it out with scissors. Fold it along a diameter, and then cut along the diameter, obtaining a semicircle. Mark the center of the circle with a V, as shown in Illus. 5.23, and then mark unit angles around the semicircle. Use the same unit as in Exercise 1.

3. a) Draw an acute angle and find its measure with the protractor.

    b) Draw an obtuse angle and find its measure with the protractor.

4. What angle unit is most often used? What are the most commonly used subunits?

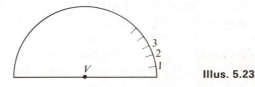

Illus. 5.23

Exploratory exercises

5. If two different angles, ∠ A and ∠ B, have the same measure,

 a) are they necessarily congruent (that is, ∠ A ≅ ∠ B)?

 b) would it be correct to say $m(∠ A) ≅ m(∠ B)$?

6. If two different angles are congruent, do they have the same measure? Why?

7. Consider the following, discuss its possible truth, and give a plausibility argument for your position: for any angles ∠ A and ∠ B, if ∠ A ≺ ∠ B (according to Definition 4.2), then $m(∠ A) < m(∠ B)$.

8. If an angle A is "twice as large" as an angle B, does $m(∠ A) = 2 · m(∠ B)$?

9. Suppose we were to define a measure for angles as follows: To find the measure of an angle ∠ ABC, choose a unit segment (Illus. 5.24). On each ray there is a (unique) point making a segment congruent to the unit segment. (Thus $AB = BC = 1$.) Now, there is a segment $\overline{AC}$, and its measure is of course a number. Assign this number as the measure of the angle.

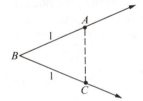

Illus. 5.24

 a) Would an angle measured in this manner have a unique number for its measure? [*Hint:* Consider what would happen if a different unit segment were used.]

 b) Would a larger angle have a greater measure than a smaller angle?

 c) What would be the measure of a straight angle?

 d) What would be the measure of a right angle?

 e) Would congruent angles have the same measure? Would angles with the same measure be congruent?

 f) If an angle ∠ A were "twice as large" as angle ∠ B, would $m(∠ A) = 2 · m(∠ B)$?

 g) Discuss the advantages or disadvantages of this kind of a measurement process for angles.

 h) If instead of using the length of segment $\overline{AC}$, we were to use the measure of a circular arc, radius of length 1, from A to C, how would that kind of measurement process compare with the one just described?

7. SOME PROPERTIES OF ANGLE MEASURES

The properties of angle measures somewhat parallel those of measures of one-dimensional figures. For example, if two angles are congruent, then they have the same measure, and conversely. If one angle is less than another, then its measure is a smaller number. We shall state these theorems without proof.

Theorem 5.4 For any angles $\angle A$ and $\angle B$, $\angle A \cong \angle B$ if and only if $m(\angle A) = m(\angle B)$.

Theorem 5.5 For any angles $\angle A$ and $\angle B$, $\angle A \prec \angle B$ if and only if $m(\angle A) < m(\angle B)$.

There is a limit to the size of an angle, since the largest angle we have defined is a straight angle. There is no largest segment, however, since we may always "add" a segment to any segment, however large, to obtain a larger one. Thus, in this respect, angle measures and segment measures differ.

Measures of segments are additive. That is, the measure of the union of several segments is their sum, provided, of course, that the intersection of any two of them has measure zero. The additive property of measures has its parallel for angle measure. Consider, for example, the figure in Illus. 5.25, consisting of two adjacent angles, $\angle ABD$ and $\angle DBC$. The measure of $\angle ABC$ is the sum of the measures of the other two angles. In this case, the large angle is not the *union* of the two small angles; hence we cannot verbalize the additive property as we did for segments. Nonetheless it is clear that measures for angles have an additive property completely analogous to that of measures for segments.*

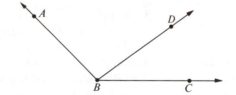

Illus. 5.25

8. TWO-DIMENSIONAL MEASURE OR AREA

One-dimensional measure, called *length*, or *distance*, applies to curves or unions of curves. The interior of a simple closed curve is a different kind of figure, and we shall devise another kind of measure for it. This kind of measure can also be devised for a figure consisting of a simple closed curve, together with its interior, as well as for unions of such figures. This kind of figure is referred to as *two-dimensional*, and the measure of a two-dimensional region will be known as its *area*. Note that the words *region* and *area* are not synonyms. A region is a set of points and an area is a measure, or number.

To answer the question "How large is a two-dimensional region?" we shall need something with which to compare it. We shall, as for segments and angles,

---

* Actually, the large angle is the union of the two small ones, "minus" the half-line $\overset{\circ}{BD}$. If one wishes to carry the analogy further, he might say that the measure of an angle formed by two adjacent angles is the sum of the measures *minus* the measure of their intersection, where of course the intersection is a ray, whose angle measure we would define to be zero.

choose some figure of the type being considered as a *unit*, assigning it a measure
of 1. We might, for example, choose as a unit the interior of a simple quadrilateral,
or a right isosceles triangle together with its interior, or in fact any figure of the
proper type, as shown in Illus. 5.26.

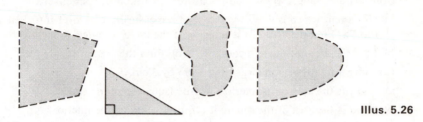

Illus. 5.26

To find the measure (area) of a two-dimensional region, we "fill it up," so to
speak, with figures congruent to the unit to determine how many are required.
This is shown in Illus. 5.27. In this example, an isosceles right triangular region
(the triangle and its interior) is being used as a unit, and the larger region is being
"filled in" with these units.

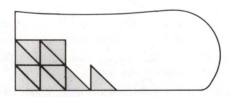

Illus. 5.27

It is easy to see that certain difficulties may arise. If an irregularly shaped unit
is used, for example, it will be quite difficult to fill the region being measured without leaving gaps or overlapping units. On the other hand, if a regularly shaped
unit, such as the triangular region, is used to measure a region that is irregularly
shaped, then the difficulty will be almost as great. Thus it becomes apparent
immediately that two-dimensional measures are more complicated than one-
dimensional measures. The basic principle is the same in both cases, however.
We try to "fill in" the region to be measured with units and subunits.

EXERCISES

1. a) Draw an isosceles right triangle whose congruent sides each measure 5 cm.

   b) For a unit, use the interior of an isosceles right triangle whose congruent sides
      measure 1 cm. Fill in the large triangular region with units. What is the area of
      the large triangular region?

2. a) Draw a rectangle whose sides measure 4 cm and 5 cm, respectively. Fill in the interior with units, and find its area, using the same unit as for Exercise 1.

   b) Use a unit which is the interior of a square, 2 cm by 2 cm. Draw a rectangle like that of part (a) and find the area of its interior.

3. Consider a rectangle whose sides measure 4 in. and 5 ft, respectively.

   a) Use a unit which is a rectangular region measuring 1 in. by 1 ft (called an "inch-foot"). Find the area of the interior of the large rectangle in inch-feet.

   b) Find the area in square inches.   c) Find the area in square feet.

4. Consider a square region measuring 3 cm by 3 cm.

   a) What is the area of its interior, in cm² (square centimeters)?

   b) What is the area of the square itself, not including the interior?

   c) What is the area of the square together with its interior?

Exploratory exercises

5. In Illus. 5.28 an isosceles right triangular region is being used as a unit to measure the interior of a polygon. The angle of the polygon at A is a right angle. One unit has been placed with its right angle at A, and we know that the sides of the triangle lie on the sides of the polygon here.

   a) How do we know this?

   b) How could you, using compass and straightedge only, find point D? What theorems would you cite to prove that you are correct?

   c) Angle $\angle ABD$ appears to be a right angle. Can you prove that it is, using only the theorems developed previously in this text?

   d) If a right angle $\angle DBE$ is now constructed, will A, B, and E be collinear if (i) angle $\angle ABD$ is a right angle (why?), (ii) angle $\angle ABD$ is not a right angle (why?)?

   e) If we had proved that the sum of the measures of the angles of a triangle is 180°, could you then prove that $\angle ABD$ is a right angle (having measure 90°)?

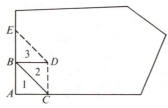

Illus. 5.28

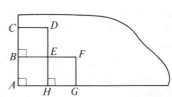

Illus. 5.29

6. We have spoken of the possibility of using a square region as a two-dimensional unit. We would expect to be able to place squares together like this, so we would hope that the sides of adjacent squares lie on a line (Illus. 5.29). For example, we would want to know if A, B, and C are collinear. This will be true when $\angle ABE$ is a right angle.

   a) Draw a square $ABCD$, with a right angle at A. Remember that a square is defined to be a parallelogram with four congruent sides (that is, rhombus) and one right angle.

b) Draw $\overline{DB}$, and consider triangles ABD and CDB. Prove that they are congruent.

c) What can you now conclude about $\angle C$?

d) What can you now conclude about $\angle D$ and $\angle B$?

e) If we had proved that the sum of the measures of angles of a triangle is 180°, could you prove that angles $\angle D$ and $\angle B$ are right angles? If so, do that.

Some of the questions raised in the preceding set of exercises point up the need for further formal development of our geometry before we can continue with the theoretical development of two-dimensional measures. We wish, of course, to be able to establish rigorously the nature of the so-called "filling in" of a region by unit regions. We cannot do this experimentally by cutting and pasting but must use the theorems, etc., developed so far.

The principal question seems to be whether unit regions such as isosceles right triangular regions will actually fit together as expected. When square regions are considered as units, a similar question arises. If, as in Illus. 5.30(a), we begin placing isosceles right triangular regions in the interior of an angle of a polygon which is a right angle, we know that units 1 and 2 will fit as shown. Unit 3, however, might not fit as expected; $\overline{BE}$ will not lie on $\overrightarrow{AP}$ unless $\angle ABD$ is a right angle. We cannot prove that $\angle ABD$ is a right angle with the geometry developed thus far but can easily do so once we have proved that the sum of the measures of the angles of a triangle is 180°.

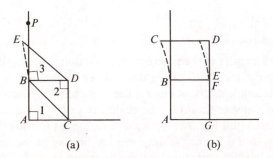

(a)   (b)   **Illus. 5.30**

Similarly, in Illus. 5.30(b) we can place a square unit in the corner, as shown. We can place a second one as shown, and will know that $E = F$. We will not, however, know whether A, B, and C are collinear, as we would hope is the case. If we could establish that the four angles of a square are all congruent, this troublesome point would be eliminated, but we have not yet done this. It can be done once we know that the sum of the measures of the angles of a triangle is 180°.

To prove this, we shall need a new assumption here, as well as some of its consequences. These will be discussed further in Chapter 6.

The properties we shall need are as follows:

1. For any line *l* and any point *P* not on *l*, there is one and only one line in the plane of *l* and *P* that contains *P* and is parallel to *l*.
2. The sum of the measures of the angles of any triangle is 180°.
3. The opposite sides and opposite angles of a parallelogram are congruent.

We shall now proceed to develop a theory of two-dimensional measures and shall use these properties, postponing what proof is needed until Chapter 6.

Area of a Rectangular Region

From the properties listed in the preceding section, it follows that any rectangle is a parallelogram with opposite sides congruent and with four right angles. Let us now consider finding the area of a rectangular region, using a rectangular region as a unit, as in Illus. 5.31(a).

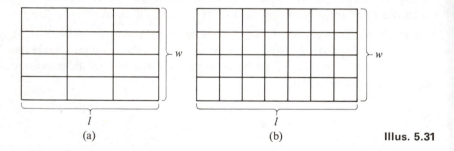

(a)   (b)   **Illus. 5.31**

In this case, the length *l* of the rectangle is three times the length of the unit and the width is four times the width of the unit. Since all the angles of the units and also of the large rectangle are right angles, the units fit nicely into the region being measured, and the measure (area) of the region is 12 (of the units).

A commonly used unit for area is a square region, as in Illus. 5.31(b). The use of such a unit often (but not always) simplifies the finding of measures, particularly if the square is of such size that its sides have linear measure 1. If the *inch* is being used as a linear unit, for example, then a convenient unit is a square region whose sides all have a measure of 1 in. It is clear that in this case the measure of the region shown in Illus. 5.31(b) is 28 units, and this may be obtained by multiplying the length by the width.

It may happen, of course, that the length or the width, or both, are such that a whole number of units does not exactly fit the rectangular region being measured. This is shown in Illus. 5.32. The length of the rectangle is $m + p$, where m is a whole number and p is some other number between 0 and 1. The width is $n + q$, where n is a whole number and q is a number between 0 and 1. There are $m \cdot n$ units that fit into the shaded region, as was the case in Illus. 5.31(b). We now consider the rest of the region being measured, consisting of regions I, II, and III. We can think of region A as a subunit. If q is $\frac{1}{2}$, then it is a half unit, with area $\frac{1}{2}$. If q is $\frac{1}{4}$, then region A is a quarter unit, with area $\frac{1}{4}$, and so on. Whatever q may be, the area of the subunit A is q. Similarly, the area of the subunit B is p. Now, m of the subunits, of area q, will fill region I; n of the subunits of area p will fill region II. Hence the area of region III may be seen to be $p \cdot q$. For example, if $q = \frac{1}{2}$ and $p = \frac{1}{4}$, then region III is half of a quarter-unit, or $\frac{1}{4}$ of a half-unit. Thus its area would be $\frac{1}{8}$.

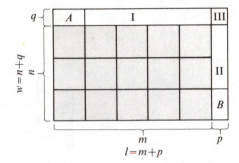

Illus. 5.32

The total area of the rectangular region has now been shown to be $m \cdot n + m \cdot q + n \cdot p + p \cdot q$. The question to which we seek an answer is this: Can we find the area of the region by multiplying length by width, as we did before? The answer seems to be yes, for

$$l \cdot w = (m + p)(n + q) = m \cdot n + m \cdot q + n \cdot p + p \cdot q,$$

which is the same as obtained above.

From the preceding discussion it seems plausible that the area of a rectangular region may be obtained by multiplying its length by its width. Actually, there are other logical questions to be considered before one can rigorously conclude that this is the case, but we shall accept the fact now and proceed to attack the problem of finding areas of other kinds of regions.

Sweeps

Suppose we consider a segment to move or *sweep* across a rectangular region, as in Illus. 5.33. The measure of the segment is w, which is the width of the rectangle. If the sweeping segment starts on one side of the rectangle and sweeps to the other side, it will have moved a distance l, which is the length of the rectangle. The area of the region over which it sweeps is, of course, $l \cdot w$. In other words, it is the product of the length of the sweeping segment and the distance it sweeps. Of course, the distance swept is measured along a line perpendicular to the sweeping segment.

There is little difficulty involved in finding areas of rectangular regions by sweeps. The same notion can be used to find areas of other regions. In Illus. 5.34, for example, the region being swept is not rectangular. The sweeping segment maintains a constant measure w, and sweeps a distance l, measured along a line perpendicular to the sweeping segment. The area of the region is then the product of the measure of the sweeping segment and the distance it sweeps, $w \cdot l$. The fact that this works can be made more plausible by cutting the region and rearranging it to form a rectangular region, as shown in Illus. 5.35.

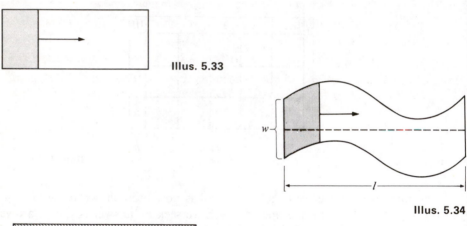

Illus. 5.33

Illus. 5.34

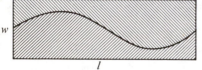

Illus. 5.35

If a figure is such that a sweeping segment varies in length as it sweeps across the figure, the method above may still be used with some adaptation. It will be necessary to find some sort of average length of the sweeping segment.

EXERCISES

Exploratory exercises

1. a) Draw a trapezoid with altitude 8 in. and bases measuring 1 in. and 9 in., respectively, as shown in Illus. 5.36. Angle M is a right angle.

   b) Mark points A, B, and C, dividing the side into four congruent segments, and draw segments $\overline{AP}$, $\overline{BQ}$, and $\overline{CR}$ perpendicular to $\overline{MN}$.

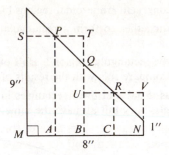

Illus. 5.36

c) Complete rectangles *BMST* and *BNVU*. Approximate the area of the trapezoidal region by finding the sum of the areas of the rectangular regions.

d) Average the heights of the two rectangular regions. Multiply this average by 8 (the length of side $\overline{MN}$).

e) Average the lengths of the two bases of the trapezoid. Multiply this average by 8.

2. a) Draw another trapezoid with dimensions as shown in Illus. 5.37. Mark points *A, B, C, D, E, F,* and *G*, dividing the side into eight congruent segments. Construct four rectangles, as shown.

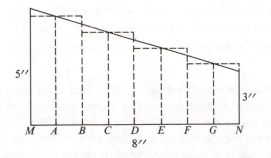

Illus. 5.37

b) Approximate the area of the trapezoidal region by finding the sum of the areas of the rectangular regions.

c) Average the heights of the four rectangular regions. Multiply this average by 8 (the length of the side $\overline{MN}$).

d) Average the lengths of the two bases of the trapezoid. Multiply this average by 8.

3. Repeat the procedure with the trapezoid in Exercise 2, this time using eight rectangular regions to approximate the area. If you were to use 500 rectangular regions, how would you expect the result to compare?

4. Consider a segment sweeping the trapezoidal region of Exercise 1, left to right. If the area is to be the product of the *average* length of the sweeping segment and the distance it sweeps, what is the simplest kind of average to take?

5. Suppose you have two congruent trapezoidal regions like that of Exercise 1.

    a) Can you place them together to form a rectangular region? Make a sketch to show how.

    b) What is the area of the rectangular region? Half of it should be the area of the trapezoidal region. Compare this with the results of Exercises 1, 2, and 3.

6. a) Copy this figure (Illus. 5.38) and draw rectangles in it as shown. Note that the heights of the rectangles vary but all have the same width w.

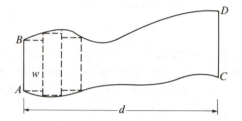

                                                              Illus. 5.38

    b) Measure the height of each; also measure the common width. Calculate the area of each rectangular region. Add, and thus approximate the area of the region.

    c) Find the average height. Multiply it by the distance d.

    d) Average the lengths AB and CD. Multiply this average by the distance d.

7. a) Repeat Exercise 6, this time using twice as many rectangles.

    b) Is your approximation of the total area better this time?

    c) If you were to use 100 rectangles, would you expect to obtain a better approximation?

Activities outlined in the preceding exercises lead us to make some conclusions about the use of sweeps to find areas. If a segment changes its length uniformly, that is, keeping its endpoints on two lines as it sweeps, then the average length of the sweeping segment may be found by merely averaging its initial and final length.

If the sweeping segment does not change its length uniformly, then there seems to be no easy way in general to find the proper average. In such cases an approximation to the desired average may be found graphically by drawing a number of equally spaced segments, representing various positions of the sweeping segment, and averaging the lengths of them. This is shown in Illus. 5.39. To see this, let us suppose that there are n of these segments and the lengths are $h_1, h_2, h_3, \ldots, h_n$, respectively. Let us further suppose that the distance between these segments is a. An approximation to the area of the first region, starting at the left, is $a \cdot h_1$. The area of the second region is about $a \cdot h_2$, that of the third is $a \cdot h_3$, and so on. Thus an approximation to the total area is

$$a \cdot h_1 + a \cdot h_2 + a \cdot h_3 + \cdots + ah_n,$$

or

$$a(h_1 + h_2 + h_3 + \cdots + h_n).$$

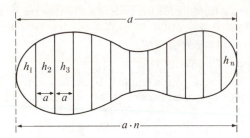

Illus. 5.39

Now, multiplying by n/n, we obtain the equivalent expression

$$(a \cdot n) \cdot \left(\frac{h_1 + h_2 + h_3 + \cdots + h_n}{n} \right).$$

We note that $a \cdot n$ is the distance d, and

$$\frac{h_1 + h_2 + h_3 + \cdots + h_n}{n}$$

is the average length of the n segments.

Now that the validity of the principle of sweeps has been made clearer, we shall use it to find areas of several other kinds of regions.

EXERCISES

1. Using the concept of a sweep, find a formula for the area of any trapezoidal region.

2. Using the concept of a sweep, find a formula for the area of a right triangular region.

3. Using the concept of a sweep, find a formula for the area of any triangular region.

4. Using the concept of a sweep, find a formula for the area of any parallelogramic region.

5. Using sweeps, find a formula for the area of an obtuse triangular region ABC, as shown in Illus. 5.40, by:

   a) finding the area of region ADC,

   b) finding the area of region BDC, and

   c) subtracting. Show that your formula is equivalent to that obtained in Exercise 3.

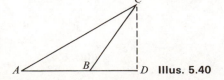 **Illus. 5.40**

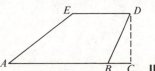 **Illus. 5.41**

6. Using sweeps, find a formula for the area of an obtuse trapezoidal region, as shown in Illus. 5.41, by:

   a) finding the area of region $ACDE$,

   b) finding the area of region BCD, and

   c) subtracting. Show that your formula is equivalent to that obtained in Exercise 1.
7. The area of a right trapezoidal region, as shown in Illus. 5.42, has been found to be

$$\frac{m_1 + m_2}{2} \cdot h.$$

Assume this, and without using sweeps, find a formula for the area of a trapezoidal region as shown at the right above. The bases have length b_1 and b_2, respectively, and the height is h.

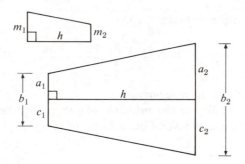

Illus. 5.42

9. PROPERTIES OF TWO-DIMENSIONAL MEASURES

One-dimensional measures and angle measures, as we have developed them, are additive. From our discussions and the exercises, it should be clear intuitively that areas are also additive. A region may not have an area if its boundary is very complicated. But if each of two regions has an area (is measurable), then their union and intersection are measurable and the area of the union is the sum of their measures, minus the area of their intersection. For example, the area of the region shown in Illus. 5.43 would be the area of region X plus the area of region Y minus the area of the intersection, which is the segment $\overline{AB}$. In order for this to be sen-

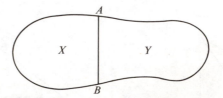

Illus. 5.43

sible, it appears that we should agree to assign an area of zero to any one-dimensional figure. We shall agree to do this, but it is still difficult to prove that we have the additive property for two-dimensional measures. In fact, this must actually be an axiom, although we shall not state it formally.

Circular regions   The concept of a *sweep* can also be used to find a formula for the area of a circular region, but the sweep concept must be modified somewhat.

EXERCISES

Exploratory exercises

1. a) Draw a circular region with a 10-cm diameter, and use the principle of sweeps, as already developed, to approximate the area. To do this, draw 10 or 15 equally spaced chords, as shown in Illus. 5.44, measure them, and find the average length. Then multiply by the distance the sweeping segment travels.

Illus. 5.44

   b) Can you find a simple way to compute the average length of the sweeping segment in this case?

2. a) Draw a circle and two perpendicular diameters. Connect the ends of the diameters to form an inscribed square.

   b) Find the area of the square region. Note that it consists of two triangular regions having a diameter as a common base and with altitudes $D/2$.

   c) Is the area of the square region greater or less than that of the circular region?

3. a) Draw a circle and two perpendicular diameters. Draw tangents to the circle at the ends of these diameters to form a circumscribed square.

   b) Find the area of the square region. Is it greater or less than that of the circular region?

4. a) Draw a circle with a 10-cm diameter, and draw a central angle of 30° (that is, vertex at the center of the circle). The angle intersects the circle at two points. Call them A and B. Draw the segment $\overline{AB}$.

   b) Measure $\overline{AB}$. Calculate the length of the arc AB. (It is 30/360 of the circumference.)

   c) Find the area of the triangular region OAB, where O is the center of the circle.

5. Repeat Exercise 4, using a 3°-angle. Compare the results and discuss.

The concept of a *sweep* may be used to find the area of a circular region. In this case, we shall consider a radius to do the sweeping, keeping one end fixed at the center of the circle like a spoke in a wheel, as shown in Illus. 5.45. It is clear that the sweeping segment (or radius) does not change in length as it sweeps the interior of the circle. The distance traveled by each point of the radius is different, however. For example, the distance traveled by the endpoint at the center of the circle is zero, while the distance traveled by the other endpoint is $2\pi r$ (which is the circumference of the circle). Two questions now arise: Can we find the average distance swept by all of these points, and does the distance traveled by each point change uniformly as we move from one endpoint to the other?

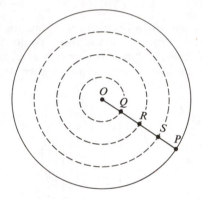

Illus. 5.45

It is not difficult, of course, to see intuitively that the average distance could be obtained by averaging the distances traveled by each endpoint. This average would be $0 + 2\pi r/2 = \pi r$. Since the length of the sweeping segment is r, the area would be $\pi r \cdot r$ or πr^2.

To determine whether the distance traveled by each point changes uniformly, we consider some points on the sweeping radius, as shown. Here, $OQ = QR = RS = SP = \frac{1}{4}r$. The distance traveled by the points Q, R, and S is the circumference of the concentric circles (shown dashed). Since the radius of the circle containing Q is $OQ = \frac{1}{4}r$, Q travels $2\pi(\frac{1}{4}r)$ or $\frac{1}{2}\pi r$. In a similar manner, R travels $2\pi(\frac{1}{2}r)$ or πr, and S travels $2\pi(\frac{3}{4}r)$ or $\frac{3}{2}\pi r$.

We now note that the *change* in distance traveled is the same as we move along the sweeping segment from O to Q to R, etc.

Note also, that the average distance traveled by the five points O, Q, R, S, and P is

$$\frac{0 + \frac{1}{2}\pi r + \pi r + \frac{3}{2}\pi r + 2\pi r}{5} = \pi r.$$

Hence the area swept is the product of the length of the sweeping segment (r) and the average distance swept by all of the points (πr) or πr^2. A formula for the area of the interior of a circle is therefore $A = \pi r^2$.

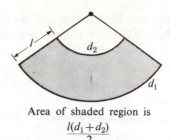

Area of shaded region is

$$\frac{l(d_1+d_2)}{2}$$

Illus. 5.46

The principle of sweeps may also be applied in situations such as that of Illus. 5.46, where the sweeping segment is only part of a radius. The area of the region swept (shaded area in the figure) is the product of the length of the sweeping segment and the average of d_1 and d_2, that is, $l(d_1 + d_2)/2$.

EXERCISES

1. Find the area of a circular region having an 18-in. radius
   a) in square inches,
   b) in square feet,
   c) in inch-feet. [*Hint:* In the formula πr^2, r appears twice. Consider the formula to be $\pi r \cdot r$ and substitute for r twice, once in inches, and once in feet.]

2. Find the area of a right circular cylinder whose altitude is 10 cm and whose radii measure 5 cm. Draw a picture and label the dimensions.

3. Find the lateral area of a right circular cone, if a radius of the base measures 10 cm and the slant height (length of an element of the cone) measures 20 cm. Draw a picture and label the dimensions.

4. Find the lateral area of a right pentagonal prism whose height is 16 cm, where the perimeter of the bases is 27 cm. Draw a picture and label the dimensions.

5. Find the lateral area of a right hexagonal pyramid, if the base is a regular hexagon, 2 in. on a side, and the altitude of each triangular lateral face is 8 in. Draw a picture and label the dimensions.

6. Find the area of a circular ring (annulus) as shown in Illus. 5.47.

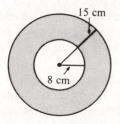

Illus. 5.47

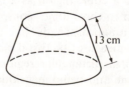

Illus. 5.48

7. Find the lateral area of the figure in Illus. 5.48. (It is called a "frustrum" of a cone.) A radius of the bottom base is 10 cm and a radius of the upper base is 7 cm. The slant height is 13 cm.

8. Show that the sweep principle illustrated in Illus. 5.46 is correct, by considering the region swept to be

   a) approximately trapezoidal,

   b) the "difference" between two regions swept by radii of different lengths.

9. If the area of a circle is equal to its circumference, what is the measure of its radius?

10. Find the length of b, where $a = 2\sqrt{2}$ and the area of the annulus is equal to the area of the inner circular region (Illus. 5.49).

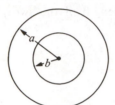

Illus. 5.49

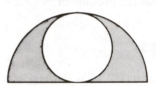

Illus. 5.50

11. In Illus. 5.50 the diameter of the small circle has the same measure as the radius of the semicircle. How does the area of the shaded region compare with the area of the small circular region?

10. THREE-DIMENSIONAL MEASURE OR VOLUME

In Chapters 3 and 4 we defined various kinds of surfaces such as cylinders, cones, spheres, and polyhedrons. The interior of such a figure (or the union of the interior and the surface) is a three-dimensional region and the measure of such a region is known as its *volume*. A completely rigorous development of volume is beyond the scope of this book, and we shall therefore treat the subject of volume on a more or less intuitive basis.

To determine the volume of a three-dimensional region we shall, as before, choose as a unit some figure of the general type being considered and assign it a measure of 1. Various three-dimensional regions might be used, but we shall choose a cubic region in analogy with our choice of a square region as an area unit. A cube is a rectangular parallelepiped with congruent faces and congruent edges. The use of a cubic region as a unit provides for a rather natural extension from one to two to three dimensions, since each face of the cube is a unit of area and each edge a unit of length.

When measures of two-dimensional regions were discussed, the formula for the area of a rectangular region, $A = l \cdot w$, was taken as the basic relationship between one and two-dimensional measures. The value of such a relationship is that it provides us with a method of determining area indirectly by means of linear measure. Formulas for determining the area of other kinds of two-dimensional regions were then deduced from this basic relationship. We shall proceed in a similar way for three-dimensional regions. That is, we shall choose a basic relationship (formula) which shows how to calculate volume in terms of area and length.

Volumes of Rectangular Parallelepipeds*

Consider a rectangular parallelepiped with length, width, and height, 7, 4, and 3, respectively, as shown in Illus. 5.51(a). If we "fill up" the interior with cubic regions, it is intuitively clear that the number of units in the first layer is the same as the number of two-dimensional units (square regions) in the base, which is $4 \cdot 7$, or 28. The third linear measure then determines the number of these layers and the total number of units is the product $3 \cdot 4 \cdot 7$, or 84 (Illus. 5.51b).

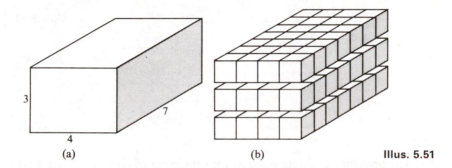

(a) (b) **Illus. 5.51**

This example makes it plausible that the volume of a rectangular parallelepiped may be obtained by multiplying its length by its width by its height. In other words, we multiply the area of the base by the height. We shall proceed, using the formula

$$V = lwh$$

for the volume of a rectangular parallelepiped.

Volumes of Prisms

A rectangular parallelepiped is a rectangular prism having rectangles for all faces. The volume of such a prism may be found by multiplying the area of the base by the height. We shall now compare the volume of such a prism with the volume of a prism whose base is rectangular but whose lateral faces are not all rectangles, as

---

* When we speak of the volume of a rectangular parallelepiped, we, of course, mean to include the interior, since the parallelepiped is a surface and therefore has zero volume. Since the meaning is clear, this should cause no problem and we will speak of the volumes of cones, cylinders, spheres, etc., in the same manner.

shown in Illus. 5.52. The second prism shown is like the first, except that it appears to "lean" to one side. Let us imagine that the rectangular prism is composed of a stack of rectangular cards and that the second one is obtained from the first by pushing the stack of cards to the side, as shown in Illus. 5.53. Since the volume of the rectangular prism is the sum of the volumes of the cards and pushing them to one side would not change the volume, the two prisms must have the same volume. The volume of the rectangular prism is the product of the height and the area of the base, and therefore the volume of the second is also the product of the height (measured perpendicular to the bases) and the area of the base.

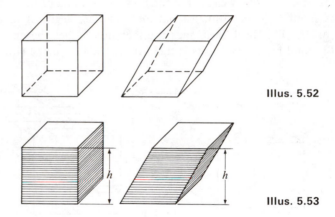

Illus. 5.52

Illus. 5.53

   The preceding example suggests that the concept of a *sweep* might be used as an aid in determining volume formulas, as it was for area formulas. In the case of areas we imagined a segment sweeping across a region. In the case of volumes we imagine some two-dimensional region sweeping through a three-dimensional region. In the case of prisms, we may imagine that a base of the prism sweeps through the figure. The volume is obtained by finding the product of the sweeping region (the area of a base) and the distance it sweeps, measured perpendicular to the sweeping region. The volume is thus the product of the area of a base and the height. If we use the sweep concept, it is then easy to see that the volume of any prism is the product of its height and the area of a base, whether that base be rectangular, triangular, or of any shape. In fact, the volume of any cylinder would be the product of the height and the area of a base.

   The volume of any cylinder is the product of its height and the area of a base.

   The concept of a sweep would also enable us to find the volume of a figure like that shown in Illus. 5.54, which we might consider to be a "bent" prism or cylinder. That is, if the sweeping base does not change size or shape as it sweeps but moves "sideways" at times, the volume of the figure may still be found by multiplying the height by the area of a base.

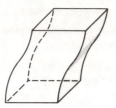

Illus. 5.54

EXERCISES

Rigorous exercises

1. Find a formula for the volume of a circular cylinder in terms of its height and the length of a radius of the base.

2. In the preceding discussion we assumed that for a prism, the polygonal base does not change its area as it sweeps. In fact, we seemed to suppose that it does not change size or shape. In other words, we assumed that any cross section parallel to the bases is congruent to the bases. Prove that any cross section of a prism parallel to a base is congruent to that base. (Assume that opposite sides of a parallelogram are congruent.)

3. Show, using the principle of sweeps for finding area, that congruent triangles have the same area.

4. Using the result of Exercise 3, show that any two congruent simple polygons have the same area.

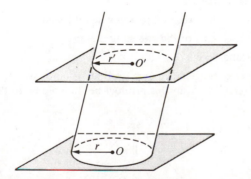

Illus. 5.55

5. Consider a circular cylinder like the one shown in Illus. 5.55. Prove that every cross section parallel to the (circular) base is a circular region and that this region has the same area as the base.

Exploratory exercises

6. Consider the triangular prism shown in Illus. 5.56 and imagine that the square region $MNPQ$ sweeps to $\overline{RS}$. As it sweeps, its area decreases.

   a) What is the area of the region $MNPQ$?

b) What is the area of the sweeping region when it reaches $\overline{RS}$?

c) What is the average area of the sweeping region?

d) What is the distance swept?

e) What is the product of the average area of the sweeping region and the distance swept? Is this the volume of the prism?

f) How else could sweeps be used to find the volume?

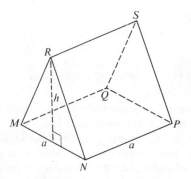

Illus. 5.56

7. Consider the pyramid in Illus. 5.57. Its base $MNPQ$ is the same as the correspondingly named face of the prism of Exercise 6, and its height h is the same as h for the prism. Imagine that the square region $MNPQ$ sweeps to the vertex R. As it sweeps, its area decreases.

a) What is the area of $MNPQ$?

b) What is the area of the sweeping region when it reaches R?

c) What is the average area of the sweeping region?

d) What is the distance swept?

e) Find the product of the average area and the distance swept. Compare this with the result of Exercise 6. Can this product be the volume of the pyramid? Why?

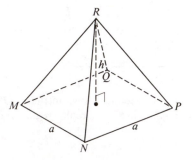

Illus. 5.57

Volumes of Pyramids and Cones

The concept of sweeps that has been useful in finding formulas for volumes of prisms and cylinders becomes difficult to apply to pyramids or cones, as was indicated in Exercises 6 and 7 above. If we apply the principle to finding the volume of the prism in Illus. 5.58(a), we might think of the rectangular base as sweeping upward and shrinking to finally become the segment $\overline{RS}$, with zero area. The average area of the sweeping rectangle would than be half the area of the face $MNPQ$, or $ab/2$. The distance swept is h, and we obtain $\frac{1}{2}ah \cdot b$ for the volume. This result is correct, as we may verify by imagining the triangular base MNR to sweep to the other base and calculating accordingly. If, however, we try to find the volume of the pyramid in Illus. 5.58(b) by a similar procedure, we find that we are in difficulty. If we again imagine the rectangular base to sweep upward, we find that it shrinks to the point R, with area zero. The average area of the sweeping rectangle would seem to be the same as before, as would the distance swept, but the volume of the pyramid is clearly less than that of the prism.

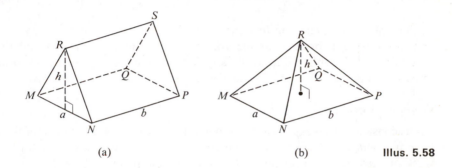

(a)   (b)   **Illus. 5.58**

The reason the difficulty just described arises is not that the concept of *sweeps* is not valid or breaks down. Rather, it is that we do not now have a simple or convenient way of finding the proper average area of the sweeping polygonal region. We shall therefore use a different concept to find formulas for volumes of cones and pyramids.

EXERCISES

1. a) Find the average of 1, 2, 3, 4, 5, and 6 by adding all the numbers and then dividing by 6.

   b) Add 1 and 6 and divide by 2, and see if this result is the average found in part (a).

2. a) Find the average of 1^2, 2^2, 3^2, 4^2, 5^2, and 6^2 by adding all the numbers and then dividing by 6.

   b) Add 1^2 and 6^2 and divide by 2, and see if this result is the average found in part (a).

Rigorous exercise

3. a) From stiff paper or thin cardboard, make a model of a triangular pyramid, using a pattern as shown in Illus. 5.59. The model can be held together with masking tape. Note that the base is an equilateral triangle.

b) Make another pyramid exactly like that of part (a).

c) Make a third pyramid as shown in Illus. 5.60. The lengths of the segments correspond to those of the first pyramid, as shown in the drawings.

d) See if you can fit the three pyramids together to form a triangular prism.

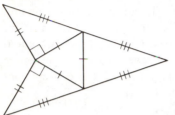

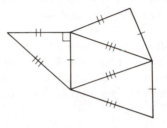

Illus. 5.59   **Illus. 5.60**

The plan for developing formulas for the volumes of cones and pyramids is first to establish a formula for the volume of a triangular pyramid and then to generalize the formula to apply to any kind of cone or pyramid. A principle somewhat like the principle of sweeps will be helpful.

Cavalieri's Principle

Let us consider the stacks of cards shown in Illus. 5.53. In this case, all of the cards have the same area as well as volume, and it is intuitively clear that shifting the stack of cards does not change the volume of the prisms. Now let us consider other three-dimensional figures, and suppose they consist of stacks of cards, as in Illus. 5.61. In this case the two figures have the same height h and the bases have the same area A. If we take a corresponding pair of cards, one from each stack, we find that they have the same area A' and, hence, the same volume. That is, any card in one stack has the same area (and volume) as the card in the other stack which is the same height above the base. Since this is true for every pair of corresponding cards in the stacks, the two figures must have the same volume. In essence this is the principle that we shall find useful.

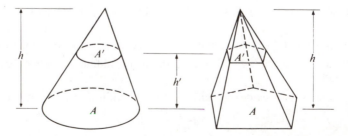

Illus. 5.61

Cavalieri's Principle Consider two three-dimensional figures and some fixed plane. Suppose that the two figures can be oriented so that a plane parallel to the given plane which intersects one of the figures also intersects the other. If for each such plane the cross sections of the figures determined by that plane have the same area, then the two figures have the same volume.

Let us now consider two pyramids which are not congruent but have the same height and the same base area (Illus. 5.62). If we take cross sections at a height b above the base, we find that these cross sections each have area

$$\frac{(h - b)^2}{h^2} \cdot A,$$

where A is the area of a base. We shall not prove this fact here, but it provides, together with Cavalieri's Principle, a means of proving the following theorem.

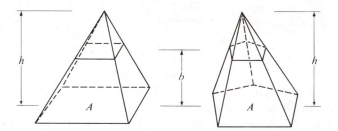

Illus. 5.62

Theorem 5.6 If two pyramids have the same height and the same base area, then they have the same volume.

We are now in a position to derive a formula for the volume of a triangular pyramid. The proof is essentially that given by Euclid, loc. cit., Book XII, Proposition 7.

Theorem 5.7 The volume of a triangular pyramid is one-third the product of the height and the base area.

Consider a triangular pyramid and a triangular prism with the same altitude and base area (Illus. 5.63). We wish to show that the triangular prism is the union of three triangular pyramids, each having the same volume as the original pyramid.

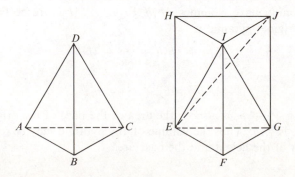

Illus. 5.63

Consider pyramids *EHIJ* and *EGIJ*, with bases *EHJ* and *EGJ*, respectively (Illus. 5.64). The pyramids have the common vertex *I*. We note first that their bases have the same area since $\triangle EHJ \cong \triangle JGE$. Further, since they have a common vertex *I*, they have the same height. Thus they have the same volume.

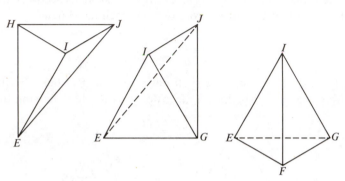

Illus. 5.64

Finally, consider pyramids *EFGI* and *EGIJ*, with bases *IFG* and *IGJ*. These pyramids have the common vertex *E*. Since $\triangle IFG \cong \triangle GJI$, their bases have the same area. Since they have a common vertex *E*, they have the same height. There-fore they have the same volume.

We have shown that all three pyramids have the same volume as the original pyramid. Now suppose that triangular base *EFG* has area *A* and that the height of the prism is *h*. Then the volume of the prism is *Ah*, and the volume of each of the pyramids is $\frac{1}{3}Ah$. This completes the proof.

The volume of any pyramid is now easy to find. Consider any pyramid, as in Illus. 5.65. The base is, of course, a polygonal region. If we select one vertex of the base and consider the diagonals from that vertex, we see that the base is the union of several triangular regions. In Illus. 5.65 the diagonals from *B* divide the base into three triangular regions, for example. We now consider the several pyramids having these triangular regions as bases and having the common vertex *V*. The original pyramid is thus seen to be the union of several triangular pyramids. If the bases of these triangular pyramids have areas $b_1, b_2, \ldots, b_n$, respectively, the triangular pyramids have volumes $\frac{1}{3}b_1h, \frac{1}{3}b_2h, \ldots, \frac{1}{3}b_nh$, respectively, and the volume of the original pyramid is

$$\tfrac{1}{3}b_1h + \tfrac{1}{3}b_2h + \cdots + \tfrac{1}{3}b_nh$$

or

$$\tfrac{1}{3}h(b_1 + b_2 + \cdots + b_n).$$

Since $b_1 + b_2 + \cdots + b_n$ is also *A*, the area of the base of the original pyramid, the total volume is $\frac{1}{3}Ah$. Thus the volume of any pyramid may be found by taking $\frac{1}{3}$ of the product of the height and the base area.

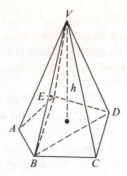

Illus. 5.65

EXERCISES

1. Find the volume of a pyramid whose height and base area are as follows.
   a) $h = 13$ in. $A = 47$ in.^{2}
   b) $h = 41$ ft, $A = 75$ ft^2
   c) $h = 34.2$ ft, $A = 19$ yd^2

2. Find the volumes of the cones with the following properties.
   a) $h = 17$ cm, $A = 31.2$ cm^2
   b) $h = 35$ in., $r = 7$ in. (circular cone)

Rigorous exercises

3. a) Show that the cross sectional area of a pyramid whose height is h and whose base has area A is

$$\frac{(h - b)^2}{h^2} \cdot A,$$

   where b is the height of the section above the base. The cross section is, of course, assumed to be parallel to the base.

   b) For any cross section of a circular cone parallel to the base, the area of that cross section is

$$\frac{(h - b)^2}{h^2} \cdot A,$$

   where h is the height, A the area of the base, and b the height of the section above the base. Assuming that this is true, prove that if a pyramid and a cone have the same height and the same base area, then they have the same volume.

   c) Prove that the volume of a circular cone is one-third of the product of its height and its base area.

   d) Find a formula for the volume of a circular cone whose height is h and whose base has a radius of length r.

4. Show that the volume formula we obtained for circular cylinders could also be obtained using Cavalieri's Principle.

Volumes of Spheres

Determining the volume of a sphere presents a rather interesting problem because it may involve such a variety of geometric concepts as well as a number of seemingly unrelated figures. It also demonstrates the usefulness of Cavalieri's Principle in determining volume relationships.

Our plan of attack will be as follows. We shall look for another three-dimensional region, whose volume we can determine and whose cross sectional areas are the same as those of the sphere. In other words, if a sphere and our other figure are placed "side by side," any plane parallel to a given plane and intersecting both figures will produce cross sections with the same area. Our first task, therefore, is to see how to determine the areas of the cross sections of the sphere.

Consider a sphere with radius r and choose an arbitrary cross section as shown in Illus. 5.66. The cross section is a circular region whose radii have length x and is a distance y from the center of the sphere. Since r is constant, the area of a cross section will vary according to the distance y, and by the Pythagorean Theorem this relationship is

$$x^2 = r^2 - y^2.$$

(Note that when $y = r$, $x = 0$. This would represent a plane tangent to the sphere, and the cross section would be the point of tangency. Note also that when $y = 0$, $x = r$. This would represent a cross section bounded by a great circle.)

Now, the area of the cross section of radius x is

$$\pi x^2 = \pi(r^2 - y^2) = \pi r^2 - \pi y^2.$$

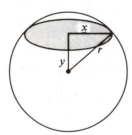

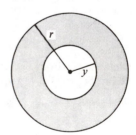

Illus. 5.66   Illus. 5.67

It will be recalled that the area of a circular ring (annulus) with outer radius r and inner radius y is also $\pi r^2 - \pi y^2$ (cf. Exercise 6, p. 165). Such a figure is shown in Illus. 5.67.

We now form a three-dimensional region having an annulus such as this for its cross sections. To do this, consider a right circular cylinder with radius of length r and altitude $2r$. Let O be the midpoint of the axis of the cylinder, and then consider two cones whose bases are the bases of the cylinder and whose common vertex is the point O. Finally, consider a plane that is tangent to the sphere and contains one base of the cylinder. Such an arrangement is shown in Illus. 5.68.

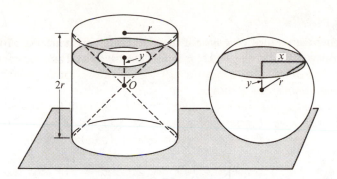

Illus. 5.68

It will be noted that the three-dimensional region lying inside the cylinder and outside the cones is exactly the kind of region we need. First, each of its cross sections is an annulus, and second, at any distance y from O, the area of the cross section (annulus) is $\pi r^2 - \pi y^2$, which is the same as the area of a cross section of the sphere at a distance y from the center of the sphere. Thus, by Cavalieri's Principle, the volume of this region is the same as the volume of the sphere. Now we need to find the volume of this region, which is the volume of the cylinder minus the volumes of the two cones, or

$$\pi r^2 \cdot 2r - 2 \cdot \tfrac{1}{3}\pi r^2 r = 2\pi r^3 - \tfrac{2}{3}\pi r^3 = \tfrac{4}{3}\pi r^3.$$

This completes the proof of the following theorem.

Theorem 5.8 The volume of a sphere with radius r is $\tfrac{4}{3}\pi r^3$.

EXERCISES

1. Find the volume of a sphere when the length of a radius is
    a) 3 in.,   b) 6 ft,   c) 1 cm,   d) 2 yd.

2. Find the volume of a sphere with radius
    a) r,   b) $2r$,   c) $\tfrac{1}{2}r$.

3. How does a change in the radius affect the volume of a sphere?

4. The diameter of the earth is approximately 4 times the diameter of the moon.
    a) Compare the radii of the earth and the moon.
    b) Compare their volumes.

5. A can contains three tennis balls, each having radius of length r. The inner radius of the base of the container has length r, and the altitude is $6r$. Find the total volume of the tennis balls and the volume of the container. Compare the volumes.

6. The volume of one sphere is twice the volume of another sphere. Compare their radii.

7. If d represents the diameter of a sphere, show that $\tfrac{1}{6}\pi d^3$ represents the volume of the sphere.

Exploratory exercise

8. Two spheres are called *concentric* if they have the same center. Consider two concentric spheres with radii r and $r + h$, as shown in Illus. 5.69.

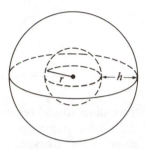

Illus. 5.69

a) Find the volume of the inner sphere.

b) Find the volume of the outer sphere.

c) The three-dimensional region between the spheres is called a *spherical shell*. Find its volume.

d) Let V represent the volume of the spherical shell. Suppose we let h become smaller and smaller. What happens to V?

e) Let A represent the surface area of the inner sphere. If h is very small (for example, 1/1000 in.), how would Ah compare with V?

f) If h is very small, how would V/h compare with A?

g) Using the equation in (c) above, find V/h.

h) Look at the equation in (g) above. As h approaches zero (that is, becomes smaller and smaller), V/h approaches what value?

i) Make a conjecture as to the surface area of a sphere with radius r.

11. SURFACE AREA

While a formula for the lateral area of a cone or a cylinder is relatively easy to determine, the surface area of a sphere is somewhat more challenging. For one thing, a sphere cannot be "flattened" in a manner similar to that of a cone or cylinder, thus making it difficult to compare its area with that of a two-dimensional region in a plane.

An approach to this problem was presented in Exercise 8 above, and although the method rests on a kind of "limit concept," it is not difficult to accept on an intuitive level. The basic idea is that if V is the volume of a spherical shell of inner radius r and shell thickness h, then V/h approaches A, the area of the sphere, as h approaches zero. At the same time it can be shown that V/h approaches $4\pi r^2$ as h

approaches zero. We therefore conclude that A and $4\pi r^2$ represent the same number. We state this conclusion in the following theorem.

Theorem 5.9 The surface area of a sphere with radius r is $4\pi r^2$.

EXERCISES

1. Find the area of a sphere when the length of a radius is
   a) 1,   b) 2,   c) 4,   d) 8.
2. How does a change in the radius affect the area of a sphere?
3. Compare the surface area of the moon and the earth. (Use 4000 miles as the radius of the earth.)
4. How is the area of a sphere related to the area of the interior of a great circle of the sphere?
5. Express the area of a sphere in terms of its diameter d.
6. A sphere of radius r is enclosed in a right circular cylinder with radius r and altitude $2r$. Compare the total surface (lateral and bases) area of the cylinder with the area of the sphere. How does your answer compare with that of Exercise 5 of the previous exercise set?

JUST FOR FUN

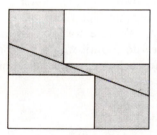

Illus. 5.70

1. What is the area of the large rectangle in Illus. 5.70?
2. What is the area of the shaded portion?
3. Trace the figure on another piece of paper and cut it into six pieces as indicated.
4. Arrange the four shaded pieces to form a square. What is the area of the square?
5. Now rearrange the four shaded pieces to form a rectangle. What is the area of the rectangle?
6. Can you explain why the area changes?

CHAPTER 6

PARALLELISM AND SIMILARITY

1. INTRODUCTION

In the preceding chapter certain theorems were temporarily assumed because they could not be proved until an additional assumption about parallels had been made. This important assumption will be made shortly and the consequences of it, which are far-reaching and profound, will be discussed. In this chapter and the one to follow, it will become clearer how the axioms we choose determine the kind of geometry we obtain.

A definition of parallel lines was given in Chapter 2, and we have been speaking of parallel lines, tacitly assuming that such things exist. It is, of course, possible to define things that do not exist. We have not formally assumed that parallels exist, nor have we proved that they do. Until we do one or the other, we do not, of course, know whether there are such things as parallel lines. Let us begin, therefore, by proving that parallels do exist.

2. EXISTENCE OF PARALLELS

The first theorem involves the notion of lines "cut by a transversal" and of "alternate interior angles," to use traditional parlance. In Illus. 6.1 we may consider lines l and m to be cut by the "transversal" $\overleftrightarrow{EF}$, in which case a pair of angles such as $\angle AEF$ and $\angle DFE$ are "alternate and interior" (the same being true of $\angle BEF$ and $\angle CFE$). Before proceeding, let us make the following definition.

Definition 6.1 Given a line n which meets two distinct lines l and m in a plane at distinct points E and F, respectively, and points A on l and D on m, on opposite half-planes relative to n, line n is called a *transversal* with respect to lines l and m, and $\angle AEF$ and $\angle DFE$ are called *alternate interior angles*.

Theorem 6.1 If lines in a plane cut by a transversal are such that alternate interior angles are congruent, then the lines are parallel.

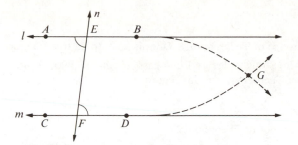

Illus. 6.1

Proof. Consider any two distinct lines, *l* and *m*, in a plane and a transversal *n*, meeting *l* and *m* at distinct points *E* and *F*, respectively. Choose a point *A* on *l* and a point *D* on *m* (Illus. 6.1). Then ∠*AEF* and ∠*DFE* are alternate interior angles. By hypothesis, we take them to be congruent. If lines *l* and *m* were to meet, it would have to be on one side of *n* or the other. Suppose that they meet at a point *G* on the same side of *n* as point *D*. This gives us a triangle △*EFG* with exterior angle ∠*AEF*. By Theorem 4.14, ∠*AEF* ≻ ∠*DFE*, contradicting our hypothesis that ∠*AEF* ≅ ∠*DFE*. Therefore it is impossible for *l* and *m* to intersect and thus the lines are parallel.*

We can now prove easily that parallels exist. More specifically, we will show that for any line *l* and any point *P* off the line there is *at least one* line parallel to *l* and containing *P*.

Theorem 6.2 For any line *l* and any point *P* not on *l*, there is (at least) one line *m*, in the plane of *l* and *P*, which contains *P* and does not meet *l*.

Outline of proof. Consider a line *l* and a point *P* not on *l*. Call some point of *l* point *Q*, thus determining line $\overleftrightarrow{PQ}$ (Illus. 6.2). There are points *A* and *B* on *l* and on opposite half-planes relative to $\overleftrightarrow{PQ}$. Now there is a half-line $\overrightarrow{PD}$ in $\overset{\circ}{PQ}/A$ such that ∠*QPD* ≅ ∠*PQB*. We thus conclude that $\overleftrightarrow{PD}$ is parallel to line *l*.

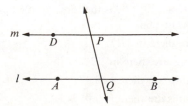

Illus. 6.2

---

* There is nothing special about assuming that *l* and *m* meet on the "right" side of *n* (for example, look at the drawing upside down). The proof is the same if we assume that *l* and *m* meet on the "left" side of *n*.

EXERCISES

1. State the converse of Theorem 6.1. Does it seem to be true? Discuss.
2. Construct, with straightedge and compass, a pair of parallel lines, and discuss the significance of Theorem 6.1 in this regard.

Rigorous exercises

3. Complete the proof of Theorem 6.2.
4. Prove that lines perpendicular to the same line are parallel.
5. Suppose that there were another (different) line containing P and not meeting l (see Illus. 6.2). We would then have two lines parallel to the same line. Would those lines be parallel to each other? For the second line and line l, would alternate interior angles be congruent? Why? (Prove your answer.)
6. Here is a "proof" that for any point P not on a line l, there is only one line m, in the plane of l and P, which contains P and does not meet l.

   From P drop a perpendicular to l at Q (which is unique by Theorem 4.20). Through P there is a unique perpendicular, m, to $\overline{PQ}$ by Theorem 4.13 as shown in Illus. 6.3.

   By Exercise 4 above, m is parallel to l. Since the perpendiculars are unique, m is the only parallel to l through P.

   What is wrong with the above argument?

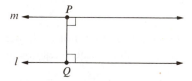

Illus. 6.3

REVIEW PRACTICE EXERCISES

1. Name each number with a fractional numeral whose denominator is a power of ten.

   Example. $15.78 = \frac{1578}{100}$

   a) 1.06   b) 2.3   c) 7.003   d) 0.175   e) 100.02   f) 0.03429

2. Since $1.06 = \frac{106}{100} = 106 \times \frac{1}{100}$, 1.06 is the same as 106 hundredths, or we might say there are 106 hundredths in 1.06.

   a) How many tenths are there in 3.7?
   b) How many hundredths are there in 3.7?
   c) How many thousandths are there in 3.7?
   d) How many thousandths are there in 7.02?
   e) How many ten-thousandths are there in 13.2459?

3. a) How many halves are there in $\frac{7}{2}$?
   b) How many fourths are there in $\frac{7}{2}$?

    c) How many sixteenths are there in $\frac{3}{4}$?

    d) How many fifteenths are there in $\frac{17}{3}$?

    e) How many fifteenths are there in $\frac{22}{5}$?

4. The prime factorization of 60 is "$2 \times 2 \times 3 \times 5$." It is a prime (or complete) factorization because every factor is a prime number. Write prime factorizations for these numbers.

    a) 12   b) 210   c) 63   d) 231

5. Write prime factorizations for these numbers.

    a) 36   b) 100   c) 196   d) 900   e) 225   f) 484

6. a) For each of the factorizations in Exercise 5, did any prime factor occur only once? Three times? An odd number of times?

    b) If a number is a perfect square, what is true of its prime factors?

3. THE PARALLEL AXIOM

We have proved that for any line l and a point P not on l there exists *at least one* line containing P and parallel to l. We have not proved that there is *only one* such line. We now question whether, as in Illus. 6.4, more than one parallel may exist. Of course, we cannot answer this question experimentally because we regard lines as being of unlimited extent. To travel to the unlimited outermost reaches of space to see whether lines meet there is impossible.

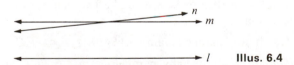

                                                                Illus. 6.4

    Is it possible that we might *prove* the uniqueness of the line parallel to l and containing P? The answer, although not obvious, is negative. Mathematicians tried this over a period of several hundred years, without success. The question was finally settled when a kind of geometry was devised in which parallels were not unique but which was consistent with the rest of ordinary Euclidean geometry. If it were possible to prove the uniqueness of parallels from the rest of geometry, then it would not be possible to find a geometry not having uniqueness of parallels and at the same time being consistent with the rest of geometry. If, therefore, we wish our geometry to have uniqueness of parallels, we shall have to assume it.

Axiom 6.1 (Parallel Axiom) For any line l and a point P not on l there is no more than one line parallel to l and containing the point P.

    Combining Axiom 6.1 and Theorem 6.2, we know that for any line l and any point P not on l there is *one and only one* line parallel to l and containing P. This is of fundamental importance to much of the development of geometry. As we shall see, it is this assumption about uniqueness of parallels that gives Euclidean geometry its unique characteristics. Many important theorems can now be proved which would not follow without the parallel axiom. The following theorem, which is the converse of Theorem 6.1, is an example.

Theorem 6.3 If two parallel lines are cut by a transversal, the alternate interior angles are congruent.

Outline of proof. Consider two parallel lines *l* and *m*, and a point *P* on *m* (Illus. 6.5). By Theorem 6.2 there is a line containing *P* which is parallel to *l* and for which alternate interior angles are congruent. By Axiom 6.1 there is only one such line, hence it must be *m*. Thus, for any two parallel lines cut by a transversal, alternate interior angles are congruent.

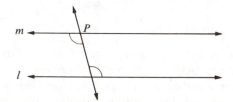

Illus. 6.5

Theorem 6.3 allows us to prove fairly easily that the opposite angles of a parallelogram are congruent and then that opposite sides are congruent. These facts are given in the following theorems, which, like Theorem 6.3, would not be true without the parallel axiom.

Theorem 6.4 In any parallelogram $\square ABCD$, $\angle A \cong \angle C$ and $\angle B \cong \angle D$.

Outline of proof. Consider any parallelogram $\square ABCD$ and angles $\angle ADB$ and $\angle CBD$ (Illus. 6.6). They are alternate interior angles, and $\overline{BC} \parallel \overline{AD}$. The angles are therefore congruent. Similarly $\angle ABD \cong \angle CDB$ because $\overline{AB} \parallel \overline{CD}$. It follows that $\angle B \cong \angle D$. Similar reasoning allows deduction of the fact that $\angle A \cong \angle C$.

Corollary 6.5 All of the angles of a rectangle are right angles.

Theorem 6.6 In any parallelogram $\square ABCD$, $\overline{AB} \cong \overline{CD}$ and $\overline{AD} \cong \overline{BC}$.

The proofs of Corollary 6.5 and Theorem 6.6 are left as an exercise.

The foregoing theorems are well-known theorems from elementary geometry.

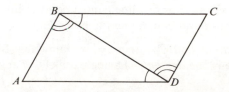

Illus. 6.6

Although they are important theorems, our main concern at the moment is in noting how many important theorems depend on the parallel axiom. It is also well known in Euclidean geometry that in any triangle the sum of the measures of the angles is 180°, and it is also true that this theorem depends on the parallel axiom.

Theorem 6.7 The sum of the measures of the angles of any triangle is 180°.

Outline of proof. Consider any triangle $\triangle ABC$ (Illus. 6.7). There is one and only one line $\overleftrightarrow{RS}$ containing C and parallel to $\overleftrightarrow{AB}$. The sum of the measures of $\angle RCA$, $\angle ACB$, and $\angle SCB$ is 180° because $\angle RCS$ is straight. But $m\angle A = m\angle RCA$ because they are alternate interior angles. Similarly $m\angle B = m\angle SCB$. Thus $m\angle A + m\angle B + m\angle C = 180°$.

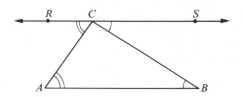

Illus. 6.7

One of the most famous theorems of geometry is the theorem of Pythagoras, concerning right triangles. There are many ways of proving this interesting theorem, but they all depend on the parallel axiom.

Theorem 6.8 (Pythagorean Theorem) In any right triangle $\triangle ABC$, where $\angle C$ is a right angle, $[m(\overline{AB})]^2 = [m(\overline{BC})]^2 + [m(\overline{AC})]^2$. (Or, if the measures of the sides opposite angles $\angle A$, $\angle B$, and $\angle C$ are respectively a, b, and c, then $a^2 + b^2 = c^2$.)

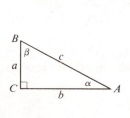
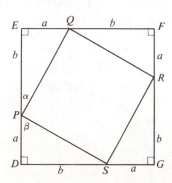

Illus. 6.8

Outline of proof. Consider any right triangle $\triangle ACB$, with right angle $\angle C$ (Illus. 6.8). The measures of the sides opposite A, B, and C will be called "a," "b," and "c," respectively, and the measures of angles $\angle A$ and $\angle B$ will be called "α" and "β," respectively. Now consider a square $\square DEFG$, each of whose sides has measure $a + b$. On $\overline{DE}$ there is a point P with D-P-E and for which $m(\overline{DP}) = a$. It follows that $m(\overline{PE}) = b$. Similarly there are points Q, R, and S on the other sides of the square, as shown, for which $m(\overline{EQ}) = m(\overline{FR}) = m(\overline{GS}) = a$, and $m(\overline{QF}) = m(\overline{RG}) = m(\overline{SD}) = b$. All of the triangles $\triangle SDP$, $\triangle PEQ$, $\triangle QFR$, and $\triangle RGS$ are congruent to $\triangle ABC$. Since $\angle DPE$ is a straight angle and $\alpha + \beta = 90°$, we know that $\angle SPQ$ is a right angle. Similarly, the other angles of $PQRS$ are right angles, so that the quadrilateral is a square.

The area of square $\square DEFG$ is $(a + b)^2$ and that of square $\square PQRS$ is c^2. Since the larger square region is composed of square $\square PQRS$ and four triangles, each having area $\frac{1}{2}ab$, the area of the large square is also $c^2 + 4(\frac{1}{2}ab)$. Thus we have

$$(a + b)^2 = c^2 + 4(\tfrac{1}{2}ab),$$

which simplifies to

$$a^2 + b^2 = c^2,$$

and this is what we set out to prove.

The converse of the Pythagorean theorem is also true. That is, if the measures of the three sides of a triangle, a, b, and c, are related by the equation $a^2 + b^2 = c^2$, then the triangle is a right triangle, the angle opposite the longest side being a right angle.

Theorem 6.9 (Converse of Pythagorean Theorem) Consider any triangle $\triangle ABC$, in which the measures of the sides opposite angles $\angle A$, $\angle B$, and $\angle C$, respectively, are a, b, and c. If $a^2 + b^2 = c^2$, then $\angle C$ is a right angle.

Outline of proof. We shall show that there exists a right triangle congruent to the given triangle. Consider any ray with endpoint P (Illus. 6.9). On this ray there is a point Q such that $\overline{PQ} \cong \overline{CB}$. There is then a ray with endpoint P and perpendicular to $\overrightarrow{PQ}$, and on it a point R such that $\overline{RP} \cong \overline{AC}$. Now $\triangle QPR$ is a right triangle, hence $RQ^2 = a^2 + b^2$, or c^2. Then $\triangle ABC \cong \triangle RQP$, and since $\angle P$ is a right angle, $\angle C$ also is a right angle.

This theorem gives us a way of determining whether an angle is a right angle, by comparing measures of segments. It is an important theorem, especially in

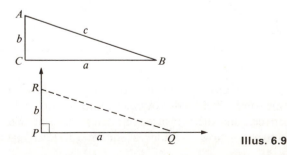

Illus. 6.9

applications of geometry. For some reason, in elementary mathematics courses it is often ignored or worse, tacitly assumed. The latter amounts to a bad mistake in reasoning—confusing a theorem with its converse. That is, one proves the Pythagorean theorem and then proceeds as though both it and its converse have been proved.

EXERCISES

1. What other angles are congruent in the drawing for Theorem 6.3 (Illus. 6.5)?

2. It can be proved that the diagonals of a rectangle are congruent. Using the converse of the above, describe how you could brace the rectangular frame of a screen door to keep it rectangular.

3. It can be proved that if the opposite sides of a simple quadrilateral are congruent, then it is a parallelogram. Using this fact, show how one could construct a parallelogram with a straightedge and compass.

4. A rectangular field is 30 yd wide and 40 yd long. How long are the diagonals? On what theorem does your answer depend?

5. An ancient surveying method of laying out a square corner was to tie knots in a rope, equally spaced, then stretch the rope into a 3–4–5 triangle as shown in Illus. 6.10. On what theorem does this method depend?

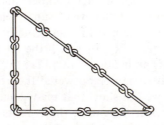

Illus. 6.10

6. A baseball diamond is a square 90 ft on a side. How far is it from home plate to second base?

Rigorous exercises

7. Complete the proof of Theorem 6.3.

8. Complete the proof of Theorem 6.4.

9. Write a complete proof of Corollary 6.5.

10. Write a complete proof of Theorem 6.6.

11. Complete the proof of Theorem 6.7.

12. Complete the proof of Theorem 6.9.

13. Prove that the diagonals of a parallelogram bisect each other.

14. Prove that any two consecutive angles of a parallelogram are supplementary.

15. Prove that the diagonals of a rectangle are congruent.

16. Prove that if the diagonals of a parallelogram are perpendicular, then it is a rhombus

17. Prove that if the opposite sides of a simple quadrilateral are congruent, then it is a parallelogram.

18. Prove that the diagonals of a rhombus are perpendicular.

19. Prove that if the opposite angles of a simple quadrilateral are congruent, then it is a parallelogram.

20. Prove that if two triangles have two pairs of corresponding angles congruent, then the other angles are congruent.

4. MORE ABOUT MEASURES

In Chapter 5, in order to develop the theory of two-dimensional measure, or area, we found it necessary to make some assumptions. In particular, we found it necessary to assume that the sum of the measures of the angles of a triangle is 180° and that all the angles of a rectangle are right angles. It would not have been possible to prove these facts then, without having the parallel axiom. Now we have that axiom and in fact have already proved the essential theorems. Thus the gaps in the prior development are filled.

When we discussed one-dimensional measures in Chapter 5, it will be recalled, we devised a method for assigning a number to every segment. In some cases the measure of a segment is a rational number; in others, irrational. Irrational measures arise when, in using subunits, we obtain an endless nonrepeating pattern. There is some question as to whether this can actually occur. In other words, are there any segments in our geometry for which the measure (with respect to some particular unit, of course) is an irrational number? While such a question cannot be answered experimentally, it can be answered theoretically. Now that we have the Pythagorean theorem, we can show that many such segments actually exist.

For any segment whatever as a unit, there exists a right triangle such as is shown in Illus. 6.11. The measure of each of the perpendicular sides is 1, and the hypotenuse has length c. By the Pythagorean theorem $c^2 = 1^2 + 1^2 = 2$. Thus $c = \sqrt{2}$, which is an irrational number. A segment with measure $\sqrt{2}$ therefore exists.

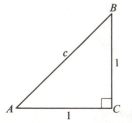

Illus. 6.11

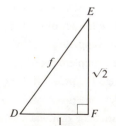

Illus. 6.12

It is similarly easy to show now that a segment exists having measure $\sqrt{3}$, as in Illus. 6.12. The triangle pictured here exists, and by the Pythagorean theorem, $f^2 = 1^2 + (\sqrt{2})^2 = 1 + 2 = 3$. Hence, $f = \sqrt{3}$, which is an irrational number. Many other segments with irrational measures may be shown to exist.

EXERCISES

1. a) Illustration 6.13 may be considered as follows. We draw a right triangle whose legs each have measure 1. Then on its hypotenuse we draw another right triangle whose legs have measure $\sqrt{2}$ and 1, and so on. What is the measure of the hypotenuse of the second triangle?

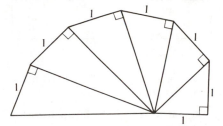

Illus. 6.13

   b) What is the measure of the hypotenuse of the third triangle? The fourth? The fifth?

   c) Generalize the results of part (b), and establish the generalization as rigorously as you can.

Exploratory exercises

2. Suppose the segments in Illus. 6.14 have measures 1 and 5.12, respectively.

   a) Would a tenth-unit fit an integral number of times in the unit? In the other segment?

   b) Would a hundredth-unit fit an integral number of times in the unit? In the other segment?

   c) Would a thousandth-unit fit an integral number of times in the unit? In the other segment?

   d) What is the largest decimal subunit that would fit an integral number of times in both segments below?

3. If two segments have measures 3.12 and 6.543, respectively, what is the largest decimal subunit that would fit an integral number of times in both segments?

————— 1

————————————————————— 5.12   **Illus. 6.14**

4. If two segments have measures 4.12_5 and 34.421_5, what is the largest quinary subunit that would fit an integral number of times in both segments?

5. If two segments have measures $\frac{13}{3}$ and $\frac{12}{5}$, respectively, what is the largest subunit that would fit an integral number of times in both segments?

6. If two segments have measures 1.6666... and 3.1111..., respectively, what is the largest subunit that would fit an integral number of times in both segments?

7. If two segments have measure 1 and $\sqrt{2}$, respectively, what is the largest subunit that would fit an integral number of times in both segments?

8. The rectangular region in Illus. 6.15 has length 5.12 and width 1. Can it be filled precisely with squares? If so, what is the largest square that could be used? [*Hint:* Refer to Exercise 2.]

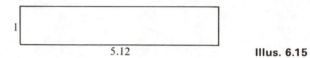

1

5.12   **Illus. 6.15**

9. Can a rectangular region which is $\frac{13}{3}$ by $\frac{12}{5}$ be filled precisely with squares? If so, what is the largest square that could be used?

10. Can a rectangular region that is 1 by $\sqrt{2}$ be filled precisely with squares? If so, what is the largest square that could be used?

11. Can the rectangular region in Illus. 6.16 be filled precisely with squares? If so, what is the largest square that could be used?

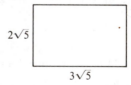

$2\sqrt{5}$

$3\sqrt{5}$ **Illus. 6.16**

👀 *Rigorous exercise*

12. Prove that there is no rational number m/n for which $(m/n)^2 = 2$. [*Hint:* This equation is equivalent to $m^2 = 2n^2$, where m and n are natural numbers. Consider prime factorizations of m^2 and $2n^2$ and the number of times factors occur on both sides.]

5. COMMENSURABILITY

We have established that for any segment we might choose as a unit, there exist many segments with irrational measures. For these segments there are no subunits, however small, for which the measuring process ends or repeats. This situation is related to the following question: Given two segments, is there a small segment which will fit an integral number of times in each? If there is, then the segments are said to be *commensurable*, and the small segment is said to *measure* both of them. Otherwise, the segments are called *incommensurable*, and of course in this case there is no segment, however small, that measures both of them.

Definition 6.2 Two segments are said to be *commensurable* if and only if there is a segment such that when it is used as a unit, each segment has a natural number for its measure.

Suppose two segments have measures $\frac{13}{5}$ and $\frac{5}{3}$, respectively. A fifth-unit will measure the first, fitting into it 13 times, and a third-unit will measure the second, fitting into it 5 times. A fifteenth-unit will measure both of them, fitting into the first segment 39 times and into the second 25 times. This segment will measure both segments, since 15 is the least common multiple (l.c.m.) of the denominators 5 and 3.

Let us similarly consider two segments with measures 4.333 . . . and 7.1111 . . . , respectively. In this case the unending numerals might seem to indicate that the segments are incommensurable. Certainly there is no decimal subunit that measures either of the segments, for if there were the decimal numeral would terminate. The segments are commensurable, however, as can be seen by finding fractional numerals for their measures. They are $\frac{13}{3}$ and $\frac{64}{9}$, respectively. The l.c.m. of the denominators is 9, and hence a ninth-unit will measure both segments. The segments are commensurable.

Let's now consider two segments with measures 1 and $\sqrt{2}$. Are they commensurable? If they were, the small segment that measures both of them would fit an integral number of times into the unit, say, m times. The small segment would also measure the other segment, fitting into it, say n times. The measure of the small segment would then be $1/m$. Since it fits into the other n times, the measure of the other is $n \cdot 1/m$, or n/m. This we see is impossible, because n/m is a rational number (n and m being natural numbers), while $\sqrt{2}$ is irrational. Segments with measures 1 and $\sqrt{2}$ are incommensurable.

We now ask the general question: Under what conditions are two segments commensurable? Before proceeding, however, let us note that this question is equivalent to the following: Under what conditions can a rectangular region be filled exactly with squares? If the sides of the rectangle are commensurable, then there is a segment that will fit into both of them an integral number of times, as shown in Illus. 6.17. If it fits m times and n times, respectively, then there will be m squares in each row and there will be n rows, making a total of $m \cdot n$ squares. If the sides are incommensurable, then there is no square, however small, that will fit into the rectangular region an integral number of times.

In the examples preceding, we might note that if the measures of two segments were rational numbers, then those segments were commensurable. If one measure was rational and the other irrational, then the segments were incommensurable. If two segments both have irrational measures, then they may or may not be commensurable, as we shall see. It is fairly easy to show that if the quotient, or

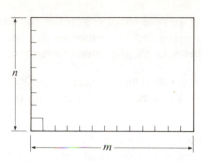

Illus. 6.17

ratio, of the measures is a rational number, the segments are commensurable, and conversely.

Theorem 6.10 Two segments with measures M and N are commensurable if and only if M/N is a rational number.

Outline of proof. First, let us suppose that segments M and N are commensurable. Then there is a segment with measure p that measures both segments. In other words,

$$M = p \cdot m \qquad \text{for some natural number } m,$$

and

$$N = p \cdot n \qquad \text{for some natural number } n.$$

From these equations it follows that $M/N = (pm/pn) = m/n$. Now, m/n is rational; hence M/N is rational. Therefore, if the segments are commensurable, then the ratio of their measures is rational. Conversely, let us suppose that the segments are such that M/N is a rational number. Then there are natural numbers m and n such that $M/N = m/n$ because every positive rational number is namable as the quotient of two natural numbers. (In this case we may restrict ourselves to natural numbers. Why?) It follows that $M/m = N/n$. For simplicity, let us symbolize M/m or N/n by "p." Now a segment of length p will measure both segments, because

$$p \cdot m = \frac{M}{m} \cdot m = M \qquad \text{and} \qquad p \cdot n = \frac{N}{n} \cdot n = N.$$

Thus if the ratio of the measures of two segments is rational, then they can both be measured by a segment of length p.

EXERCISES

1. The following are lengths of some pairs of segments. Which pairs of segments are commensurable?

   a) 4.12 and $\frac{15}{7}$ b) $\sqrt{3}$ and 5

   c) $\sqrt{4}$ and $\sqrt{9}$ d) $4\sqrt{3}$ and $17\sqrt{3}$

   e) $\sqrt{8}$ and $\sqrt{6}$

2. a) The dimensions of a rectangular region are 16.5 and 24.75. Are the sides commensurable? Show that your answer is correct.

   b) Can the region be filled with squares whose sides measure 0.01? Show that your answer is correct.

   c) Find the ratio of the lengths of the sides of the rectangle, and simplify. What is the largest square with which the rectangular region can be filled?

Rigorous exercises

3. See if you can generalize Exercise 2. That is, if a rectangular region has dimensions M and N, and the sides are commensurable, what is the largest square with which the region can be filled?

4. Complete the proof of Theorem 6.10.

5. In the second part of the proof of Theorem 6.10 it is shown that a segment of length p will measure both segments in question. How do we know that such a segment exists?

6. THE COMPLETENESS AXIOM

In the proof of Theorem 6.10 we found it necessary to make an assumption somewhat as follows: for any real number p there is a segment having measure p. We know that for every rational number r there is a segment having measure r, and that many segments exist with irrational measures such as $\sqrt{2}$, but we have not proved that for *any* irrational number there exists a segment with that number for its measure. Furthermore, we shall not prove it here for the very good reason that it would be impossible. Mathematicians have shown that there are geometries without the property in question, but which satisfy all the assumptions (axioms) we have made so far. To complete our geometric system, we shall now make an assumption about the existence of certain segments. This assumption will guarantee that any line will have as many points as there are real numbers. This last assumption is called the *completeness axiom*.

Axiom 6.2 (Completeness) For any unit segment and any positive real number p, there exists a segment having p for its measure.

EXERCISES

1. a) Using the width of your paper, draw a line and mark a point 0 at about the center. You now have two half-lines.

   b) Use a unit segment 1 cm long. Mark a unit segment on the line, having 0 as one endpoint. Mark the other endpoint "1." Mark another unit segment having 1 for an endpoint and mark the other endpoint "2." It should be on the opposite side of 1 from 0. Continue marking unit segments in this fashion on that half-line.

   c) On the other half-line mark unit segments, similarly labeling their endpoints $-1, -2, -3$, and so on.

2. Use the drawing of Exercise 1.

   a) Construct a right triangle whose legs have length 1. The hypotenuse then has measure $\sqrt{2}$. Mark a point on the positive half-line with 0 as one endpoint and having measure $\sqrt{2}$. Label the other endpoint $\sqrt{2}$.

   b) On the negative half-line mark a segment congruent to that of part (a). One endpoint should be 0, and the other endpoint should be labeled $-\sqrt{2}$.

3. a) Draw two perpendicular lines, intersecting in the middle of a sheet of paper, as shown in Illus. 6.18. Label each of them as in Exercise 1.

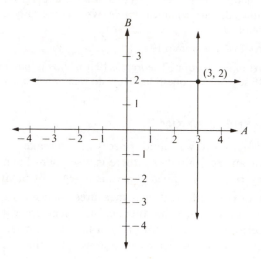

Illus. 6.18

   b) Draw a line perpendicular to line A and containing the point 3. Draw a line perpendicular to line B and containing the point 2. How many such lines are there? Do these lines meet? In how many points?

   c) Through a point P, as shown, draw a line parallel to line A and also a line parallel to line B. How many such lines are there? The first line meets line B in how many points? The second line meets line A in how many points?

⊙⊙ *Rigorous exercises*

   4. On the basis of Exercise 3, answer the following questions:

   a) Given an ordered pair of numbers such as (3, 2), is there a corresponding point of the plane? Are there more than one? Why?

   b) Given a point of the plane, is there an ordered pair of numbers? Are there more than one? Why?

   5. Is there a point of the line for every real number? Is there more than one? Why?

   6. Is there a real number for every point of the line? Is there more than one? Why?

Now that we have Axiom 6.2, we know that the real numbers correspond to the points of any line. If we consider any line and any segment as a unit (Illus. 6.19), we can mark any point of the line as O. Then on each half-line, there exist unit segments as shown. Thus far we have a point on the line for every integer. By Axiom 6.2, there exists a segment having measure p, where p is any positive real number. Let us consider such a segment $\overline{PQ}$ as in Illus 6.19. Then there exist two points, A and B, on either half-line, such that $\overline{OA} \cong \overline{PQ}$ and $\overline{OB} \cong \overline{PQ}$. The real number p corresponds to point A and the real number $-p$ corresponds to point B. In this fashion we can locate a point of the line for any real number, positive or negative. The point O, of course, corresponds to the real number zero.

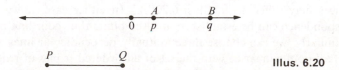

                                                 Illus. 6.19

Coordinatization of Lines and Planes

In developing measures of segments, we devised theoretically a method of assigning a real number, rational or irrational, to any segment. Thus we know that to every point of a line there corresponds a real number. Furthermore, the number corresponding to a point is unique for a given unit segment. Similarly, given any real number, the point corresponding to it is unique. In other words, there is a one-to-one correspondence between the set of real numbers and the set of points on a line. The real number corresponding to a point of a line is called the *coordinate* of the point. Obviously, the coordinate of a point P is the measure of the segment $\overline{OP}$, if P is on the positive half-line, and the additive inverse of the measure, if P is on the negative half-line.

Now consider a segment $\overline{AB}$ on the line, as shown in Illus. 6.20. Its endpoints have coordinates p and q. The measure of this segment is the absolute value of the difference of p and q, or $|p - q|$, regardless of the position of the endpoints.

                                       A B

                                    0 p q

     P Q

                                                    Illus. 6.20

Let us summarize the preceding results by saying that a line can be *coordinatized*. We shall make this a theorem, even though we shall not prove it in detail.

Theorem 6.11 For any line and for any segment as a unit, the line can be *co-ordinatized* in such a way that: (1) there is a one-to-one correspondence between the real numbers and the points of the line; (2) the length of any segment on the line is $|p - q|$, where p and q are the coordinates of the endpoints; (3) the points having coordinates 0 and 1 may be chosen arbitrarily, and the half-line to be designated positive can be chosen arbitrarily.*

It now follows easily that the points of a plane can be matched, in one-to-one correspondence, with the set of all ordered pairs of real numbers. This is referred to as the *coordinatization* of a plane. We simply choose any line on a plane and coordinatize it. Then we also coordinatize the line in the plane perpendicular to the first line at the zero point, or *origin*. This is shown in Illus. 6.21.

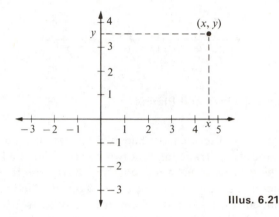

Illus. 6.21

Theorem 6.12 For any plane and any two distinct lines in the plane which intersect, the plane can be *coordinatized* in such a way that there is a one-to-one correspondence between the points in the plane and the set of all ordered pairs of real numbers.

It should be noted that while the two coordinatized lines are usually perpendicular, they do not have to be. For example, one might easily establish a one-to-one correspondence between the points in the plane and the set of all ordered pairs of real numbers, as indicated in Illus. 6.22(a). Similarly, the unit distances may differ, as in Illus. 6.22(b). In either case, however, a one-to-one correspondence can be established and the plane thus coordinatized.

Similarly, we can choose three mutually perpendicular lines in space and thus match all points in space with the set of all ordered triples of real numbers. With real numbers, or pairs, or triples of real numbers corresponding to points in this fashion, we now have an algebraic tool for the study of geometry, and the properties

---

* If distance, rather than congruence, had been undefined, the content of Theorem 6.11 could have been introduced by means of two axioms: The Ruler Axiom, which says that every line has a coordinate system and The Ruler Placement Axiom, which says that for any points p and q on a line, there is a coordinate system for the line such that p is the zero point and q is positive.

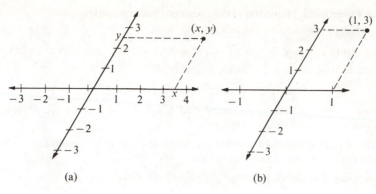

(a)   (b)

Illus. 6.22

of the real number system can be brought to bear on many geometric questions. Such a study of geometry is known as *analytic geometry*. An introduction to it is found in Chapter 8.

EXERCISES

1. Find the length of a segment $\overline{AB}$, if the coordinates of A and B, respectively, are

   a) 3 and 7,   b) -5 and 13,   c) -10 and -3,
   d) -5.3 and 0,   e) -4 and $\sqrt{2}$,   f) $-\pi$ and π.

2. Draw a pair of perpendicular coordinatized lines (axes) on a piece of graph paper, as shown in Illus. 6.23. The coordinates of points on the plane are then ordered pairs of real numbers (a, b), where $|a|$ gives the distance from the second axis and $|b|$ the distance from the first axis. Plot the points having the following coordinates.

   a) $(3, 4)$ b) $(4, 4)$ c) $(6, 4)$
   d) $(6, 8)$ e) $(3, -4)$ f) $(4, -4)$
   g) $(-3, 4)$ h) $(-4, 4)$ i) $(-4, -4)$
   j) $(-3, -4)$

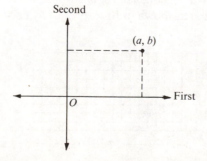

Illus. 6.23

3. Refer to Exercise 2. How long is the segment between points

a) (3, 4) and (4, 4), b) (3, 4) and (6, 4), c) (6, 4) and (6, 8),

d) (3, 4) and (3, −4), e) (−3, 4) and (−3, −4), f) 0 and (3, 4),

g) 0 and (3, −4), h) (3, 4) and (6, 8)?

4. Draw a pair of axes as in Exercise 2. Mark the point (5, 2).

a) Draw a line containing this point and parallel to the first axis. Name the co-ordinates of five points on this line.

b) Draw a line containing the point (5, 2) and parallel to the second axis. Name the coordinates of five points on this line.

c) Are parallel lines everywhere equidistant? Discuss.

Rigorous exercise

5. Prove Theorem 6.12.

7. INCOMMENSURABILITY: A PROBLEM

In developing the theory of two-dimensional measures (Chapter 5), we found a formula for the area of a rectangular region, $A = l \cdot w$. This formula was the basis of deducing formulas for finding areas of other kinds of regions. At the time we derived the formula $A = l \cdot w$ we had not yet considered the possibility that incommensurable segments exist. The argument we used, then, is essentially correct if the sides of the rectangular region are commensurable segments, but otherwise it is not. Let us now establish this formula on firmer ground, taking into account the fact that segments may be incommensurable.

First, let us consider any rectangular region for which the sides are commensurable. This means that there is some segment $\overline{AB}$ that measures all sides of the rectangle (Illus. 6.24). We may choose any unit of length, of course. To simplify the problem, let us choose $\overline{AB}$ as a unit. We may then also choose a square region whose sides measure 1 as our area unit. Then l unit segments fit into the longer side and w unit segments fit into the smaller side. The number of unit squares that fit into the rectangular region is $l \cdot w$, hence the area is $l \cdot w$.

Now let us consider a rectangular region whose sides are incommensurable. We can approximate the area shown in Illus. 6.25(a) by choosing a square as a unit

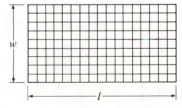

Illus. 6.24

and putting as many squares congruent to it into the region as will fit. The product $l_1 \cdot w_1$ is the area of the shaded region. This is less than the area in question but is an approximation. To obtain a better approximation we can choose a smaller square for a subunit, as in Illus. 6.25(b). The product $l_2 \cdot w_2$ is a better approximation than the product $l_1 \cdot w_1$. A still better approximation can be obtained by using a much smaller (subunit) square and successively better approximations can be made, as long as we please.

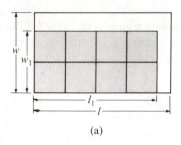
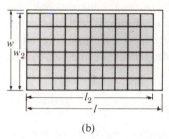

(a)   (b)

Illus. 6.25

As we take smaller and smaller squares, we shall never find one small enough to fill the region exactly, because the sides of the rectangular region are incommensurable. We thus have an unending sequence of numbers $l_1 \cdot w_1, l_2 \cdot w_2, l_3 \cdot w_3, \ldots$ These numbers are areas of rectangular regions that become more nearly the same as the one whose measure is in question. Thus these numbers get closer and closer to the area of the region. This unending sequence of numbers gets closer and closer to the product $l \cdot w$ but never exceeds it.

Let us begin again but, this time, approximate the area of the rectangular region by taking just enough squares to cover the region, as in Illus. 6.26. This time the approximation we obtain will be too large, but as we take smaller and smaller squares as subunits the approximations will become better and better, decreasing toward the product $l \cdot w$. We now conclude with confidence that the area of the rectangular region is $l \cdot w$.

Incommensurability of segments posed rather serious problems in ancient times. The existence of incommensurable segments was discovered by a group of Greek mathematicians known as the Pythagoreans. Not only was it an important

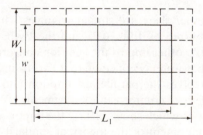
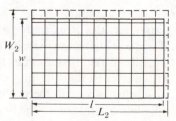

Illus. 6.26

mathematical discovery, but it was also upsetting because, in proving some of the most fundamental theorems, they had assumed that any two segments are commensurable. The problem was solved when Eudoxus developed a scheme similar to that we have just described for the area of a rectangular region.

8. SIMILARITY

Similar figures, roughly speaking, are figures that have the same shape, but not necessarily the same size. For example, the two triangles in Illus. 6.27 are similar. Notice that their corresponding angles are congruent; furthermore the ratios of corresponding sides are the same; that is,

$$\frac{AB}{A'B'} = \frac{AC}{A'C'} = \frac{BC}{B'C'} = \frac{1}{2}.$$

We shall confine our study of similar figures mainly to triangles. We first define similar triangles.

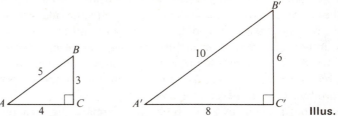

Illus. 6.27

Definition 6.3 Two triangles are *similar* if and only if they can be labeled ABC and $A'B'C'$ in such a way that $\angle A \cong \angle A'$, $\angle B \cong \angle B'$, and $\angle C \cong \angle C'$, and also such that

$$\frac{AB}{A'B'} = \frac{BC}{B'C'} = \frac{CA}{C'A'}.$$

We write "$\triangle ABC \sim \triangle A'B'C'$" to assert that the triangles are similar.

There may be a question as to whether there are similar triangles which are not also congruent; this is partially answered by Illus. 6.27 and the Pythagorean theorem. Details are left as an exercise. Another important question concerns how much information is required in order that we may conclude that two triangles are similar. The answer to this question involves the parallel assumption and, as we shall see, also brings up again the problem of commensurable segments. We now prove two preliminary theorems, the first of which shows clearly the important role the parallel assumption plays in the development of the theory of similarity.

Theorem 6.13 If three parallel lines cut off congruent segments on one transversal, then they cut off congruent segments on any transversal.

Outline of proof. Let us assume that lines *l*, *m*, and *n* are parallel, and that line *t* meets these lines at points *C*, *B*, and *A*, respectively, such that $\overline{AB} \cong \overline{BC}$ (Illus. 6.28). Let *u* be any line that intersects *l*, *m*, and *n* at points *L*, *M*, and *N*, respectively.   Now there exists a unique line *v* parallel to *u* and containing *B*, and *v* intersects *l* and *n* at *D* and *E*, respectively. Thus two triangles $\triangle DBC$ and $\triangle EBA$ exist, and are congruent.   Therefore $\overline{DB} \cong \overline{BE}$.   Now, *DLMB* and *BMNE* are parallelograms, and it therefore follows that $\overline{LM} \cong \overline{MN}$. Since *u* is an arbitrary transversal, we conclude that if congruent segments are cut off on one transversal, then congruent segments are cut off on any transversal.

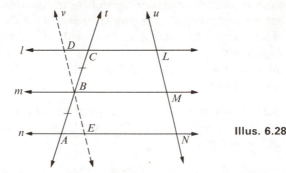

Illus. 6.28

The question of commensurability comes up in connection with the next theorem. We shall prove it only for the case in which the segments concerned are commensurable.

Theorem 6.14 If a line meeting two sides of a triangle is parallel to the third side, then it divides the two sides proportionally. In other words, if line *l* meets $\triangle ABC$ at interior points of $\overline{AC}$ and $\overline{BC}$, at points *D* and *E*, respectively, and if $l \parallel \overline{AB}$, then $AC/DC = BC/EC$.

Outline of proof for the case where $\overline{AC}$ and $\overline{DC}$ are commensurable. Choose a segment $\overline{XY}$ that measures both $\overline{AC}$ and $\overline{DC}$, and consider the points P_1, P_2, P_3, . . . , P_{m-1} on $\overline{AC}$ such that $\overline{AP_1}$, $\overline{P_1 P_2}$, $\overline{P_2 P_3}$, . . . , $\overline{P_{m-1} C}$ are all congruent to the small segment, $\overline{XY}$ (Illus. 6.29). Now $\overline{AC}$ is divided into *m* congruent segments, and $\overline{DC}$ is also divided into a number of congruent segments, say *n* of them. Suppose further that the measure of the small segment, $\overline{XY}$, is *p*. Then $AC = m \cdot p$ and $DC = n \cdot p$, and it follows that $AC/DC = m/n$.

Through each of the points P_1, P_2, . . . , P_{m-1} and *C* there is a unique line parallel to $\overline{AB}$. Since these lines cut off congruent segments on $\overline{AC}$, they also cut off congruent segments on $\overline{BC}$. Suppose each of these has measure *q*. Then

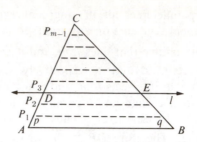

Illus. 6.29

$BC = m \cdot q$ and $EC = n \cdot q$, and it follows that $BC/EC = m/n$. Now we have $AC/DC = BC/EC$.

Although we shall not consider a proof for the incommensurable case, we may remark that a proof consists of taking smaller and smaller segments and using a limiting process somewhat similar to that used on p. 199 for the area of a rectangular region.

We are now in a position to prove the theorem of main importance to our study of similar triangles.

Theorem 6.15 If two triangles have two pairs of congruent angles, then they are similar; i.e., if in $\triangle ABC$ and $\triangle A'B'C'$ we have $\angle A \cong \angle A'$ and $\angle C \cong \angle C'$, then $\triangle ABC \sim \triangle A'B'C'$.

Outline of proof. We consider any two triangles $\triangle ABC$ and $\triangle A'B'C'$ for which $\angle A \cong \angle A'$ and $\angle C \cong \angle C'$ (Illus. 6.30). These triangles may be congruent. If so, they are similar. If not, then we proceed as follows: On $\overrightarrow{A'B'}$ there is a point D such that $\overline{A'D} \cong \overline{AB}$. On $\overrightarrow{A'C'}$ there is a point E such that $\overline{A'E} \cong \overline{AC}$. Now, $\triangle A'DE \cong \triangle ABC$. Since $\angle A'ED \cong \angle C \cong \angle C'$, we know that $\overline{DE} \parallel \overline{B'C'}$. Thus, by Theorem 6.14, $A'D/A'B' = A'E/A'C'$, and it follows that $AB/A'B' = AC/A'C'$. We may similarly show that $AC/A'C' = BC/B'C'$ by finding a triangle congruent to $\triangle ABC$ with one vertex at C'. We now have the three pairs of corresponding sides proportional. It remains to show that $\angle B \cong \angle B'$. This may be done easily with the help of Theorem 6.7.

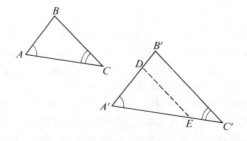

Illus. 6.30

9. A GENERAL CONCEPT OF SIMILARITY

In Chapter 4 the general concept of congruence was discussed and two figures were said to be congruent if and only if there exists a one-to-one correspondence between their points such that the segment determined by any two points of one figure is congruent to the segment determined by the corresponding pair of points of the other figure (see Definition 4.18). Let us now extend the concept of similarity to other geometric figures in a similar manner.

From the definition of similar triangles given in the preceding section it will be noted that if two triangles are congruent, then they are also similar, since the ratio of their corresponding sides (the sides being congruent and hence the same length) is always 1. The set of all congruent triangles is therefore a subset of the set of all similar triangles. The converse, however, is not true, since two triangles may be similar but not congruent.

Now suppose we consider a figure C, as in Illus. 6.31, and "construct" another figure which we feel would be called similar to it. Choose an arbitrary point P on or near figure C and consider segments determined by P and other points on C, for example, Q, R, and S. Now choose arbitrary points P' and Q', and construct points R' and S' so that $PQ/P'Q' = PR/P'R' = PS/P'S'$, and $\angle RPQ \cong \angle R'P'Q'$ and $\angle SPQ \cong \angle S'P'Q'$. If this process could be continued for every point on figure C, one would obtain a figure C' as shown in Illus. 6.32.

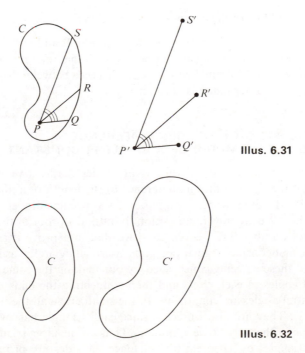

Illus. 6.31

Illus. 6.32

If two figures are related like this, we shall wish to call them similar. Note that if we choose any two points, V and W, on figure C and find the corresponding points, V' and W', on figure C', as in Illus. 6.33, they determine segments $\overline{VW}$ and $\overline{V'W'}$. Since $VP/V'P' = PW/P'W'$ and $\angle VPW \cong \angle V'P'W'$, it can be shown (see Exercise 11, p. 206) that $\triangle VPW \sim \triangle V'P'W'$, and therefore $VW/V'W' = VP/V'P' = PW/P'W'$. Similarly, for any other pairs of corresponding points, X, Y and X', Y', $XY/X'Y' = VW/V'W'$.

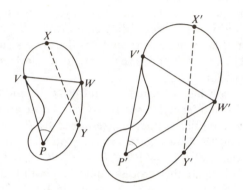

Illus. 6.33

On the basis of the observations above, we now make the following definition.

Definition 6.4 Two figures are said to be *similar* if and only if there exists a one-to-one correspondence between their points, such that the ratio of the distance between any pair of points of one figure and the distance between the corresponding pair of points of the other figure is constant.

10. THE RATIO OF THE CIRCUMFERENCE TO THE DIAMETER OF A CIRCLE IS CONSTANT

When we defined and discussed π in an earlier chapter, we made the tacit assumption that the ratio of the circumference to the length of a diameter is the same for all circles. In other words the number π is a specific number—the same number for all circles. We are now in a position to outline a proof of this important fact. We shall not number this theorem, and we shall not prove it rigorously or in detail.

We first consider regular polygons inscribed in circles, as in Illus. 6.34. We may think of these as being constructed by our making the same number of congruent central angles in each circle and then connecting the ends of the radii. Thus all the central angles are congruent. In the small circle all the triangles are congruent by SAS. They are also isosceles, since each triangle has two radii as sides. The situation in the large circle is similar. Thus all the large triangles are similar to the small triangles by Theorem 6.15. Hence the sides are proportional to the radii;

that is, $s/r = s'/r'$. It now follows that the perimeters of the polygons are proportional to the lengths of the radii. We have now established that if regular polygons with the same number of sides are inscribed in two circles, their perimeters are proportional to the lengths of the radii.

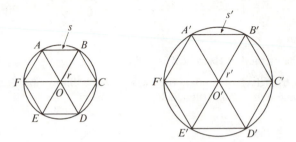

Illus. 6.34

Now consider any two circles with regular polygons inscribed. If the number of sides is very large, the perimeters approach the circumference of the circle very closely. In fact, the more sides, the closer the approximation to the circumference. If we consider inscribing successive polygons in the two circles, each with a larger number of sides, but each time putting the same kind of polygon in each circle, we can see that the ratio of perimeters to length of radii is the same every time. Since the perimeters approach the circumference, the ratio of the circumference to the length of a radius (or diameter) must be the same for both circles. Since we chose any pair of circles, this ratio is the same for all circles.

EXERCISES

1. Suppose, in $\triangle ABC$ and $\triangle A'B'C'$, that $\angle ABC \cong \angle A'B'C'$, $\overline{AB}$ is twice as long as $\overline{BC}$, and $\overline{A'B'}$ is twice as long as $\overline{B'C'}$. Are the triangles similar? Explain your answer.

2. Is similarity of triangles an equivalence relation?

3. From the discussion above, write a definition of *regular polygon* without using the circle or its central angles.

4. a) Draw a circle with a 3-inch radius and inscribe a regular hexagon in it. Measure the perimeter of the polygon. Calculate the circumference of the circle, using 3.14 for π.

   b) Repeat part (a) using an inscribed dodecagon.

5. Here is a method for dividing a segment into any number, n, of congruent segments: If $\overline{AB}$ is the given segment, draw $\overrightarrow{AC}$, and on it step off n congruent segments of any convenient length, using a compass. Call the last endpoint D. Draw $\overrightarrow{DB}$ and construct $n - 1$ segments parallel to it, as shown in Illus. 6.35. Use this method to divide a 4-inch segment into ten congruent segments.

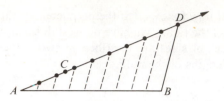

Illus. 6.35

👓 *Rigorous exercises*

6. Prove that the method in Exercise 5 is valid.

7. If these two triangles are similar, then $a/a' = b/b'$. Show (algebraically) also that $a/b = a'/b'$ and $a'/a = b'/b$ (Illus. 6.36).

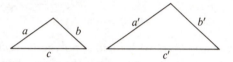

Illus. 6.36

8. Is the converse of Theorem 6.14 true? If so, prove it. (If incommensurability is a problem, consider only the commensurable case.)

9. Is the converse of Theorem 6.15 true? If so, prove it.

10. Prove that if two triangles have their sides respectively parallel, then they are similar.

11. Is the following statement true? If so, prove it.
    If in two triangles two pairs of sides are proportional and the included angles are congruent, then the triangles are similar.

12. Is the following statement true? If so, prove it.
    If in two triangles three pairs of sides are proportional, then the triangles are similar.

13. Prove the Pythagorean theorem, using the concept of similar triangles (Illus. 6.37). [*Hint:* In a right triangle, the altitude to the hypotenuse forms two right triangles, each similar to the original triangle.]

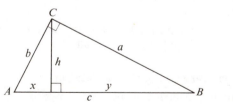

Illus. 6.37

14. Show that if two triangles are "similar" according to Definition 6.4, then they are similar according to Definition 6.3, and conversely.

15. Show that if two simple quadrilaterals have corresponding angles congruent and the ratio of corresponding sides is constant, then they are similar.

16. Prove that any two circles are similar.

11. APPLICATIONS OF SIMILAR TRIANGLES

The theory of similar triangles gives rise to a great many applications, principally those involving indirect measurement. A simple and common example is that of finding the height of a tree or flagpole without climbing. This may be done by measuring the shadow of the tree, and measuring a post and its shadow. If, as in Illus. 6.38, the post has length $B'C'$ and its shadow has length $A'C'$, then $BC/AC = B'C'/A'C'$ because triangles $\triangle ABC$ and $\triangle A'B'C'$ are similar. They are similar because the rays of the sun are (for all practical purposes) parallel. Hence $\angle A' \cong \angle A$ and $\angle C \cong C'$, the latter being both right angles. It follows that

$$ BC = AC \cdot \frac{B'C'}{A'C'}. $$

Thus we can compute the height of the tree, once we have measured the post and the two shadows.

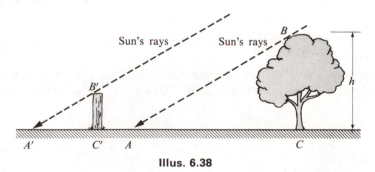

Illus. 6.38

Example. How tall is a tree that casts a 45-ft shadow at the same time that a 5-ft post casts a 9-ft shadow (Illus. 6.39)?

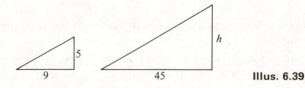

Illus. 6.39

Since these triangles are similar, $h/45 = 5/9$. Then $h = 45 \cdot \frac{5}{9}$, or 25 ft.

Another common type of problem is that of finding the width of a river without crossing it. This may be done as shown in Illus. 6.40. A landmark such as a tree is chosen at Y, on the far side of the river, and stakes are set at X and A. Another stake is set at B on a line with X and A. Angle $\angle AXY$ is measured and $\angle B$ is made the same size. Thus a point C is determined, which is also on the line $\overleftrightarrow{AY}$. Now $\triangle AXY \sim \triangle ABC$; hence $XY/AX = BC/AB$, or $XY = AX \cdot (BC/AB)$. Thus after the accessible segments $\overline{AX}$, $\overline{BC}$, and $\overline{AB}$ have been measured, the width of the river, XY, can be computed.

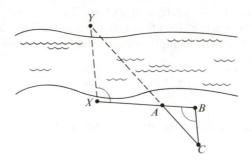

Illus. 6.40

Example. How wide is the river in Illus. 6.40 if $AX = 120$ ft, $AB = 3$ ft, and $BC = 4$ ft?

As explained above, the triangles are similar because $\angle XAY \cong \angle BAC$ (they are vertical angles), and $\angle AXY \cong \angle ABC$. Hence $XY/AX = BC/AB$, and

$$XY = AX \cdot \frac{BC}{AB}.$$

We then find

$$XY = 120 \cdot \tfrac{4}{3} \quad \text{or} \quad 160 \text{ ft.}$$

EXERCISES

1. How tall is a flagpole that casts a 36-ft shadow at the same time a yardstick casts a 4-ft shadow?

2. The width of a river is being measured, as shown in Illus. 6.41. The angles at X and B are right angles, $AB = 5$ ft, $BC = 12$ ft, and $AX = 50$ ft. Find XY. Also find AY.

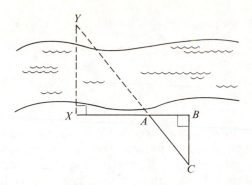

Illus. 6.41

3. A string is stretched tight from level ground to the top of a hill (Illus. 6.42). A stick placed under the string at D is $3\frac{1}{3}$ ft long and just touches the string when D is 4 ft from A. The string is then taken down and measured. Its length is 112 ft. How high is the hill? How long is $\overline{AB}$?

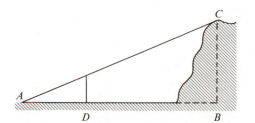

Illus. 6.42

4. Two triangular fences are as shown in Illus. 6.43, with sides parallel. If $a' = 2a$ and the perimeter of the smaller fence is 35 ft, what is the perimeter of the other?

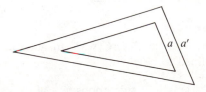

Illus. 6.43

5. A picture of the front of a building measures $1\frac{1}{4}$ in. by 4 in. The building is actually 120 ft long. How tall is it?

6. Three towns on a map are situated as in Illus. 6.44. $AB = 3$ in., $BC = 4$ in., and $AC = 5.5$ in. It is known that A and B are actually 20 miles apart. What is the distance from B to C? From A to C?

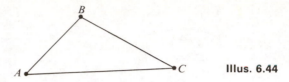

Illus. 6.44

Trigonometry

By Theorem 6.15 it follows immediately that if two right triangles have a pair of congruent acute angles, they are similar. This fact is the basis of the study of triangles known as *trigonometry*. If in right triangles $\triangle ABC$ and $\triangle A'B'C'$, $\angle A \cong \angle A'$, as in Illus. 6.45, the triangles are similar. Therefore corresponding sides have the same ratio; for example, $a/a' = c/c'$. By a little algebra we see that this is equivalent to $a/c = a'/c'$. From the latter equation we may conclude the following: For all right triangles having an acute angle of a given size, the ratio of the side opposite that angle to the hypotenuse is the same (Illus. 6.46). This ratio does not depend on the size of the triangle, but only on the size of the angle. This particular ratio is known as the *sine ratio*. Let us define it more precisely.

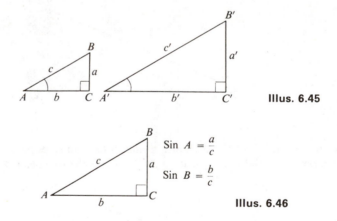

Illus. 6.45

$$\text{Sin } A = \frac{a}{c}$$

$$\text{Sin } B = \frac{b}{c}$$

Illus. 6.46

Definition 6.5 In a right triangle $\triangle ABC$ with an acute angle $\angle A$, *sine A* (abbreviated "sin A") is defined to be a/c or

$$\frac{\text{length of side opposite } \angle A}{\text{length of hypotenuse}}.$$

It is easy to see that small angles have small sines and that if $\angle P \prec \angle Q$, then $\sin P < \sin Q$. Sines of angles of various sizes have been computed and are listed in the table on page 213. The entries are approximate, being correct to three places.

There are five other ratios of sides that we might define. We shall consider only two of them here. They are as shown in Illus. 6.47 and in the following definitions.

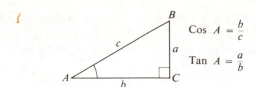

$$\text{Cos } A = \frac{b}{c}$$

$$\text{Tan } A = \frac{a}{b}$$

Illus. 6.47

Definition 6.6 In a right triangle $\triangle ABC$, with an acute angle $\angle A$, *cosine A* (abbreviated "cos A") is defined to be b/c or

$$\frac{\text{length of leg adjacent to } \angle A}{\text{length of hypotenuse}}.$$

Definition 6.7 *Tangent A* (abbreviated "tan A") is defined to be a/b or

$$\frac{\text{length of leg opposite } \angle A}{\text{length of leg adjacent to } \angle A}.$$

Note that the table of sines also contains entries for cosines and tangents, also approximate.

When tables of trigonometric ratios are available, problems of indirect measurement are simplified, as shown in the following examples.

Example 1. How long is $\overline{BC}$ in $\triangle ABC$ if $m\angle A = 31°$ and $c = 15$ m? (See Illus. 6.48.) We know that $a/c = \sin A$. Thus $a = c \sin A = 15 \cdot \sin 31° = 15 \cdot 0.515$, which is about 7.73 m.

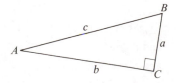

Illus. 6.48

Example 2. How far is it across this river if $AC = 50$ ft and $m\angle A = 73°$? See Illus. 6.49. We know that $a/b = \tan A$. Thus $a = b \tan A = 50 \cdot \tan 73° = 50 \cdot 3.27$, or about 163 ft.

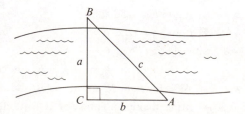

Illus. 6.49

Example 3. To find the height of a building, a man measures the angle between the horizontal and his line of sight to the top of the building, and finds it to be 13° (Illus. 6.50). He then measures the distance d, finding it to be 410 ft. He then calculates as follows:

$$h/d = \tan 13°, \quad \text{hence} \quad h = d \tan 13° = 410 \cdot 0.231,$$

or about 94.7 ft. He then adds his height h', which is 6 ft, obtaining 100.7 ft, approximately, for the height of the building.

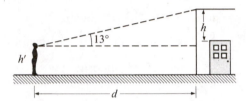

Illus. 6.50

Note how, in each of the preceding examples, we choose the appropriate trigonometric ratio. In each problem we seek to find the length of some side of a triangle. The length of some other side is known. The ratio we choose is the one that involves those two sides.

EXERCISES

1. Given triangle $\triangle ABC$ with right angle at C.
   a) Find a if $m\angle A = 34°$ and $c = 17$ ft.
   b) Find a if $m\angle A = 16°$ and $b = 35$ yd.
   c) Find b if $m\angle A = 41°$ and $c = 34$ m.
   d) Find c if $m\angle A = 26°$ and $b = 13$ cm.
   e) Find c if $m\angle A = 38°$ and $a = 28$ ft.
2. Given triangle $\triangle ABC$ with right angle at C.
   a) Find $m\angle A$ if $a = 10$ ft and $c = 20$ ft.
   b) Find $m\angle A$ if $a = 131$ ft and $b = 258$ ft.
   c) Find $m\angle A$ if $b = 171$ ft and $c = 200$ ft.
3. Given triangle $\triangle ABC$ with right angle at C.
   a) Find $m\angle B$ if $m\angle A = 23°$. Find $\sin B$, $\sin A$, $\cos B$, and $\cos A$.
   b) Find $m\angle A$ if $m\angle B = 43°$. Find $\sin B$, $\sin A$, $\cos B$, and $\cos A$.
4. A kite is flying on a 500-ft string, making an angle of 21° with the horizontal. Assuming the string is straight, how high is the kite?

TABLE 1

NATURAL TRIGONOMETRIC FUNCTIONS

Angle					Angle				
De-gree	Ra-dian	Sine	Co-sine	Tan-gent	De-gree	Ra-dian	Sine	Co-sine	Tan-gent
0°	0.000	0.000	1.000	0.000					
1°	0.017	0.017	1.000	0.017	46°	0.803	0.719	0.695	1.036
2°	0.035	0.035	0.999	0.035	47°	0.820	0.731	0.682	1.072
3°	0.052	0.052	0.999	0.052	48°	0.838	0.743	0.669	1.111
4°	0.070	0.070	0.998	0.070	49°	0.855	0.755	0.656	1.150
5°	0.087	0.087	0.996	0.087	50°	0.873	0.766	0.643	1.192
6°	0.105	0.105	0.995	0.105	51°	0.890	0.777	0.629	1.235
7°	0.122	0.122	0.993	0.123	52°	0.908	0.788	0.616	1.280
8°	0.140	0.139	0.990	0.141	53°	0.925	0.799	0.602	1.327
9°	0.157	0.156	0.988	0.158	54°	0.942	0.809	0.588	1.376
10°	0.175	0.174	0.985	0.176	55°	0.960	0.819	0.574	1.428
11°	0.192	0.191	0.982	0.194	56°	0.977	0.829	0.559	1.483
12°	0.209	0.208	0.978	0.213	57°	0.995	0.839	0.545	1.540
13°	0.227	0.225	0.974	0.231	58°	1.012	0.848	0.530	1.600
14°	0.244	0.242	0.970	0.249	59°	1.030	0.857	0.515	1.664
15°	0.262	0.259	0.966	0.268	60°	1.047	0.866	0.500	1.732
16°	0.279	0.276	0.961	0.287	61°	1.065	0.875	0.485	1.804
17°	0.297	0.292	0.956	0.306	62°	1.082	0.883	0.469	1.881
18°	0.314	0.309	0.951	0.325	63°	1.100	0.891	0.454	1.963
19°	0.332	0.326	0.946	0.344	64°	1.117	0.899	0.438	2.050
20°	0.349	0.342	0.940	0.364	65°	1.134	0.906	0.423	2.145
21°	0.367	0.358	0.934	0.384	66°	1.152	0.914	0.407	2.246
22°	0.384	0.375	0.927	0.404	67°	1.169	0.921	0.391	2.356
23°	0.401	0.391	0.921	0.424	68°	1.187	0.927	0.375	2.475
24°	0.419	0.407	0.914	0.445	69°	1.204	0.934	0.358	2.605
25°	0.436	0.423	0.906	0.466	70°	1.222	0.940	0.342	2.748
26°	0.454	0.438	0.899	0.488	71°	1.239	0.946	0.326	2.904
27°	0.471	0.454	0.891	0.510	72°	1.257	0.951	0.309	3.078
28°	0.489	0.469	0.883	0.532	73°	1.274	0.956	0.292	3.271
29°	0.506	0.485	0.875	0.554	74°	1.292	0.961	0.276	3.487
30°	0.524	0.500	0.866	0.577	75°	1.309	0.966	0.259	3.732
31°	0.541	0.515	0.857	0.601	76°	1.326	0.970	0.242	4.011
32°	0.559	0.530	0.848	0.625	77°	1.344	0.974	0.225	4.332
33°	0.576	0.545	0.839	0.649	78°	1.361	0.978	0.208	4.705
34°	0.593	0.559	0.829	0.675	79°	1.379	0.982	0.191	5.145
35°	0.611	0.574	0.819	0.700	80°	1.396	0.985	0.174	5.671
36°	0.628	0.588	0.809	0.727	81°	1.414	0.988	0.156	6.314
37°	0.646	0.602	0.799	0.754	82°	1.431	0.990	0.139	7.115
38°	0.663	0.616	0.788	0.781	83°	1.449	0.993	0.122	8.144
39°	0.681	0.629	0.777	0.810	84°	1.466	0.995	0.105	9.514
40°	0.698	0.643	0.766	0.839	85°	1.484	0.996	0.087	11.43
41°	0.716	0.656	0.755	0.869	86°	1.501	0.998	0.070	14.30
42°	0.733	0.669	0.743	0.900	87°	1.518	0.999	0.052	19.08
43°	0.750	0.682	0.731	0.933	88°	1.536	0.999	0.035	28.64
44°	0.768	0.695	0.719	0.966	89°	1.553	1.000	0.017	57.29
45°	0.785	0.707	0.707	1.000	90°	1.571	1.000	0.000	

5. A 15-ft ladder is leaning against a building. The lower end of the ladder is 7.5 ft from the building. Find the measure of the angle the ladder makes with the ground. With the building.

6. A 50-ft wire is used to guy an antenna support mast. The wire makes an angle of 52 with the ground. At what height is it attached to the mast?

7. A rafter rises 2 ft for every 3 ft of horizontal run. What size angle does it make with the horizontal?

8. On a road with 3% grade (this means the ratio of rise to run is 3%) how high does one rise in traveling a mile?

9. An airplane climbs at an angle of 7° with the horizontal. How high will it be 4 minutes after takeoff if it flies 180 mph?

10. A tunnel is dug into the earth at an angle of 8° with the horizontal. What is the vertical drop between two junctions in the tunnel that are 150 ft apart?

A GEOMETRIC FALLACY

See if you can find a flaw in the argument below.

Theorem. If a quadrilateral $ABCD$ has $\angle B \cong \angle D$, $\overline{AD} \cong \overline{BC}$, then $ABCD$ is a parallelogram.

Consider perpendiculars $\overline{AE}$ and $\overline{CF}$ as shown in Illus. 6.51.

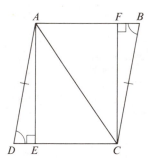

Illus. 6.51

We note that $\triangle AED \cong \triangle CFB$ by *AAS*. Thus $\overline{AE} \cong \overline{CF}$, and $\overline{DE} \cong \overline{BF}$. Now $\triangle AEC \cong \triangle CFA$ by Theorem 4.19, and therefore $\overline{EC} \cong \overline{FA}$. By Exercise 17, page 188, quadrilateral $ABCD$ is a parallelogram.

CHAPTER 7

NON-EUCLIDEAN GEOMETRIES

In the preceding chapters the importance of the parallel axiom became apparent. Some of the fundamental theorems of Euclidean geometry are not provable without the use of Axiom 6.1 or something equivalent to it. Without this axiom, for example, one cannot prove that the sum of the measures of the angles of a triangle is 180° or that the opposite sides of a parallelogram are congruent. The entire theory of similar figures also rests on this fundamental axiom.

1. EUCLID'S FIFTH POSTULATE

The assumption made in Chapter 6 about parallels (Axiom 6.1) is, at first glance, quite different from that of Euclid, but the two statements are in fact equivalent, as we shall see. Axiom 6.1 is essentially as formulated by the Scottish mathematician Playfair,* while the assumption, or *postulate*, of Euclid that is equivalent to it is his famous fifth postulate. Euclid's postulate is as follows, except for some minor rewording.

Axiom 7.1 If two lines are cut by a transversal such that the sum of the measures of the interior angles on one side of the transversal is less than 180°, then the two lines meet on that side of the transversal.

Euclid's fifth postulate says that the lines in Illus. 7.1 must meet on the right side of the transversal, since the sum of the measures of the interior angles is less than 180° on that side.

---

* Playfair's form of the parallel postulate was also suggested by Proclus, a Greek mathematician of the fifth century, as well as other mathematicians before Playfair.

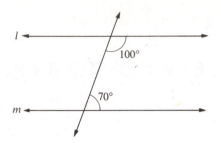

Illus. 7.1

Euclid's fifth postulate refers to lines in a plane, of course. It is this postulate that gives Euclidean geometry its unique characteristics. This postulate is *equiv-* *alent* to Axiom 6.1 in the logical sense; that is, it implies 6.1 and, conversely, 6.1 implies Euclid's postulate. This means that the two statements are true or false together and that either may be used in place of the other as we please.

To prove the equivalence of the two statements, we assume all of the previous axioms as background and then prove, first, that Euclid's postulate implies 6.1 and, second, that 6.1 implies Euclid's postulate, 7.1. We now prove this equivalence.

1. *If Euclid's fifth postulate holds, then Axiom 6.1 holds.*

Outline of proof. Suppose, as hypothesis, that Euclid's fifth postulate, 7.1, holds. Then consider a line *l* and a point *P* not on *l*, as in Illus. 7.2. We wish to show that there is not more than one line on *P* that is parallel to *l*. By Theorem 6.3 there is at least one line *m* which contains *P* and is parallel to *l*. Now choose any point Q on *l* and consider the transversal $\overleftrightarrow{PQ}$. The line *m*, proved to exist in Theorem 6.3, is such that $\angle RPQ \cong \angle PQV$. Now let us consider any other line *n* on *P*. Since this line is different from *m*, we know that $\overset{\circ}{\overrightarrow{PB}}$ is in the interior of $\angle QPS$ or in the exterior of that angle.

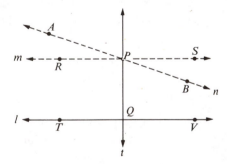

Illus. 7.2

If $\overset{\circ}{\overrightarrow{PB}}$ is in the interior of $\angle QPS$, then the sum of the measures of $\angle PQV$ and $\angle QPB$ is less than 180; then, by Euclid's postulate, lines *l* and *n* meet on the side of $\overleftrightarrow{PQ}$ containing *B*. On the other hand, if $\overset{\circ}{\overrightarrow{PB}}$ is in the exterior of $\angle QPS$, then $\overset{\circ}{\overrightarrow{PA}}$

is in the interior of $\angle RPQ$ and thus lines n and l meet on the side of $\overleftrightarrow{PQ}$ containing A. Therefore, any line on P different from m meets l; hence the parallel, m, is unique.

   2. *If Axiom 6.1 holds, then Euclid's postulate (Axiom 7.1) holds.*

Outline of proof. Suppose, as hypothesis, that 6.1 holds, and consider any line l and a point P not on l. Also consider any line $\overleftrightarrow{PQ}$, containing P and a point Q of l, as in Illus. 7.3. By Theorems 6.2 and 6.3 there exists a line m containing P such that $\angle APQ \cong \angle PQD$ or, what amounts to the same thing, $m\angle BPQ + m\angle PQD = 180°$, and this line is parallel to l. By hypothesis, this line is unique.

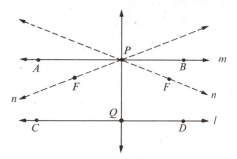

Illus. 7.3

   Now consider a line $\overleftrightarrow{PF}$, other than $\overleftrightarrow{PQ}$ or m, such that $m/FPQ + m\angle PQD < 180$. Then $\overleftrightarrow{PF}$ is interior to $\angle BPQ$ and $\overleftrightarrow{PF}$ is not parallel to l, since m is the only parallel to l which contains P. It follows that $\overleftrightarrow{PF}$ meets l at some point X and this point X must be on the same side of $\overleftrightarrow{PQ}$ as F. Otherwise, $\triangle PQX$ would have exterior angle $\angle QPF$ less than interior angle $\angle PQC$. The proof is the same if F is on the opposite side of $\overleftrightarrow{PQ}$ and $m\angle FPQ + m\angle PQC < 180$. We have now shown that if the parallel containing P is unique, then Euclid's postulate holds.

   Now that we have proved the equivalence of these two axioms, or assumptions, we can use either of them in place of the other as it suits our convenience. There are several other statements that are equivalent to Euclid's fifth postulate (and, of course, to 6.1). Of interest to us here is the following:

Axiom 7.2 For any distinct lines l and m and n, if $l \parallel n$ and $m \parallel n$, then $l \parallel m$.

In the preceding chapters we have perhaps tacitly assumed this statement, which says that if two distinct lines are parallel to the same line they are also parallel to each other. As we can now see, such an assumption is not to be made lightly, because it is equivalent to a most important and vital axiom. A proof of the equivalence of this statement and 6.1 is left as an exercise.

EXERCISES

👀 1. If in a geometry we have Axiom 7.2 as an axiom, is parallelism of lines an equivalence relation? If not, how could you alter the notion of parallelism slightly to make it so?

2. Find three other statements that are equivalent to the parallel postulates above.

👀 *Rigorous exercise*

3. a) Prove that Axiom 6.1 implies Axiom 7.2.

b) Prove that Axiom 7.2 implies Axiom 6.1.

c) What can you then conclude about Axioms 7.1 and 7.2?

2. INDEPENDENCE OF THE PARALLEL AXIOM

From early times mathematicians were apparently not quite satisfied with Euclid's fifth postulate. This axiom seemed different from the others, yet so obvious that they felt it should be a theorem rather than an axiom. There is nothing logically incorrect, of course, about using an axiom that can be proved as a theorem from the other axioms. But mathematicians habitually try, partly for aesthetic reasons, to boil down a set of axioms until they are independent of each other, and thus no one of them is a consequence of the others. Euclid himself was apparently not entirely happy with his axiom, for he listed it last and did not use it until necessary.

The attempts of many fine mathematicians to deduce the fifth postulate as a consequence of the others extended over many centuries. All of the attempts failed. Notable was the work of the Jesuit, Saccheri, in the early 18th century, and that of Bolyai, a Hungarian, Lobachevsky, a Russian, and Riemann, a German, in the 19th century. Bolyai and Lobachevsky independently discovered that it is possible to construct a mathematical system, a geometry, in which parallels are not unique but in which all of the other axioms of ordinary Euclidean geometry hold. This proved that Euclid's fifth postulate is independent of the rest and, hence, cannot be proved as a consequence of them. A bit later, Riemann showed that it is also possible to construct a geometry in which there are no parallel lines. In his kind of geometry most, but not all, of the other axioms of ordinary Euclidean geometry hold. These two kinds of geometry are known as *non-Euclidean* geometries. Had Saccheri not been in error with his proofs or had his attitude been different, he might have been the discoverer of both kinds of non-Euclidean geometry.

The discovery of non-Euclidean geometries was upsetting to the mathematical community in much the same way that the discovery of incommensurable segments had been, about 2000 years earlier. The geometry of Euclid was thought to be an infallible description of space, developed from "self-evident" axioms, and that kind of geometry had deeply influenced mathematical thinking for many centuries. The notion that a consistent kind of geometry exists, different from Euclid's, was not easy to accept. The idea that these non-Euclidean geometries might have applications to or provide descriptions of space was still more difficult to accept; in fact, the mathematical community was not ready to accept it until some years had passed. The non-Euclidean geometries were felt to be mere curiosities, and Euclidean geometry was held to be the *true* geometry. After some time had passed, these ideas were dispelled. Non-Euclidean geometries were shown to have applications of various sorts. It was also shown that they can serve to describe space. The crowning achievement in this regard was Einstein's use of a non-Euclidean geometry of the Riemann type in his general theory of relativity. That theory is now believed to provide a better description of space and the universe than any other. It may therefore be said that space is probably not Euclidean after all.

REVIEW PRACTICE EXERCISES

One method of approximating square roots is illustrated below.

Example. Find $\sqrt{46}$. Suppose the first estimate is 6.5.

Divide:

$$6.5\overline{)46.00} \quad 7.07+ \quad .$$

Thus $6.5 < \sqrt{46} < 7.07$.
The average of the estimate and the quotient,

$$\frac{6.5 + 7.07}{2} = 6.78$$

provides a better estimate.
Divide:

$$6.78\overline{)46.00} \quad 6.79+ \quad .$$

Thus $6.78 < \sqrt{46} < 6.79$, or $\sqrt{46} \approx 6.8$ (to the nearest tenth). (The process can of course be repeated as often as you wish.)

1. Approximate these square roots to the nearest tenth.
   a) $\sqrt{52}$ b) $\sqrt{91}$ c) $\sqrt{123}$
   d) $\sqrt{675}$ e) $\sqrt{253}$ f) $\sqrt{878}$

2. Approximate these square roots to the nearest unit.
   a) $\sqrt{2596}$ b) $\sqrt{8924}$ c) $\sqrt{11,496}$
   d) $\sqrt{5329}$ e) $\sqrt{4304}$ f) $\sqrt{1000}$

3. In the following problems, a, b, and c represent the measures of the two legs and the hypotenuse, respectively, of a right triangle. Given the measures of two sides, find the measure of the third side. When the measure is irrational, find a rational approximation to the nearest tenth.

Example. Find the measure of b when $a = 6$ and $c = 15$.

By the Pythagorean theorem, $a^2 + b^2 = c^2$.

Substituting, we get $36 + b^2 = 225$, or $b = \sqrt{189}$. Thus, $b \approx 13.7$.

a) Find b, where $a = 5$ and $c = 9$.

b) Find c, where $a = 1$ and $b = 2$.

c) Find a, where $b = 25$ and $c = 4225$.

d) Find c, where $a = \sqrt{7}$ and $b = \sqrt{18}$.

e) Find b, where $a = \sqrt{13}$ and $c = \sqrt{14}$.

f) Find a, where $b = 2\sqrt{5}$ and $c = 3\sqrt{13}$.

4. Show that the ratio of a leg to the hypotenuse of an isosceles right triangle is $1/\sqrt{2}$.

5. Given equilateral triangle $\triangle ABC$ (Illus. 7.4), where x is the measure of each side and $\overline{AD}$ is the perpendicular bisector of $\overline{CB}$, find the measure of $\overline{AD}$.

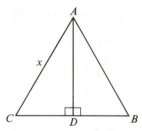

Illus. 7.4

3. RIEMANNIAN GEOMETRIES

A geometry in which there are no parallel lines and hence any two lines in a plane intersect is said to be Riemannian. There are many kinds of Riemannian geometry, all of them, of course, non-Euclidean. We shall illustrate this kind of geometry by considering a particular example, a geometry on a sphere.

A set of points on a sphere will constitute the set of points in this geometry. The lines of this geometry will be taken to be the great circles on the sphere. With this interpretation of "line," we find that the union of all the lines is a *plane*, fitting the definition of a plane given in Chapter 2. Some of the theorems of Chapter 2 do not hold here. For example, it is not true that any two lines meet in at most one point. In fact, any two lines have an intersection which contains exactly two points. These points are opposite ends of a diameter, and are called *antipodal* points (Illus. 7.5). Some of the betweenness properties developed in Chapter 2 likewise do not hold in this geometry. For example, it is no longer the case that for any three distinct points of a line exactly one is between the other two. Any three points on a great circle are in an arrangement such as that in Illus. 2.18, in which each is between the other two.

Aside from a few differences such as those just described, the geometry on the sphere has all of the properties of Euclidean geometry except for those that depend

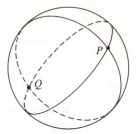

Illus. 7.5

on the parallel axiom. It is the lack of parallels in which we are primarily interested here.

Since any two lines have a nonempty intersection, there are no parallel lines in this geometry. This means that none of the theorems depending on the parallel axiom will be true here. Theorems 6.4 through 6.9 and 6.13 through 6.15 are not theorems in this geometry. Since parallels do not exist, parallelograms do not exist. The sum of the measures of the angles of a triangle is not 180° in the geometry on the sphere. For example, Illus. 7.6 shows a triangle with three right angles. Such a triangle exists on the earth, where P is the North Pole and A and B are points on the equator that are 90° of longitude apart. The sum of the measures of the angles of such a triangle is 270°. In this geometry there does not exist any triangle for which the sum of the angle measures is 180°. The sum of the angle measures is different for different triangles, but it is more than 180° for any triangle. It follows that no quadrilateral has four right angles, the sum of the angle measures being always more than 360°.

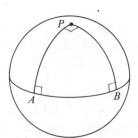

Illus. 7.6

The Pythagorean theorem does not hold in this geometry. In fact, since triangles with more than one right angle exist, the statement of the Pythagorean theorem does not even make sense if we try to apply it to this geometry. Furthermore, there are no similar triangles in this geometry except for congruent ones. That is, there are no noncongruent triangles having their corresponding angles congruent and corresponding sides proportional. The ratio of the circumference of a circle to the length of a diameter is not a constant. It is different for circles of different size.

EXERCISES

For these exercises, obtain a spherical object and something that will write on it (such as a library globe and a grease pencil, a blackboard globe and chalk, or a tennis ball and a ball pen).

1. a) Draw a "line."

    b) Draw another line that meets the first one at right angles.

    c) Draw a third line, to form a right triangle.

    d) With a protractor, measure the angles of the triangle.

2. a) Draw a triangle as small as it can be and still have its angles measured comfortably. Measure the angles.

    b) Draw as large a triangle as you can and measure its angles.

    c) Compare and generalize.

    d) What seems to be the maximum number of degrees in the angles of triangles? What seems to be the minimum?

3. Draw a quadrilateral with two right angles. Measure the other two angles, and discuss.

4. Try to draw two similar triangles, one inside the other. Discuss the difficulties.

5. a) Draw two concentric circles, one quite small and the other quite large.

    b) Measure the diameters. This should be done with a string, remembering that a diameter is a segment of a great circle.

    c) Lay string around the circles and measure their circumferences.

    d) Find the ratio of circumference to length of diameter in each case and compare.

6. Show by an example that the exterior-angle theorem does not hold in this geometry.

From several points of view, spherical geometry is an interesting example of a geometry. For one thing, it shows how we can use the Euclidean geometry we have developed to build a model of a non-Euclidean geometry. We use a sphere, as we view it in the Euclidean sense, as a plane. This set of points *is* a plane in the Riemannian geometry. Great circles as we see them in the Euclidean sense are lines in the Riemannian geometry. The fact that we use Euclidean geometry to build such a model of Riemannian geometry shows that Riemannian geometry is just as consistent as Euclidean geometry.

The example of spherical geometry also shows very clearly how important the parallel axiom is in giving Euclidean geometry its unique characteristics. It might be argued, however, that in viewing spherical geometry as we have done, we have taken the position that we know what the "true" geometry is like, in which planes are "flat" and lines "straight," that spherical geometry does not tell us what space is really like. But let us consider a very large sphere, say the size of the earth. Suppose there lives on this sphere a very small creature, such as an ant (or a man). The small creature can experience only a very small region of the "plane" on which he lives, and he thinks it is flat. It appears that there are parallels, and when triangles are measured, the angle measures sum to 180. The small creature views his world as Euclidean, whereas it is only when he can view the whole thing that he sees it is not. Compare this now to man in the universe. He may view the universe as Euclidean, while someone outside the universe would say that it is not. Since our communication with persons outside the universe is not fluent, we cannot be too sure of what the "true" picture is.

4. BOLYAI-LOBACHEVSKIAN GEOMETRIES

We have now considered Euclidean geometry, in which there is a unique parallel to a line through a point off the line. We have also considered Riemannian geometry, in which the parallel axiom is denied, by saying that through a point off a line there is no parallel. There is one other way to deny the parallel axiom, namely to say that through a point off a line there is more than one line parallel to the given line. A geometry in which this statement is used instead of the parallel axiom is known as a Bolyai-Lobachevskian geometry, also called an *hyperbolic* geometry.

In this kind of geometry it is not necessary to deny any of the earlier axioms, as it was for Riemannian geometry. Every theorem of Euclidean geometry can also be a theorem of hyperbolic geometry, provided it does not depend on the parallel axiom. In hyperbolic geometry we have the axiom that through a point off a line there is more than one parallel to the given line. An immediate consequence is that there must be an infinite number of parallels. For suppose there are two distinct lines l and m, both parallel to n, as in Illus. 7.7. Then any line $\overleftrightarrow{PC}$, where C is in the interior of $\angle APB$, cannot meet n. There is an infinite number of such lines; hence there is an infinite number of lines containing P and parallel to n.

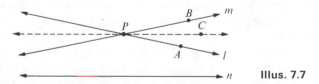

Illus. 7.7

Let us now consider a model of an hyperbolic geometry, due to F. Klein. As for the model of Riemannian geometry, we take something from Euclidean geometry to serve as a plane, this time the interior of a circle. The lines in this geometry are chords without endpoints. It is now clear that this model satisfies the condition of multiple parallels, as shown in Illus. 7.8. Lines l and m are both parallel to n, since the intersection of either of them with n is empty. It is also easy to note from this drawing that in this geometry, lines that are parallel to the same line are not necessarily parallel to each other.

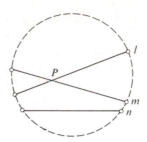

Illus. 7.8

The open segments that are the lines of this geometry satisfy all of the axioms about lines except the parallel axiom, and all of the betweenness conditions hold. This can be checked rather easily, but it is important to remember that the points of the circle are not included in the geometry, but only the interior. Thus the lines have no endpoints.

Angles are defined to be congruent if they bear a certain relationship which we shall not describe here. A scheme of assigning measures to segments is devised and then segments are defined to be congruent if they have the same measure. Under this scheme of assigning measures, segments near the center of the circular region have considerably smaller measures than comparable segments near the edge of the region. This is shown roughly in Illus. 7.9, where all of the segments have the same measure. We omit the details of this manner of assigning measures. It can be done in a precise mathematical fashion, however, and the interested reader is referred to the Bibliography.

   Illus. 7.9

Once measures are assigned to segments, we then define two segments to be congruent if they have the same measure. This means that a segment near the center of the circle that appears large to us as we take a Euclidean view may be congruent to a segment that appears very small, near the edge of the region. This means of assigning measures is done very carefully, so that all of the triangle congruence theorems hold. Archimedes' axiom also holds, as do all of the other axioms and theorems, except the parallel axiom and those properties that depend on it. Each line is of "unlimited extent" in the sense that there are no endpoints. Furthermore, lines or rays do not have finite measures. We may see this from Illus. 7.10. If we were to try to measure a ray, using any segment as a unit and, of course, placing congruent segments end to end, we would find that the measuring process is an endless one. No matter how many unit segments we place end to end, the shrinkage as we approach the circle is such that we never reach it.

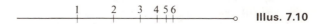   Illus. 7.10

In this kind of geometry the sum of the measures of the angles of a triangle is less than 180 for any triangle. There are no parallelograms and no quadrilaterals with four right angles. No triangles are similar unless they are congruent, and the Pythagorean theorem does not hold.

REVIEW PRACTICE EXERCISES

1. Draw a pair of coordinate axes on graph paper and plot these points.

   a) $(3, 2)$ b) $(-4, 1)$ c) $(-2, -6)$

   d) $(\frac{1}{2}, -3)$ e) $(0, -2)$ f) $(3, 0)$

   g) $(0, 1.5)$ h) $(-2, -2)$

2. Draw a pair of axes on graph paper.

   a) Plot the points $(0, 2)$, $(\frac{1}{2}, \frac{5}{2})$, $(3, 5)$, $(7, 9)$, $(-2, 0)$, and $(-\frac{1}{3}, \frac{5}{3})$.

   b) For each of the points in (a), how is the x-coordinate related to the y-coordinate?

   c) Name five other points whose x- and y-coordinates are related in this way. Plot these points. How many other such points are there?

   d) Use set-builder notation to describe the set of all points, (x, y) whose x- and y-coordinates are related in the manner described in (b) above.

3. Consider the set $\{(x, y) \mid x = 3y\}$.

   a) Name the coordinates of ten points that belong to this set.

   b) Using the same coordinate axes as in Exercise 2, plot these points.

   c) Are there any points that belong to both this set and the one described in Exercise 2? What is the intersection of the sets in Exercises 2 and 3?

4. Draw a pair of axes on graph paper.

   a) Plot these points: $(0, 0)$, $(1, 1)$, $(-1, 1)$, $(\frac{1}{2}, \frac{1}{4})$, $(-\frac{1}{2}, \frac{1}{4})$, $(2, 4)$, $(-2, 4)$, $(3, 9)$, $(-3, 9)$, $(\frac{1}{3}, \frac{1}{9})$, and $(-\frac{1}{3}, \frac{1}{9})$.

   b) For each of the points in (a), how is the x-coordinate related to the y-coordinate?

   c) Name five other points whose x- and y-coordinates are related in this way. Plot these points. How many other such points are there?

   d) Use set-builder notation to describe the set of all points, (x, y), whose x- and y-coordinates are related in the manner described above.

   e) Are there any points that belong to the set in Exercise 2 and the set described above? If so, name them.

5. AN INFINITE BUT BOUNDED UNIVERSE

The geometric model described above is of philosophical interest, since it illustrates that our notion of what is meant by "unlimited extent" is at best a vague one. It may be that a universe is infinite to an inhabitant of that universe, while it appears bounded to an outside observer. Let us consider a universe which consists of a plane as shown in Illus. 7.8. We may consider it to be very large, say as large as the earth or, perhaps, the solar system. In this universe, let us suppose, there lives a very small creature, say the size of a gnat or a man. As we view this universe from an outside vantage point, we observe that any object that moves from the center of the universe toward the edge shrinks. The nearer it approaches the edge, the more it shrinks. To the inhabitant of this universe, however, the view is quite different. As he moves toward the edge of his universe, he is not aware that he shrinks, because all of his surroundings shrink. If he were to try to measure his universe, his yardsticks would shrink as he moves toward the edge, but he would be unaware of it. His observation would be that the universe is infinite and unbounded and that any line is of unlimited, or infinite, extent. But to the outside observer the universe would be bounded.

This example may at first seem strange but, upon some reflection, it becomes clear that we observe our universe from within, as does the gnat described above. If our universe should be like that of the gnat, we would have no way to discover it.

6. EUCLIDEAN GEOMETRY AS AN APPROXIMATION

The models described earlier, of Riemannian geometry on a sphere and of hyperbolic geometry on the interior of a circle, illustrate the existence of consistent non-Euclidean geometries. At least they are just as consistent as Euclidean geometry. It is also of interest that the geometry on a sphere and the Euclidean geometry of a plane are nearly the same, so long as we confine ourselves to small portions of the sphere. This is, of course, common experience to those living on the earth. Even surveyors, who lay out large tracts of land, find Euclidean plane geometry generally satisfactory, even though they are working on a sphere. Hyperbolic geometry, too, coincides very closely with Euclidean if we confine ourselves to small regions. Very small triangles in any of these geometries have angle measure sums that are very nearly 180°, for example, and the Pythagorean theorem is approximately true.

Thus, even if a non-Euclidean geometry provides the best description of the universe, any discrepancies between that geometry and Euclidean geometry are too small to be noticeable or measurable in most circumstances, since we experience only a small portion of our universe.

REVIEW PRACTICE EXERCISES

The absolute value of a real number x, symbolized $|x|$, is defined as follows:

$$|x| = x, \text{ when } x \text{ is positive or zero,}$$

and

$$|x| = -x \text{ (the additive inverse of } x) \text{ when } x \text{ is negative.}$$

Example.

$$|\tfrac{1}{2}| = \tfrac{1}{2}, \qquad |-\tfrac{1}{2}| = \tfrac{1}{2}, \qquad |7| = 7, \qquad \text{and} \qquad |-3| = 3.$$

1. Find the absolute value.

    a) $|-5|$ b) $|1.5|$ c) $|-1.5|$ d) $|-\pi|$

    e) $|(-2)(-3|$ f) $|(2)(-3)|$ g) $\left|\dfrac{-6}{2}\right|$ h) $\left|\dfrac{-6}{-2}\right|$

2. Find the absolute value.

    a) $|3 - 1|$ b) $|1 - 3|$ c) $|4 - 5|$ d) $|5 - 4|$

    e) $|1 - \tfrac{1}{2}|$ f) $|\tfrac{1}{2} - 1|$ g) $|2 - (-5)|$ h) $|-5 - 2|$

3. What relationship is illustrated in Exercise 2?

4. Find the absolute value of the following.

    a) $|2 - 3|$ b) $|2 - 7|$ c) $|1 - 9|$

    d) $|-1 - 5|$ e) $|-3 - (-1)|$ f) $|\tfrac{1}{4} - 2|$

5. Simplify.

    a) $-(2 - 3)$ b) $-(2 - 7)$ c) $-(1 - 9)$

    d) $-(-1 - 5)$ e) $-(-3 - (-1))$ f) $-(\tfrac{1}{4} - 2)$

6. What relationship is illustrated in Exercises 4 and 5?

JUST FOR FUN

The figure in Illus. 7.11 shows the largest tetrahedron that can be contained in a cube. Assuming that the cube measures 1 unit on each edge, (a) find the measure of each edge of the tetrahedron, and (b) find the volume of the tetrahedron.

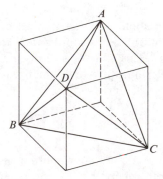

                                                               Illus. 7.11

CHAPTER 8

COORDINATE GEOMETRY

In Chapter 6 we established a one-to-one correspondence between the points of a line and the set of real numbers. In other words, any line may be *coordinatized* as described in Theorem 6.11. In a similar manner, a plane may be coordinatized, as described in Theorem 6.12, so that there is a one-to-one correspondence between the points of the plane and the set of all ordered pairs of real numbers.* When the system of real numbers is used in studying the structure and properties of geometry, it is often referred to as *coordinate geometry*, or *analytic geometry*.

1. DISTANCE

Recall that in coordinatizing a line we arbitrarily select some point on the line as the origin and assign the number 0 as its coordinate. Any other point may then be chosen to have coordinate 1. This gives us a unit segment and the coordinates of all other points are then determined. A coordinatized line is shown in Illus. 8.1.

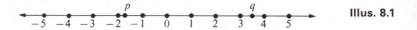

Illus. 8.1

The distance between two points with coordinates p and q is $|p - q|$, the absolute value of the difference between the coordinates. For example, if $p = -1.8$ and $q = 3.5$, the distance between these points is $|-1.8 - 3.5|$, or 5.3.

We usually picture a coordinatized plane as in Illus. 8.2. The two coordinatized lines are called *coordinate axes*. They may be referred to as the *first axis* and *second axis*. We often assign letters to these axes, usually in alphabetical order.

---

* Space may also be coordinatized, so that there is a one-to-one correspondence between the points of space and all ordered triples of real numbers. We shall confine our attention here, however, to geometry on a plane.

Most often we use x and y. The two dashed lines containing P are parallel, respectively, to the axes, and intersect the first axis at 3 and the second axis at 2. The ordered pair (3, 2) identifies the point P, where 3 is called the *first coordinate* and 2 is called the *second coordinate*. We can denote the point by "P," "(3, 2)," or "$P(3, 2)$." The latter is read, "The point P whose coordinates are 3, 2."

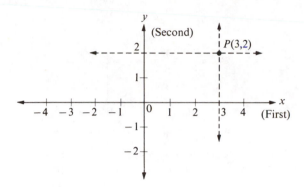

Illus. 8.2

We know how to find the distance between any two points on a coordinate axis. For lines that are parallel to either of the coordinate axes, distance is determined in a similar manner. In Illus. 8.3(a), A and B are points on a line l which is parallel to the x-axis. From the previous paragraph, it is intuitively clear that A and B have the same second coordinate, 4. In Illus. 8.3(b), lines m and n are parallel to the y-axis and perpendicular to the x-axis. All of this adds up to quadrilateral $ABCD$ being a rectangle, which means that $\overline{AD} \cong \overline{BC}$ (hence A and B have the same second coordinate), and that $\overline{AB} \cong \overline{DC}$. This last congruence tells us that the distance between A and B is the same as between D and C, or, simply stated, $m(\overline{AB}) = |5 - 2| = 3$.

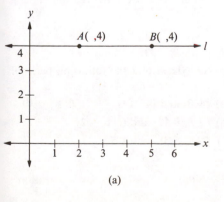

(a)

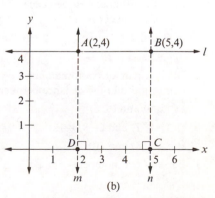

(b)

Illus. 8.3

The notation, $A(x, y_1)$ and $B(x, y_2)$, in Illus. 8.4 means that points A and B have the same first coordinate, x. Again, lines m and n are parallel, this time, to the x-axis, and again, quadrilateral $ABCD$ turns out to be a rectangle. Conclusion: Line l is parallel to the y-axis and $\overline{AB} \cong \overline{DC}$. Thus the distance between points A and B is the same as between D and C, or, $m(\overline{AB}) = |y_1 - y_2|$.

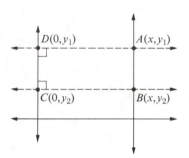

Illus. 8.4

The preceding discussion may be summarized with the following theorems.

Theorem 8.1   Two points have the same first (second) coordinate if and only if the line containing them is the second (first) axis or parallel to it.

Theorem 8.2   If two points have coordinates (x, y_1) and (x, y_2), then the distance between these points is $|y_1 - y_2|$. If two points have coordinates (x_1, y) and (x_2, y), then the distance between these points is $|x_1 - x_2|$.

EXERCISES

1. Find the distance between each of the pairs of points of a coordinatized line having the following coordinates.

   a) 9 and 2 b) 8 and -5 c) -5 and 7
   d) -2 and -19 e) -10 and 16 f) -14 and -12

2. Draw a pair of coordinate axes on graph paper. Then plot the following pairs of points and find the distance between them.

   a) $(1, 5)$ and $(6, 5)$ b) $(4, 8)$ and $(4, -1)$
   c) $(-8, 7)$ and $(-8, -12)$ d) $(-6, -12)$ and $(-6, -2)$
   e) $(-3, 8)$ and $(10, 8)$ f) $(-5, -5)$ and $(-5, -1)$

3. For any real numbers a, b,

$$|a - b| = |b - a|.$$

   Illustrate with five examples.

4. For any real numbers a, b,

$$|a - b|^2 = (a - b)^2.$$

Illustrate with five examples.

5. a) If $A(4, -2)$ and B are points on a line that is parallel to the x-axis, what do we know about the coordinates of B?

   b) The points $P(0, -5)$ and $Q(0, 7)$ belong to what line?

Exploratory exercises

6. Use the Pythagorean theorem to find the length of the hypotenuse.

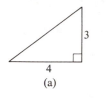 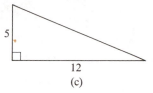

| (a) | (b) | (c) |

7. a) Draw a pair of coordinate axes on graph paper, and plot the following points: $A(2, 1)$, $B(5, 1)$, and $C(5, 5)$.

   b) Connect the three points. What type of triangle is $\triangle ABC$?

   c) What is $m(\overline{AB})$? $m(\overline{BC})$? $m(\overline{CA})$?

8. Find the coordinates of B. The dashed lines are parallel, respectively, to the coordinate axes.

(a)   (b)   (c)

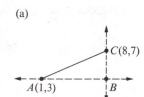 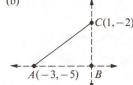 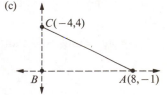

(d)

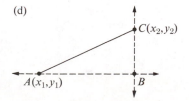

The previous exercises may have provided you with some ideas on how to find the distance between any two points in a plane, even if they are not on a line parallel to one of the axes. Let us consider a pair of points, A and B, with coordinates (x_1, y_1) and (x_2, y_2), respectively, as in Illus. 8.5. As shown, line l

contains A and is parallel to the x-axis. Similarly, line m contains B and is parallel to the y-axis. Lines l and m intersect at a point $C(x_2, y_1)$ and $\triangle ABC$ is a right triangle. The legs $\overline{AC}$ and $\overline{BC}$ have respective lengths $|x_1 - x_2|$ and $|y_1 - y_2|$.

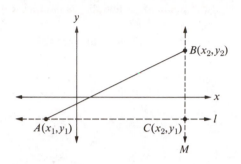

Illus. 8.5

Thus, by the Pythagorean theorem, we now have

$$[m(\overline{AB})]^2 = |x_1 - x_2|^2 + |y_1 - y_2|^2.$$

Note that the absolute value signs are not really necessary here, since squares of real numbers are never negative. The discussion above really outlines a proof of the following theorem.

Theorem 8.3 (Distance Formula) For any two points $A(x_1, y_1)$ and $B(x_2, y_2)$,

$$m(\overline{AB}) = \sqrt{(x_1 - x_2)^2 + (y_1 - y_2)^2}.$$

As the word "any" implies, this distance formula also holds for points on an axis or on a line parallel to an axis. The reader should have no difficulty verifying this fact.

Example. Find the distance between the points $A(-2, 3)$ and $B(1, 7)$.

$$\begin{aligned} m(\overline{AB}) &= \sqrt{(-2 - 1)^2 + (3 - 7)^2} \\ &= \sqrt{(-3)^2 + (-4)^2} \\ &= \sqrt{9 + 16} \\ &= \sqrt{25} \\ &= 5. \end{aligned}$$

EXERCISES

Use the distance formula in Exercises 1 through 6 to find the distance between the pairs of points.

1. $(1, 1)$ and $(5, 4)$
2. $(-1, 0)$ and $(4, 12)$
3. $(-3, 0)$ and $(0, 4)$
4. $(3, 7)$ and $(-2, -1)$

5. $(-2, -2)$ and $(4, 4)$ 6. $(-2, 5)$ and $(5, -2)$

7. Find the perimeter of a triangle having vertices $(1, 2)$, $(4, 6)$, and $(6, 3)$.

8. Find the perimeter of a quadrilateral having vertices $(0, -2)$, $(-3, 2)$, $(0, 4)$, and $(3, 0)$.

9. a) Find the lengths of the diagonals of the quadrilateral in Exercise 8.

    b) What type of quadrilateral is it?

Exploratory exercise

10. For each set of ordered pairs below, tell what they have in common. Then plot them on graph paper.

    a) $(4, 3)$, $(1, 3)$, $(-3, 3)$, $(\frac{1}{2}, 3)$

    b) $(-1, 6)$, $(-1, 0)$, $(-1, 2\frac{1}{2})$, $(-1, 4)$

    c) $(6, 0)$, $(-2, 0)$, $(7, 0)$, $(-2\frac{1}{2}, 0)$

REVIEW PRACTICE EXERCISES

1. Find the solution set for each inequality. The replacement set is the set of whole numbers (cf. Appendix, Section 5, Sentences and Sets, p. 343).

    a) $x < 9$ b) $x + 3 < 12$ c) $x - 2 < 7$

    d) $2x < 18$ e) $\frac{1}{3}x < 3$ f) $-x > -9$

2. The inequalities in Exercise 1 are all equivalent because they have the same solution set. Tell which inequalities in Exercise 1 illustrate the following statements.

    a) If $a < b$, then $a + c < b + c$.

    b) If $a < b$ and $c < 0$, then $ac > bc$.

    c) If $a < b$ and $c > 0$, then $ac < bc$.

3. For each inequality in Exercise 1, interchange the order relation (that is, $x > 9$ and $x + 3 > 12$, etc.). Using the same replacement set, find the solution set of each inequality.

4. How do the solution sets in Exercise 3 compare? Are the inequalities in Exercise 3 all equivalent?

5. a) What is the intersection of the solution sets of $x < 9$ and $x > 9$?

    b) Is there any whole number which is not in the solution set of either $x < 9$ or $x > 9$?

    c) If the replacement set is the set of whole numbers, what is the solution set of the disjunction $x < 9$ or $x = 9$ or $x > 9$? What is the solution set of the conjunction $x < 9$ and $x = 9$ and $x > 9$?

6. a) Consider the set $S = \{0, 1, 2, 3, 4, 5\}$ and the set of all ordered pairs $\{(x, y) \mid x \in S$ and $y \in S\}$ [for example, $(5, 4)$, $(3, 3)$, $(0, 2)$, etc.]. How many such ordered pairs are there?

    b) Draw a pair of coordinate axes on graph paper and plot the points having these ordered pairs as coordinates. (Such an array of points is called a *lattice* of points.)

    c) Circle the points belonging to $\{(x, y) \mid x = y\}$.

d) Use set-builder notation to describe the set of points in the region of the lattice "above" the points in (c).

e) Use set-builder notation to describe the set of points "below" the points in (c).

2. LINES

When we coordinatize a plane, we establish a one-to-one correspondence between the set of points in the plane and the set of all ordered pairs of real numbers. Given a point P in the plane, we are able to find the coordinates of P, and if we are given a pair of coordinates, we can locate the point having these coordinates. Further, we now have a way to use the coordinates of two points to find the distance between the points.

Let us now turn to the problem of describing a line in terms of ordered pairs of real numbers. We might begin by noting that every line is a set of points, and that this set of points is associated with a set of coordinates (ordered pairs of real numbers). Thus, if we can find a way to distinguish these coordinates from the coordinates of points *not* on the line, we will have described the line.

Look at the ordered pairs below. What do they have in common? To which line do they belong?

$(3, 0)$, $(-1, 0)$, $(\frac{1}{2}, 0)$, $(-\frac{1}{4}, 0)$, $(23, 0)$, $(-142, 0)$, $(0, 0)$

The first axis contains all of the points above,* and may be described as the set of all points (x, y) such that $y = 0$. Using set-builder notation, we can describe the first axis as

$$\{(x, y) \mid y = 0\}.$$

Since set notation can be cumbersome at times, we often use the defining equation to identify the line. Thus, when we refer to the x-axis as "the line $y = 0$," we understand that it is short for $\{(x, y) \mid y = 0\}$. The *graph* of a set of points is a picture of that set. When a point or points are represented on a drawing, we say we have *plotted points*.

EXERCISES

👀 *Exploratory exercises*

1. a) Use graph paper to plot these points:

$$(-3, 1), (0, 1), (1, 1), (5, 1).$$

b) How is the line containing these points related to the x-axis? The y-axis?

c) Use set-builder notation to describe the line.

---

* Since there is a one-to-one correspondence, it is often easier to refer to a point by naming its coordinates as, for example, "the point (2, 3)."

2. Follow the instructions in Exercise 1 for the points:

$$(-1, 2), (-1, 5), (-1, 0), (-1, -2).$$

3. a) Using graph paper, draw a line that is parallel to the second axis and contains the point $(2, 0)$.

   b) Name the coordinates of three other points on this line.

   c) Use set-builder notation to describe the line.

4. a) Using graph paper, draw a line that is perpendicular to the second axis and contains the point $(-2, -3)$.

   b) Name the coordinates of two other points on this line.

   c) Name the coordinates of two points not on this line.

   d) Describe this line using set-builder notation.

5. Describe the y-axis using set-builder notation.

The previous exercises suggest a way of describing any line that is parallel (or perpendicular) to either of the axes. For example, if a line l is parallel to the x-axis as in Illus. 8.6, all points on l have the same second coordinate -2, while different points have different first coordinates. Thus a defining equation for this line is $y = -2$.

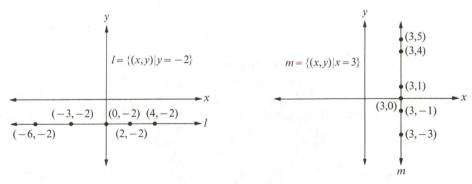

Illus. 8.6

Similarly, line m is parallel to the y-axis, and all points on m have the same first coordinate, 3. Note that no point off m has a first coordinate 3. A defining equation for m is $x = 3$. Let us now summarize these remarks in a theorem.

Theorem 8.4 Any line parallel to the x-axis may be described as

$$\{(x, y) \mid y \text{ is constant}\}.$$

The first axis is $\{(x, y) \mid y = 0\}$. Any line parallel to the y-axis may be described as

$$\{(x, y) \mid x \text{ is constant}\}.$$

The second axis is $\{(x, y) \mid x = 0\}$.

👓 **EXERCISES**

1. Use set-builder notation to describe a line which is parallel
    a) to the y-axis and 3 units to the right,
    b) to the x-axis and 2 units below it,
    c) to the y-axis and intersects the x-axis at $(-4, 0)$,
    d) to the x-axis and intersects the y-axis at $(0, 6.5)$.

2. Write an equation to describe a line which is
    a) parallel to the x-axis and intersects the y-axis at $(0, -\frac{4}{3})$,
    b) perpendicular to the x-axis at $(0, 0)$,
    c) perpendicular to the y-axis at $(0, -8)$,
    d) perpendicular to the y-axis at the origin.

3. Use graph paper to graph these lines.
    a) $\{(x, y) \mid y = 2)\}$ b) $\{(x, y) \mid x = -3.5\}$ c) $\{(x, y) \mid y = -0.5\}$
    d) $x = \frac{5}{2}$ e) $y = 0$ f) $2x = 6$

4. a) Find the coordinates of the point of intersection of the lines $x = 3$ and $y = 4$.
    b) Find the distance between this point and the origin.

5. a) Using graph paper, plot the points $A(1, 2)$ and $B(5, 4)$. Draw the line $\overleftrightarrow{AB}$.
    b) Write an equation of the line which is parallel to the x-axis and contains point A. Draw the line.
    c) Write an equation of the line which is parallel to the y-axis and contains point B. Draw the line.
    d) The lines in (b) and (c) meet at a point C. Name the coordinates of C.
    e) What type of triangle is determined by points A, B, and C?

6. The interiors of the four right angles determined by the axes are called *quadrants*. They are numbered I, II, III, and IV, as shown in Illus. 8.7.

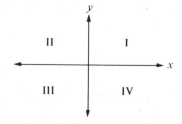

Illus. 8.7

    a) If a line, different from either axis, contains the origin, in how many quadrants can it lie?*

---

* Of course, a line does not "lie in a quadrant" in the sense of being contained in that quadrant. If *some part* of the line is contained in a quadrant, however, it is customary to say that the line "lies" in the quadrant, and it is in this sense that we use the expression here.

b) If a line contains the origin and a point in the first quadrant, in what other quadrant must it lie?

c) If a line contains the origin and a point in the third quadrant, in what other quadrant must it lie?

d) If a line, different from and not parallel to either axis, does not contain the origin, in how many quadrants does it lie?

7. In which quadrant do each of the following points lie?

a) $(4, 3)$ b) $(-5, 6)$ c) $(6, -5)$ d) $(-3, -2)$

e) $(-4, -3)$ f) $(-6, 5)$ g) $(5, -6)$ h) $(3, 4)$

8. In which quadrant is $P(x, y)$ if

a) $x < 0$ and $y > 0$,   b) $x > 0$ and $y > 0$,

c) $x < 0$ and $y < 0$,   d) $x > 0$ and $y < 0$.

3. THE GENERAL EQUATION OF A LINE

If a line l is parallel to the x-axis and intersects the y-axis at $(0, 2)$, a defining equation for l is $y = 2$. By Axiom 7.1 and Theorem 8.1, l is the only line defined by $y = 2$. It is also worth noting that the coordinates of one point and the fact that it is parallel to the x-axis is enough to distinguish l from all other lines.

We now turn to the problem of describing lines that are not parallel to either axis. Let us consider a pair of lines l and p as in Illus. 8.8. In order to describe l, we need to find a number property or relationship that is shared by the coordinates of every point on l, but does not hold for the coordinates of points not on l.

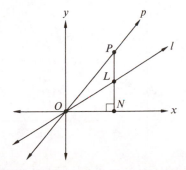

Illus. 8.8

Oddly enough, one way to discover what the points on l have in common is to ask how l differs from another line p which also passes through the origin. The most obvious difference of course is that p slants more steeply upward than does l, as is indicated by the fact that $\angle PON > \angle LON$ or that $\overline{PN} > \overline{LN}$.

EXERCISES

Exploratory exercises

1. Consider lines *l*, *n*, and *p* in Illus. 8.9.

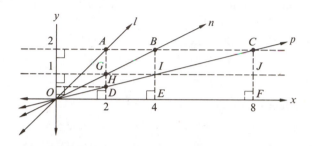

Illus. 8.9

   a) Name the coordinates of *A*, *B*, and *C*.

   b) We call the vertical distance, *AD*, the *rise* between *O* and *A*. The horizontal distance, *OD*, is called the *run* between *O* and *A*. Find the rise and run for *A*, *B*, and *C*.

   c) Find the rise/run ratio between *O* and each of the points *A*, *B*, and *C*.

   d) Find this ratio for *G* and compare with the ratio for *B*.

   e) Find this ratio for *H* and *I*. How do the ratios compare? For what other point is the ratio the same?

   f) Find the distances *IJ* and *CJ*. What is the ratio *CJ/IJ*?

2. Consider line *l* in Illus. 8.10. Lines $\overleftrightarrow{AE}$, $\overleftrightarrow{BD}$, and $\overleftrightarrow{CE}$ are parallel to the axes as indicated by the coordinates.

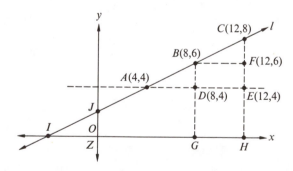

Illus. 8.10

   a) Find the ratios *BD/AD*, *CE/AE*, and *CF/BF* and compare.

   b) Triangles △*ADB*, △*BFC*, and △*AEC* are all similar. Why?

   c) Name three other triangles that are similar to the triangles in (b).

   d) Find numbers for the ratios *JO/IO*, *BG/IG*, and *CH/IH*.

   e) Use the ratios in (d) to name the coordinates of *I* and *J*.

3. Suppose that $A(-1, -3)$ and $B(2, 3)$ are points on a line l and let $P(x, y)$ be an arbitrary point on l as in Illus. 8.11.

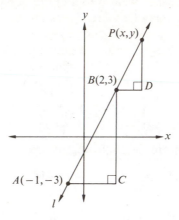

Illus. 8.11

a) Name the coordinates of C and D.

b) What is the rise/run ratio between A and B?

c) What is this ratio between B and P? How should it compare with the ratio in (b)?

d) Solve for y: $\dfrac{(y - 3)}{(x - 2)} = 2$.

e) Complete: The line l is $\{(x, y) \mid$ _____ $\}$.

The rise/run ratio between two points is the *slope* of a line. More precisely, it is defined as follows.

Definition 8.1 Given a line l, different from and not parallel to the y-axis, and two distinct points $A(x_1, y_1)$ and $B(x_2, y_2)$, on l, the real number

$$\frac{y_2 - y_1}{x_2 - x_1}$$

is called the *slope* of the line l.

Note that the slope is not defined for the y-axis or any line parallel to it since for such a line, $x_2 - x_1 = 0$. Note also that absolute value is not used. As illustrated below, lines which slant downward to the right have negative slopes—a rather convenient way to distinguish them from lines that slant upward to the right. The only caution to be observed is to subtract the x- and y-coordinates in the same order.

The method of describing lines that was developed in the previous exercises rests squarely upon the fact that the slope of a line (if it exists) is unique; that is, the slope is the same for any two points on the line. Indeed, it is this characteristic that distinguishes lines from all other types of curves. The proof of this theorem is not difficult and is left as an exercise.

Theorem 8.5 If a line has a slope, it is unique.

We have now accomplished what we set out to do, which was to find a way to describe lines in terms of sets of ordered pairs of real numbers. Essentially, we begin with two points $A(x_1, y_1)$ and $B(x_2, y_2)$ which uniquely determine the line we are interested in describing. These two points provide us with the slope of the line, $(y_2 - y_1)/(x_2 - x_1)$. Now, by Theorem 8.5, the slope between any point $P(x, y)$ on l and either of the given points, A or B, will be the same. That is,

$$\frac{y - y_2}{x - x_2} = \frac{y_2 - y_1}{x_2 - x_1}.$$

We can summarize this in the following theorem.*

Theorem 8.6 Given the points $A(x_1, y_1)$ and $B(x_2, y_2)$, where $x_1 \neq x_2$, the line $\overleftrightarrow{AB}$ is

$$\left\{ (x, y) \mid y - y_2 = \frac{y_2 - y_1}{x_2 - x_1} (x - x_2) \right\}, \tag{8.1}$$

or

$$\{(x, y) \mid y - y_2 = m(x - x_2)\}, \tag{8.2}$$

where m is the slope of the line.

The defining equation in (8.1) is referred to as the *two-point* equation of a line. The equation in (8.2) is referred to as the *point-slope* equation of a line.

Example 1. Find an equation of the line containing the points $(-4, 0)$ and $(2, -2)$, and graph the equation. (See Illus. 8.12.)

$$y - 0 = \frac{0 - (-2)}{-4 - 2} [x - (-4)],$$

$$y = \frac{2}{-6} (x + 4),$$

$$y = -\tfrac{1}{3}(x + 4).$$

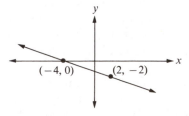

(−4, 0)   (2, −2)

Illus. 8.12

---

* In the previous exercises we outlined an argument that if $P(x, y)$ is on l, then it belongs to (8.1). To complete the argument for Theorem 8.6, one would have to prove that if $P(x, y)$ belongs to (8.1), then $P(x, y)$ is on the line l.

Example 2.   Find an equation of the line containing the points $(1, 5)$ and $(-2, -4)$, and graph the equation.   (See Illus. 8.13.)

Substituting in the two-point equation, we get the following.

$$y - (-4) = \frac{-4 - 5}{-2 - 1} [x - (-2)],$$

$$y + 4 = \frac{-9}{-3} (x + 2),$$

$$y + 4 = 3x + 6.$$

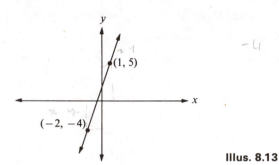

Illus. 8.13

It is intuitively clear that a nonvertical line must intersect the y-axis at some point whose coordinates are $(0, b)$, where b is some real number. This point is referred to as the *y-intercept*. In the point-slope equation, if the point (x_2, y_2) is the y-intercept, then $x_2 = 0$, $y_2 = b$, and from $y - y_2 = m(x - x_2)$ we get

$$y - b = mx - 0$$

or

$$y = mx + b. \tag{8.3}$$

Equation (8.3) is called the *slope-intercept* equation of a line.

Example 3.   Find the equation of a line which has a slope of $-\frac{1}{2}$ and intersects the y-axis at $y = 1$.   (See Illus. 8.14.)

By the slope-intercept equation, we get $y = -\frac{1}{2}x + 1$.

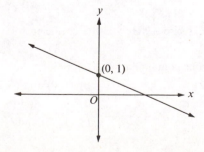

Illus. 8.14

Example 4. Find the equation of a line containing the point (2, 3) and having slope 2.

By the point-slope equation, we get $y - 3 = 2(x - 2)$ or $y = 2x - 1$.

The last equation shows that the line intersects the y-axis at the point $(0, -1)$.

EXERCISES

1. Find an equation of the line containing the following points.
   a) $(-4, 2)$ and $(5, 2)$ b) $(0, 0)$ and $(4, 2)$
   c) $(-3, -3)$ and $(2, 2)$ b) $(5, -1)$ and $(-1, -3)$

2. Find an equation of the line containing the following points.
   a) $(-2, 5)$ and $(1, -4)$ b) $(1, 4)$ and $(4, 1)$
   c) $(-1, 1)$ and $(\frac{1}{2}, -\frac{1}{2})$ d) $(-\frac{1}{2}, 4)$ and $(0, 0)$

3. In each of the following, decide whether the three points are collinear.
   a) $(3, -2), (2, -1), (-1, -2)$ b) $(-1, 1), (1, 5), (-\frac{1}{3}, 2)$

4. Consider the line $\{(x, y) \mid y = x + 1\}$.
   a) Is $(3, 4)$ a member of this set? Is $(17, 18)$?
   b) If -6 is the first coordinate of a point on this line, what is the second coordinate?
   c) Find the coordinates of five more points on this line.
   d) Which quadrant does not contain any points of this line?
   e) Find the coordinates of the point of intersection of this line and the y-axis.
   f) Find the coordinates of the point of intersection of this line and the x-axis.

5. Graph the following lines, using the same axes.
   a) $y = \frac{1}{4}x$ b) $y = \frac{1}{2}x$ c) $y = x$
   d) $y = 2x$ e) $y = 5x$ f) $y = 10x$

6. Study the graphs of Exercise 5. How is the coefficient of x related to the graph of each equation?

7. Graph the following lines, using the same axes for all of them.
   a) $y = -\frac{1}{4}x$ b) $y = -\frac{1}{2}x$ c) $y = -x$
   d) $y = -2x$ e) $y = -5x$ f) $y = -10x$

8. Study the graphs of Exercise 7. How is the coefficient of x related to the graph of each equation?

9. Find the slope of the line containing the given points.
   a) $(-3, 4)$ and $(1, -2)$ b) $(5, 0)$ and $(-2, -2)$
   c) $(1, -6)$ and the origin   d) (m, n) and (n, m)
   e) $(-b, -a)$ and (a, b) f) $(1, 73)$ and $(-2, -7)$

10. a) Find an equation of a line having a slope of $-\frac{4}{3}$ and containing the point $(1, -1)$.

    b) Find three other points on this line.

11. a) Suppose that a line containing the origin were rotated counterclockwise. What would happen to the slope of the line as it approached the y-axis?

    b) What would happen to the line in (a) as it approached the y-axis if it were rotated in a clockwise direction?

Rigorous exercises

12. Prove Theorem 8.5.

13. Prove that if $a \neq 0$ and $b \neq 0$, the line having x intercept a and y intercept b can be described by the equation

$$\frac{x}{a} + \frac{y}{b} = 1.$$

14. a) If line l is also not parallel to the x-axis, it must meet the x-axis at some point P. What do we know about the coordinates of P?

    b) Prove that for any real number a, there is a real number b such that $P(a, b)$ is a point on line l. [*Hint:* Consider a real number a and the line $x = a$ parallel to the y-axis.]

Exploratory exercises

15. Graph the following lines, using the same axes for all of them.

    a) $y = 2x$ b) $y = 2x + 1$ c) $y = 2x + 2$

    d) $y = 2x + 3$ e) $y = 2x - 4$ f) $y = 2x - 2$

16. Consider the graphs in Exercise 15.

    a) How are the graphs related to each other?

    b) How are their slopes related?

17. a) Graph the equations in Exercises 1(d) and 2(a), using the same axes.

    b) How are the graphs related?

    c) How are their slopes related?

Parallel and Perpendicular Lines

The graphs in Exercise 12 of the previous exercise set illustrate rather clearly the relationship between the slopes of parallel lines. In particular it seems fairly obvious that if two or more lines have the same slope, they are parallel. The converse is also true; that is, if two or more lines are parallel, then they have the same slope. A proof of this is outlined below.

Theorem 8.7 Two distinct nonvertical lines are parallel if and only if they have the same slope.

Outline of proof. Consider two distinct lines with the same slope m, described by the equations

$$y = mx + b$$

and

$$y = mx + c.$$

Let us suppose they are not parallel. Then they intersect at a point whose co-ordinates (x_1, y_1) satisfy both equations. That is,

$$y_1 - mx_1 = b$$

and

$$y_1 - mx_1 = c.$$

This means that $b = c$ and the lines are not distinct, contrary to hypothesis. Thus, if two distinct nonvertical lines have the same slope, they are parallel. Now suppose lines l and m are parallel, and consider a line n, parallel to the x-axis and inter-secting l and m at A and A'. Choose arbitrary points B and B' on l and m, as shown in Illus. 8.15. It can be shown that $\triangle ABC \sim \triangle A'B'C'$, which means that

$$\frac{CB}{AC} = \frac{C'B'}{A'C'}.$$

It follows that the slope of l is the same as the slope of m. Thus, if two lines are parallel, they have the same slope. If two lines are perpendicular, one might expect that their slopes are related in some special way. Such is the case as is illustrated in the exercises below.

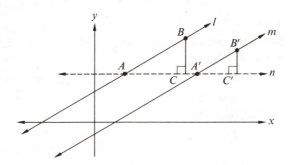

Illus. 8.15

EXERCISES

Exploratory exercises

   1. Use graph paper to graph each pair of lines.

      a) $y = x$ and $y = -x$

      b) $y = \frac{1}{2}x$ and $y = -2x$

      c) $y = 4x + 1$ and $y = -\frac{1}{4}x$

2. Which of the lines in Illus. 8.16 have
   a) positive slopes?
   b) negative slopes?
   c) no slope?
   d) slope zero?

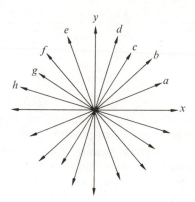

Illus. 8.16

3. If a line has a positive slope, any line perpendicular to it will have what kind of slope?
4. If a line has a negative slope, any line perpendicular to it will have what kind of slope?
5. Draw some perpendicular lines on graph paper, find their slopes, and then find the product of the slopes.

In working the exercises above, you probably discovered that any pair of perpendicular nonvertical lines has one positive slope and one negative slope. Further, the product of the slopes is -1.

Theorem 8.8 Two nonvertical lines l_1 and l_2 with slopes m_1 and m_2 are perpendicular if and only if $m_1 m_2 = -1$.

Outline of proof. It will suffice to consider lines which intersect at the origin since any other pair of perpendiculars in the plane will be pairwise parallel to them and, by Theorem 8.7, have corresponding slopes.

Now suppose $l_1 \perp l_2$ and $m_1 > 0$.* Choose points P and Q on l_1 and l_2 such that $\overline{PO} \cong \overline{QO}$. Let $\overline{QR}$ and $\overline{PS}$ be perpendiculars to the x-axis, thus producing right triangles $\triangle QRO$ and $\triangle OSP$ (Illus. 8.17). Thus, $\triangle QRO \cong \triangle OSP$, and hence $QR = OS$, $RO = SP$, and $SP/OS = RO/QR$. However, $SP/OS = m_1$ and $m_2 = -(QR/RO)$. Therefore $-(1/m_2) = RO/QR$, and $m_1 = -(1/m_2)$.

This proves that if the lines are perpendicular, then the product of their slopes is -1.

---

* For completely rigorous argument, one would need to prove that lines which "slant upward" have positive slope, that lines which "slant downward" have negative slope, and finally that nonvertical perpendicular lines have one positive and one negative slope. Since they are easily accepted on an intuitive basis, we omit the proof here.

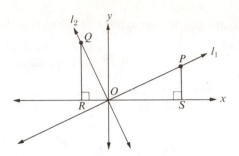

Illus. 8.17

Conversely, if $m_1 m_2 = -1$, or $m_1 = -(1/m_2)$, where $m_1 = SP/OS$ and $m_2 = -(QR/RO)$, then $SP/OS = RO/QR$. This means that right triangles $\triangle QRO$ and $\triangle OSP$ are similar and corresponding angles are congruent. It is now easy to prove that $\angle QOP$ is a right angle, and l_1 and l_2 are therefore perpendicular.

Hence we conclude that two nonvertical lines l_1 and l_2 with slopes m_1 and m_2 are perpendicular if and only if $m_1 m_2 = -1$.

Example 1. Find an equation of the line that is parallel to the line $y = 3x - 5$ and contains the point $(1, 2)$.

The slope of the line $y = 3x - 5$ is 3 and therefore any line parallel to it must also have slope 3. Using the point-slope equation $y - y_1 = m(x - x_1)$, we obtain

$$y - 2 = 3(x - 1)$$

or

$$y = 3x - 1,$$

which is an equation of the line described above.

Example 2. Find an equation of the line that is perpendicular to the line $y = 2x + 3$ and contains the point $(-2, 5)$.

The slope of the line $y = 2x + 3$ is 2 and therefore any line perpendicular to it must have slope $-\frac{1}{2}$. Using the point-slope equation $y - y_1 = m(x - x_1)$, we obtain

$$y - 5 = -\tfrac{1}{2}(x + 2)$$

or

$$y = -\tfrac{1}{2}x + 4,$$

which is an equation of the line described above.

Graphs of First-Degree Equations

An equation with one or more variables having no terms of degree greater than 1 is called a first-degree equation. Some examples of first-degree equations with one or two variables are

$$y = 3x + 5, \qquad 2x - 3y = -8, \qquad \tfrac{1}{2}y + 9x - 4 = 0,$$

$$y = -5, \qquad x - 3 = 0.$$

Any first-degree equation with one or two variables is equivalent to an equation of the form

$$ax + by + c = 0,$$

where a, b, and c represent real numbers, and a and b are not both zero. An equation such as this is said to be in *standard form*. For example, $x - y = 2$ is equivalent to $1x + (-1)y + (-2) = 0$, where $a = 1$, $b = -1$, and $c = -2$. Similarly, $x = -5$ is equivalent to $1x + 0y + 5 = 0$, where $a = 1$, $b = 0$, and $c = 5$.

In the preceding sections we determined various equations for lines, and although we have not verified it, the equations obtained are, in each case, first-degree equations.

For example, any line different from and not parallel to the y-axis has a slope m, and a y-intercept b, and can be defined by an equation $y = mx + b$, which is clearly a first-degree equation. In a similar manner, any line parallel to (or the same as) the y-axis is defined by an equation $x = k$ for some real number k, and this too is a first-degree equation. All of this can be summarized in the following theorem.

Theorem 8.9   Any line in a plane can be defined by a first-degree equation.

Approaching this from the opposite direction, we now ask whether the graph of every first-degree equation represents a line. That is, given the equation $ax + by + c = 0$, where a and b are not both zero, does the graph of such an equation represent a line? It is easy to show that it does. For suppose that $a \neq 0$ and $b \neq 0$. Then $ax + by + c = 0$ is equivalent to

$$y = -\frac{a}{b}x - \frac{c}{b},$$

which is an equation of a line with slope $-(a/b)$ and y-intercept $-(c/b)$. If $a = 0$ and $b \neq 0$, the equation is equivalent to $y = -(c/b)$, which is an equation of a line parallel to the x-axis with y-intercept $-(c/b)$. And finally, if $a \neq 0$ and $b = 0$, the equation is equivalent to $x = -(c/a)$, which is an equation of a line parallel to the y-axis with x-intercept $-(c/a)$. We have thus proved the following theorem.

Theorem 8.10 (Converse of Theorem 8.9)   The graph of every first-degree equation is a line.

Because of this correspondence, first-degree equations are often referred to as *linear* equations.

EXERCISES

In Exercises 1 through 4, find an equation of the line parallel to the given line and containing the given point.

1. $y = 4x - 5$; $(3, 2)$

2. $y + 3x = 1$; $(-1, -2)$

3. $5x + 2y - 3 = 0$; $(0, -1)$

4. $\frac{1}{4}x - y = -3$; $(-2, \frac{1}{2})$

In Exercises 5 through 8, find an equation of the line perpendicular to the given line and containing the given point.

5. $5x + y + 6 = 0$; $(2, 5)$

6. $3y - 7x - 4 = 0$; $(3, -7)$

7. $\frac{2}{3}x - y = 0$; $(0, 0)$

8. $3y - 2x = 0$; $(3, 2)$

9. Prove that each of the following sets of points are the vertices of a right triangle, and name the point which is the vertex of the right angle.

   a) $(2, -3), (1, 5), (-6, -4)$

   b) $(-3, 5), (2, -5), (1, 7)$

10. Find equations of the lines containing the altitudes of a triangle with vertices $(1, 1)$, $(3, 9)$, and $(5, 2)$.

11. a) Using graph paper, plot the points $A(2, 1)$, $B(4, 6)$, and $C(6, 1)$, and draw segments $\overline{AB}$, $\overline{BC}$, and $\overline{CA}$.

    b) Find the slope of each segment.

    c) What type of triangle is $\triangle ABC$? Why?

    d) How are $\angle BAC$ and $\angle BCA$ related? Why?

    e) How are the slopes of $\overline{AB}$ and $\overline{BC}$ related?

    f) Does the relationship in (e) hold when the base is parallel to the y-axis? When the base is not parallel to either axis?

12. Consider the linear equation $y = 3x + 5$.

    a) What is the slope of a line described by this equation?

    b) What is the y-intercept of a line described by this equation?

    c) How many lines in a plane have this slope and this y-intercept?

    d) How many lines are described by this equation?

13. Find the slope and y-intercept of the line described by each equation.

    a) $2x - y + 1 = 0$

    b) $\frac{2}{3}x = \frac{1}{3}y - \frac{1}{3}$

    c) $2y = 4x + 2$

    d) $5y - 10x = 5$

14. What is the relationship of the lines described by each of the equations in Exercise 13?

15. For any given line in a plane, how many linear equations are there that describe this line?

Rigorous exercises

16. Without using the distance formula, prove that a quadrilateral having vertices $(4, 7)$, $(0, 4)$, $(7, 3)$, and $(3, 0)$ is a square.

17. Write a complete proof of Theorem 8.8.

18. Given real numbers a, b, and c, where a and b are not both zero, prove that the linear equation $ax + by + c = 0$ describes one and only one line.

4. HALF-PLANES

Let us recall that a line l in a plane separates the plane into three sets: the line l and two half-planes, each of which is a convex set. Further, if A is in one half-plane and B is in the other, then $\overline{AB}$ intersects l at a point P (Illus. 8.18).

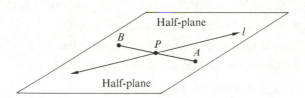

Illus. 8.18

We shall now look for a way to describe half-planes, as well as certain subsets of half-planes, in terms of ordered pairs of real numbers. Let us begin by exploring some ideas in the exercises below.

EXERCISES

Exploratory exercises

1. a) Using graph paper, graph the line $x = 2$.
   b) Plot five points to the right of $x = 2$ and find their coordinates.
   c) Plot five points to the left of $x = 2$ and find their coordinates.
   d) Does the y-coordinate tell us anything about the location of a point? If so, how?
   e) Does the x-coordinate tell us anything about the location of a point? If so, how?
   f) Label the three sets as follows: L, the line $x = 2$; H_1, the half-plane to the left; and H_2, the half-plane to the right. Where do each of these points belong?
      - (1) $(2, 7)$ (2) $(-3, -9)$ (3) $(6, 0)$
      - (4) $(14, 36)$ (5) $(1, \frac{1}{2})$ (6) $(0, -2)$
      - (7) $(2.1, -\frac{4}{3})$ (8) $(1.99, 17)$ (9) $(2, 63)$
      - (10) $(-1, -637)$ (11) $(2, -637)$ (12) $(4, -637)$
2. a) Use set notation to describe the set of points (x, y) to the *right* of $x = 2$.
   b) Use set notation to describe the set of points (x, y) to the *left* of $x = 2$.
3. Consider a line $x = a$, where a is any real number.
   a) Use set notation to describe the set of points (x, y) to the *right* of $x = a$.
   b) Use set notation to describe the set of points (x, y) to the *left* of $x = a$.
   c) What is the union of the sets in (a) and (b) above and the set (x, y) $x = a$?
   d) What is the intersection of these three sets?

4. a) Using graph paper, graph the line $y = -3$.

   b) Plot five points *below* the line $y = -3$.

   c) Plot five points *above* the line $y = -3$.

   d) Which of these points lie on $y = -3$? *Above* $y = -3$? *Below* $y = -3$?

   (1) $(2, 3)$ (2) $(-2, -3)$ (3) $(-5, -16)$

   (4) $(-\frac{1}{4}, -2.9)$ (5) $(6200, -11)$ (6) $(6200, 0)$

   (7) $(6200, 38)$ (8) $(6, -3.01)$ (9) $(-5, 1)$

   e) Which coordinate determined the answers in (d) above?

   f) Use set notation to describe the set of points *below* the line $y = -3$.

   g) Use set notation to describe the set of points *above* the line $y = -3$.

5. Consider a line $y = b$, where b is any real number.

   a) Use set notation to describe the set of points *above* $y = b$.

   b) Use set notation to describe the set of points *below* $y = b$.

   c) What is the union of the sets in (a) and (b) above and the set $y = b$?

   d) What is the intersection of these three sets?

6. a) Use set notation to describe the set of points to the right of the y-axis.

   b) Use set notation to describe the set of points above the x-axis.

   c) Use set notation to describe the intersection of the sets in (a) and (b) above. What do we call this set of points?

   In the previous exercises, the reader probably had little difficulty deciding that the regions on each side of a line $x = a$ or $y = b$ are half-planes. The graph of a half-plane may be shown by shading the appropriate region, as in Illus. 8.19. All of this is summarized in the following theorem.

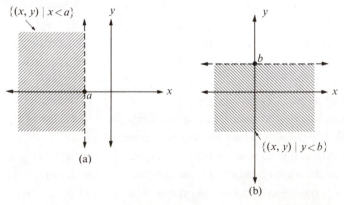

(a)

(b)

Illus. 8.19

Theorem 8.11 For any real number a, the two half-planes determined by the line $x = a$ are the sets $\{(x, y) \mid x > a\}$ and $\{(x, y) \mid x < a\}$. For any real number b, the two half-planes determined by the line $y = b$ are the sets $\{(x, y) \mid y > b\}$ and $\{(x, y) \mid y < b\}$.

It follows, of course, from the Plane Separation Axiom (3.1) that the half-planes, along with the line determining them, are disjoint sets, and that the union of these three sets is the plane. This is also easily seen when the properties of the real-number system are available as they are in a coordinatized plane. For example, if we begin with any real number a and then consider any real number x, the algebra of the real numbers tells us that *exactly* one of the following is true: $x < a$, $x = a$, or $x > a$.* From this we obtain the geometric counterpart, which is:

Given a line $x = a$ and any point $P(x, y)$ in the plane, P belongs to *exactly* one of the sets,

$$\{(x, y) \mid x < a\}, \{(x, y) \mid x = a\}, \text{ or } \{(x, y) \mid x > a\}.$$

In Exercise 6 of the preceding set the reader was asked to use set notation to describe the intersection of two half-planes. The graphs of these half-planes and their intersection are shown in Illus. 8.20. It will be observed that the x- and y-axes are drawn dashed, because they are not part of either half-plane. Since every point in the intersection must satisfy both inequalities, the intersection is

$$\{(x, y) \mid x > 0\} \cap \{(x, y) \mid y > 0\} = \{(x, y) \mid x > 0 \text{ and } y > 0\}.$$

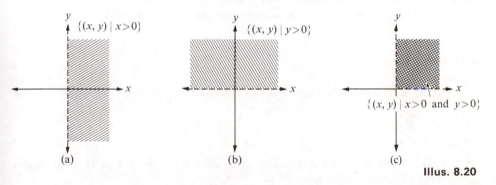

(a)   (b)   (c)

Illus. 8.20

In other words, for a point to belong to the *intersection*, its coordinates must satisfy the *conjunction* of two inequalities, $x > 0$ and $y > 0$. The reader will, of course, recognize this intersection as the first quadrant.

The union of these two half-planes will consist of the set of all points $P(x, y)$ where either $x > 0$ *or* $y > 0$. Thus

$$\{(x, y) \mid x > 0\} \cup \{(x, y) \mid y > 0\} = \{(x, y) \mid x > 0 \text{ or } y > 0\}.$$

---

* The property referred to here is the *trichotomy* property for real numbers.

The sentence "$x > 0$ or $y > 0$" is called the *disjunction* of the two inequalities. The graph of the union is shown in Illus. 8.21(a).

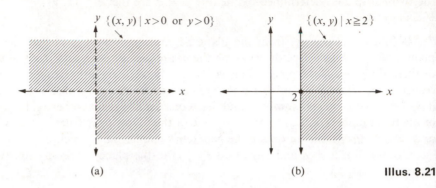

(a)   (b)   **Illus. 8.21**

It is usually convenient to describe a half-plane by using only the defining condition. Thus the two half-planes mentioned above, their intersection, and their union, could be described, respectively, as $x > 0$, $y > 0$, the conjunction $x > 0$ and $y > 0$, and the disjunction $x > 0$ or $y > 0$.

Sometimes we wish to consider the union of a half-plane and its boundary line. For example, the union of the half-plane $x > 2$ and its boundary $x = 2$ can be described by the disjunction $x > 2$ or $x = 2$. It is customary to write such a disjunction as $x \geq 2$. To distinguish between the half-plane $x > 2$ and the union $x \geq 2$, the line $x = 2$ is shown as a solid line, as in Illus. 8.21(b).

EXERCISES

In Exercises 1 through 4 use inequalities to describe the half-planes determined by the given line.

1. $y = -4$ 2. $x = \frac{2}{3}$

3. $x - 2 = 0$ 4. $y - \frac{1}{2} = 0$

In Exercises 5 through 8 use inequalities to describe the half-plane determined by the given line and containing the given point.

5. $x = \frac{3}{2}$; $(1, -5)$ 6. $y + 1 = 0$; $(0, 0)$

7. $2x + 8 = 0$; $(-\frac{1}{2}, 3)$ 8. $\frac{1}{2}y - 1 = 0$; $(1, -1)$

9. Graph the half-planes of Exercises 5 through 8.

10. a) Graph the intersection of the half-planes of Exercises 5 and 8, and use set notation to describe this intersection.

    b) Graph the intersection of the half-planes of Exercises 6 and 7, and use set notation to describe this intersection.

11. a) Graph the intersection of the four half-planes of Exercises 5 through 8.

    b) This region is the interior of what kind of figure?

12. a) Describe the second, third, and fourth quadrants, each as the intersection of half-planes.

    b) Each intersection (quadrant) is the interior of what kind of figure?

13. Graph the union of the half-planes of Exercises 6 and 7.

14. a) Graph the union of half-planes $x > 3$ or $x < -3$.

    b) Use set notation to describe the set of points in the plane that do not belong to the union in (a).

15. a) Graph the union $x \geq 2$ or $x \leq -2$.

    b) Use set notation to describe the set of points in the plane that do not belong to the union in (a).

16. a) Graph the line $y = x$.

    b) The half-plane determined by $y = x$ and containing the point $(2, 3)$ will be designated the *upper half-plane*. Find the coordinates of five other points in the upper half-plane.

    c) For each of the points in (b), how are their x- and y-coordinates related?

    d) The point $(3, 2)$ lies in the lower half-plane. Find the coordinates of five other points in the lower half-plane.

    e) For each of the points in (d), how are their x- and y-coordinates related?

    f) How might each of these half-planes be described?

17. a) Graph the line $y = 2x + 1$.

    b) Using the same axes, plot the points $(1, 1)$, $(1, 2)$, $(1, 3)$, $(1, 4)$, and $(1, 5)$. Which are in the upper half-plane? Which are in the lower half-plane?

    c) The point $(1, 3)$ is on the line because it satisfies the equation $y = 2x + 1$; that is, $3 = 2 \cdot 1 + 1$. The point $(1, 5)$ is not on the line because it does not satisfy the equation $y = 2x + 1$; that is, $5 \neq 2 \cdot 1 + 1$. Similarly, $(1, 2)$ is not on the line because $2 \neq 2 \cdot 1 + 1$. How is 5 related to $2 \cdot 1 + 1$? How is 2 related to $2 \cdot 1 + 1$? How do we know that the points $(1, 1)$ and $(1, 4)$ are not on the line $y = 2x + 1$?

    d) Using the same axes, graph the line $x = -1$. Plot two points on this line that lie in the lower half-plane. Plot two points on this line that lie in the upper half-plane. For each of these points, how is the y-coordinate related to $2 \cdot (-1) + 1$?

    e) If $P(x, y)$ is a point in the upper half-plane, how is y related to $2x + 1$? What is the relationship if P is in the lower half-plane?

18. a) Graph the line $y = -3x - 1$.

    b) Describe the half-plane determined by this line and containing the point $(1, 1)$.

    c) Describe the opposite half-plane.

Exploratory exercises

19. Consider the line $y = 2x + 1$ shown in Illus. 8.22.

    a) The line $x = 2$ intersects the line $y = 2x + 1$ at R. How do we know that $(2, 5)$ are the coordinates of R?

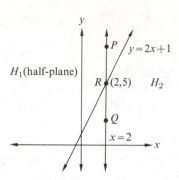

Illus. 8.22

b) Point P is on $x = 2$ and in half-plane H_1. What do we know about the y-coordinate of P?

c) What do we know about the y-coordinate of Q?

d) What is the y-coordinate of point S?

e) If a point is on the line $x = -1$ and also in H_1, then the y-coordinate is greater than what number?

20. For each point below, tell whether it is on the line $y = 2x + 1$, in H_1, or in H_2 (Illus. 8.22).

a) $(2, 3)$ b) $(1, 3)$ c) $(1, 0)$

d) $(1, 5)$ e) $(0, 1)$ f) $(0, 0)$

g) $(3, 2 \cdot 3 + 1)$ h) $(3, 2 \cdot 3 + 4)$ i) $(3, 2 \cdot 3 - 1)$

j) $(a, 2a + 1)$ k) $(a, 2a - 7)$ l) $(a, 2a + 8)$

General Description of Half-Planes

The description of a half-plane determined by a line not parallel to either axis follows much the same pattern as when the line is parallel to an axis. In each case, the half-plane can be described by an inequality.

Let's consider a line $y = mx + b$ as shown in Illus. 8.23. Of the points P, Q, and R, which one has the largest y-coordinate? Which one has the smallest y-coordinate? Since $mx_1 + b$ represents the y-coordinate of Q, we know that $y_1 > mx_1 + b$, and $y_2 < mx_1 + b$.

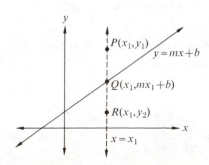

Illus. 8.23

If a line $y = mx + b$ has a negative slope, as in Illus. 8.24, one reasons the same way, concluding that the upper half-plane is described by the inequality $y > mx + b$ while the lower half-plane is described by $y < mx + b$. All of this is summarized in the following theorem.

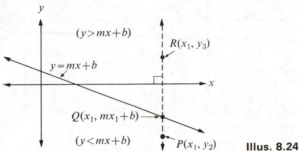

Illus. 8.24

Theorem 8.12 For any real numbers m and b, the two half-planes determined by the line $y = mx + b$ are the sets $\{(x, y) \mid y > mx + b\}$ and $\{(x, y) \mid y < mx + b\}$.

We now have a way of describing every half-plane. The results are summarized in Illus. 8.25.

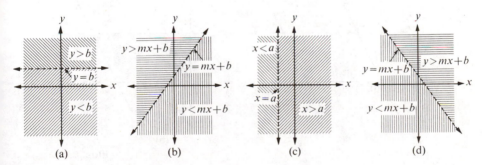

Illus. 8.25

Other regions of the plane may now be described as the intersection or union of two or more half-planes. For example, the interior of an angle was defined in Chapter 3 as the intersection of two half-planes. Such a region can now be described in terms of ordered pairs of real numbers.

Example 1. Describe the interior of an angle having vertex $(0, 0)$ and containing the points $(4, 4)$ and $(4, -4)$.

The lines containing the sides of the angle are $y = x$ and $y = -x$, as shown in Illus. 8.26(a). The intersection of the two half-planes shown in Illus. 8.26(b) is the set $\{(x, y) \mid y < x \text{ and } y > -x\}$.

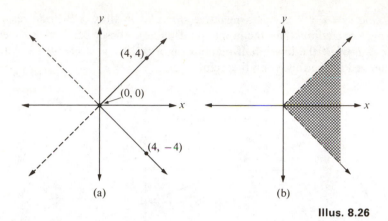

Illus. 8.26

Using the concepts of intersection and union, we can now describe a convex polygon and its interior.

Example 2. The vertices of a parallelogram are $(4, 2)$, $(1, 2)$, $(0, 0)$, and $(3, 0)$. Describe the union of the parallelogram and its interior.

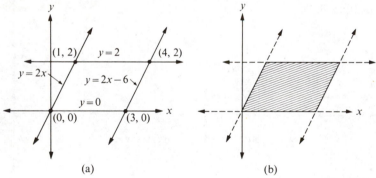

Illus. 8.27

The equations of the lines containing the sides of the parallelogram and their graphs are shown in Illus. 8.27(a). The unions of the half-planes and their respective boundaries are $y \le 2$, $y \le 2x$, $y \ge 0$, and $y \ge 2x - 6$. The intersection of these regions is described by the conjunction $y \le 2$ and $y \le 2x$ and $y \ge 0$ and $y \ge 2x - 6$. Its graph is shown in Illus. 8.27(b).

EXERCISES

In Exercises 1 through 6, write an inequality describing the half-plane determined by the given line and containing the given point.

1. $y = 5x + 2$; $(2, 3)$ 2. $y = -3x - 6$; $(-3, 4)$

3. $y - 4x = 0; (-2, -5)$ 4. $2x - y = -\frac{1}{2}; (0, 0)$

5. $3x + 2y - 5 = 0; (3, -7)$ 6. $2y - 4x = 1; (-1, 0)$

7. Graph the inequalities in Exercises 1 through 6.

8. Which of the inequalities in Exercises 1 through 6 are those of opposite half-planes?

In Exercises 9 through 14, write a sentence (inequality, conjunction, or disjunction of inequalities) to describe the region shown.

9.

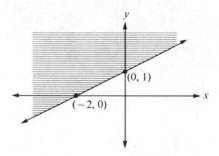

Illus. 8.28

10.

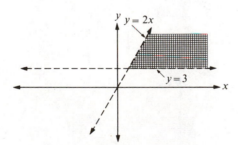

Illus. 8.29

11.

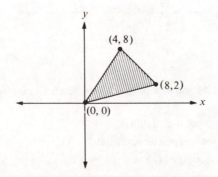

Illus. 8.30

12.

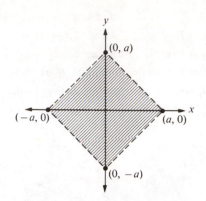

Illus. 8.31

13.

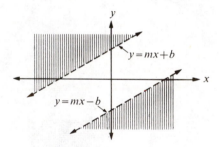

Illus. 8.32

14.

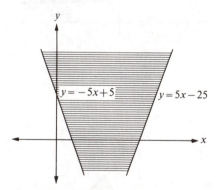

Illus. 8.33

15. The graph in Exercise 14 shows a part of an angle and its interior. What is the vertex of this angle?

16. a) Graph the conjunction $x \geq 2$ and $x \leq 4$.

b) Using the same axes, graph the equation $y = 1$.

c) Describe the intersection of (a) and (b) by a conjunction.

d) What kind of figure is the intersection?

17. a) Graph the inequality $x > 0$.
    b) Using the same axes, graph the equation $y = 3$.
    c) Describe the intersection of (a) and (b) by a conjunction.
    d) What kind of figure is the intersection?

18. a) Graph the conjunction $y < 5$ and $y > 1$.
    b) Using the same axes, graph the equation $x = -2$.
    c) Describe the intersection of (a) and (b) by a conjunction.
    d) What kind of figure is the intersection?

19. a) Graph the disjunction $x \leq 0$.
    b) Using the same axes, graph the equation $y = -4$.
    c) Describe the intersection of (a) and (b) by a conjunction.
    d) What kind of figure is the intersection?

5. SUBSETS OF LINES

We have used our knowledge of the system of real numbers to describe and find properties of geometric figures such as lines and half-planes. We have also seen how various subsets of half-planes could be described as intersections or unions of half-planes and their boundaries.

It is natural therefore that we consider ways of describing subsets of lines in terms of ordered pairs of real numbers. This is a relatively simple task and the reader perhaps recognized the intersections in the previous Exercises 16 through 19 as a segment, half-line, open segment, and ray, respectively. Essentially all we need to do is determine the equation of the line containing the subset and then find the limiting conditions that identify the particular subset in question. The following examples illustrate this.

Example 1. Describe the segment whose endpoints are $(1, 2)$ and $(3, 6)$.

The equation of the line containing this segment is $y = 2x$. In this example we have two choices of limiting conditions, $x \geq 1$ and $x \leq 3$, as shown in Illus. 8.34(a), or $y \geq 2$ and $y \leq 6$, as shown in Illus. 8.34(b).

It is customary for us to abbreviate a conjunction such as $x \geq 1$ and $x \leq 3$ as $1 \leq x \leq 3$. Similarly, the conjunction $y \geq 2$ and $y \leq 6$ can be abbreviated $2 \leq y \leq 6$. Thus a segment having endpoints $(1, 2)$ and $(3, 6)$ can be described by

$$y = 2x \quad \text{and} \quad 1 \leq x \leq 3$$

or

$$y = 2x \quad \text{and} \quad 2 \leq y \leq 6.$$

The open segment determined by the points $(1, 2)$ and $(3, 6)$ would, of course, be described by the conjunction

$$y = 2x \quad \text{and} \quad 1 < x < 3$$

or

$$y = 2x \quad \text{and} \quad 2 < y < 6.$$

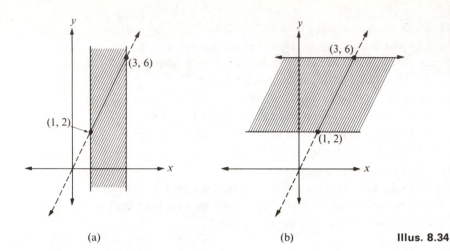

(a)   (b)   **Illus. 8.34**

Example 2. Describe the ray having endpoint $(-2, 1)$ and containing the point $(3, -2)$.

The equation of the line containing this ray is $3x + 5y + 1 = 0$. (The reader should verify this, using the two-point equation for a line.) Again, we have two choices of limiting conditions, $x \geq -2$ or $y \leq 1$ (see Illus. 8.35).

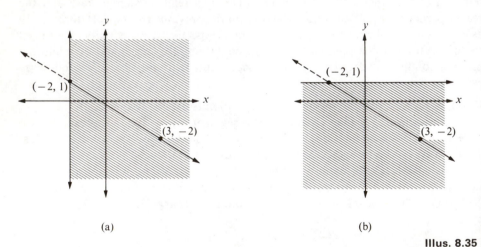

(a)   (b)

                                              Illus. 8.35

The ray shown above can therefore be described by the conjunction

$$3x + 5y + 1 = 0 \quad \text{and} \quad x \geq -2$$

or

$$3x + 5y + 1 = 0 \quad \text{and} \quad y \leq 1.$$

The half-line determined by the endpoint $(-2, 1)$ and containing the point $(3, -2)$ would of course be described by the conjunction

$$3x + 5y + 1 = 0 \quad \text{and} \quad x > -2$$

or

$$3x + 5y + 1 = 0 \quad \text{and} \quad y < 1.$$

Midpoints of Segments

In Chapter 4 we proved that every segment has exactly one midpoint (Theorem 4.15). Thus, for any two points $A(x_1, y_1)$ and $B(x_2, y_2)$, there is a unique point $P(x, y)$ such that $A\text{-}P\text{-}B$ and $AP = PB$. As we should then expect, the coordinates of P are determined by the coordinates of A and B. This can be seen as follows.

Consider a segment determined by endpoints $A(x_1, y_1)$ and $B(x_2, y_2)$ and its midpoint $P(x, y)$. There exist right triangles $\triangle ACP$ and $\triangle PDB$, as shown in Illus. 8.36. Since $\triangle ACP \sim \triangle PDB$, $AC/PD = AP/PB$. Furthermore, P is the

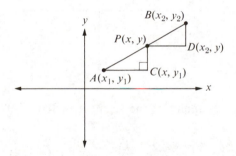

Illus. 8.36

midpoint of $\overline{AB}$, hence $AP = PB$ and $AP/PB = 1$. Expressing this in coordinates, we have

$$\frac{x - x_1}{x_2 - x} = 1,$$

from which we obtain

$$x = \frac{x_1 + x_2}{2},$$

which is the coordinate of the midpoint P.

To find the y-coordinate of P, we observe that

$$\frac{PC}{BD} = \frac{AP}{PB} = 1 \quad \text{or} \quad \frac{y - y_1}{y_2 - y} = 1,$$

from which we obtain $y = \dfrac{y_1 + y_2}{2}$.

We have now outlined a proof of the following theorem.

Theorem 8.13 The midpoint of a segment having endpoints (x_1, y_1) and (x_2, y_2) is the point

$$\left(\frac{x_1 + x_2}{2}, \frac{y_1 + y_2}{2}\right).$$

It is of interest to note that the x- and y-coordinates of the midpoint are the averages of the x- and y-coordinates, respectively, of the endpoints. Although a segment with positive slope was considered in proving Theorem 8.13, the theorem holds for any segment in the plane. The reader should verify this.

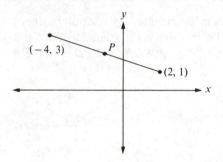

Illus. 8.37

Example 1. Find the midpoint of the segment in Illus. 8.37. By Theorem 8.13, the midpoint is

$$\left(\frac{-4 + 2}{2}, \frac{3 + 1}{2}\right) \qquad \text{or} \qquad (-1, 2).$$

Example 2. Find the midpoint of the segment in Illus. 8.38. The midpoint is

$$\left(\frac{-6 + 3}{2}, \frac{-2 + (-2)}{2}\right) \qquad \text{or} \qquad (-\tfrac{3}{2}, -2).$$

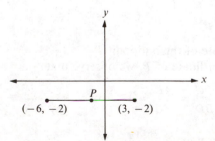

Illus. 8.38

EXERCISES

In Exercises 1 through 4, write a conjunction describing the segment having the given endpoints.

1. (0, 0) and (8, 6)   2. (2, 1) and (2, 9)

3. (0, 5) and (5, 0)   4. $(-\frac{1}{2}, -3)$ and $(\frac{9}{2}, -3)$

5. Write a conjunction describing the open segment determined by the endpoints in Exercise 2.

In Exercises 6 through 9, write a conjunction describing the ray containing the given points. The first point given is the endpoint.

6. (6, 3) and $(-1, -4)$ 7. $(-2, 3)$ and $(4, -1)$

8. (2, 3) and (3, 8)   9. (4, 4) and (4, 0)

10. Write a conjunction describing the half-line determined by the endpoint $(4, -7)$ and containing the point (5, 1).

11. Find the midpoints of the segments in Exercises 1 through 4.

Rigorous exercises

12. The following statement is equivalent to Theorem 8.13. "Given the points $A(x_1, y_1)$ and $B(x_2, y_2)$, the point

$$P\left(\frac{x_1 + x_2}{2}, \frac{y_1 + y_2}{2}\right)$$

is the midpoint of segment $\overline{AB}$." Prove it, without using the concept of similar triangles. [*Hint:* Show that P is on $\overline{AB}$ and that $AP = PB$.]

13. Consider two points, $A(x_1, y_1)$ and $B(x_2, y_2)$, and a point $P(x, y)$ such that A-P-B and $2 \cdot AP = PB$. Express the coordinates of P in terms of the coordinates of A and B.

6. APPLICATION OF ANALYTIC METHODS TO GEOMETRIC PROOFS

Earlier it was pointed out that one of the primary advantages of establishing a correspondence between the real numbers and geometric concepts is to enable us to utilize the properties of the real-number system in studying geometric relationships. Using the concepts which have been developed in this chapter, we shall now illustrate this by proving some theorems. The approach may be described as *analytical* and the proofs we construct are called *analytical* proofs meaning that we use the properties of the real numbers. In contrast, the proofs we constructed in earlier chapters are usually referred to as *synthetic* proofs.

In proving a theorem analytically, it is often helpful to choose the coordinate axes so that the algebraic computations are reduced to a minimum. If, for example, we wish to consider a triangle in a plane, then at any point in the plane we may place a pair of perpendicular lines and coordinatize the plane without changing any of the properties of the triangle. In Illus. 8.39, axes have been placed so that the origin is at a vertex and the x-axis contains one side of the triangle. This placement makes some of the coordinates zero and the slope of one of the sides zero, thus simplifying the algebra that may be required in a proof.

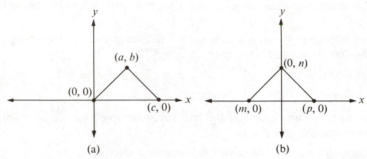

Illus. 8.39

In placing the axes and choosing the coordinates, we must be careful not to introduce any special conditions not stated in the hypothesis, however. For example, if two sides of a triangle lie on the axes, then the triangle is a right triangle, and unless the hypothesis specifies this, a special condition has been added (Illus. 8.40a). Similarly, if the coordinates are chosen as in Illus. 8.40(b), the triangle would be isosceles. If the coordinates are chosen as in Illus. 8.40(c), then the triangle would be a right isosceles triangle.

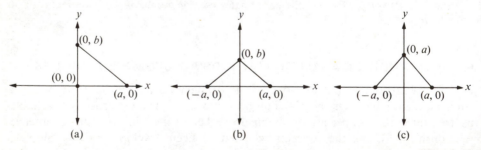

Illus. 8.40

The following example concerns a theorem presented earlier in the text. A comparison of this proof and the earlier proof provides an example to contrast analytic and synthetic proofs.

Example 1 (Theorem 6.6). In any parallelogram $\square ABCD$,

$$\overline{AB} \cong \overline{CD} \quad \text{and} \quad \overline{AD} \cong \overline{BC}.$$

Outline of proof. Consider any parallelogram $\square ABCD$ not a rectangle. Place coordinate axes so that the origin is at A, the positive half of the x-axis contains $\overline{AD}$, and at least one vertex is in the first quadrant, as in Illus. 8.41. We now know that D has second coordinate 0, but we do not know what the first coordinate is. If we call it "c," then point D has coordinates $(c, 0)$.

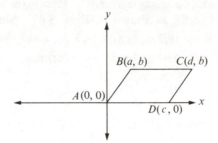

Illus. 8.41

Now, $\overline{BC} \parallel \overline{AD}$, hence we know that B and C have the same second coordinate, say b. We do not know what the first coordinates are but may call them "a" and "d." Thus B has coordinates (a, b) and C has coordinates (d, b).

By the definition of a parallelogram, $\overline{AB} \parallel \overline{CD}$ and $\overline{BC} \parallel \overline{AB}$, and by Theorem 8.7, $\overleftrightarrow{AB}$ and $\overleftrightarrow{CD}$ have the same slope; also $\overleftrightarrow{BC}$ and $\overleftrightarrow{AD}$ have the same slope. Thus,

$$\frac{b}{a} = \frac{b}{d - c}, \quad \text{(equal slopes)} \tag{8.4}$$

which means that

$$a = d - c. \tag{8.5}$$

By the distance formula, $AD = c$, and $BC = d - a$. From Eq. (8.5) we obtain $c = d - a$ and therefore $AD = BC$, which is equivalent to $\overline{AD} \cong \overline{BC}$. Similarly, $AB = \sqrt{a^2 + b^2}$ and $CD = \sqrt{(c - d)^2 + b^2}$, and from Eq. (8.5) we conclude that $AB = CD$ or $\overline{AB} \cong \overline{CD}$.

Now if parallelogram $\square ABCD$ is a rectangle, the axes may be placed so that the vertices are $A(0, 0)$, $B(0, b)$, $C(c, b)$, and $D(c, 0)$, and it is evident from the distance formula that $AB = CD$ and $AD = BC$. Thus, for any parallelogram $\square ABCD$, $\overline{AB} \cong \overline{CD}$ and $\overline{AD} \cong \overline{BC}$.

Example 2. A segment that joins the midpoints of two sides of a triangle is parallel to the third side and its measure is equal to one-half the measure of the third side.

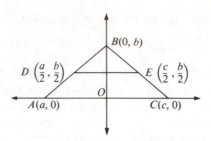

Illus. 8.42

Outline of proof. Consider a triangle $\triangle ABC$. Place axes so that the vertices are $A(a, 0)$, $B(0, b)$, and $C(c, 0)$, as shown in Illus. 8.42. Since the coordinates of the midpoints D and E are $(a/2, b/2)$ and $(c/2, b/2)$, the slope of $\overline{DE}$ is zero and, by Theorem 8.7, $\overline{DE} \parallel \overline{AC}$. Using the distance formula,

$$DE = \left| \frac{c}{2} - \frac{a}{2} \right| - \tfrac{1}{2}|c - a|$$

and $AC = |c - a|$. Therefore $DE = \tfrac{1}{2}AC$. This proves the theorem.

In contrast, one might construct a synthetic proof as follows.

Outline of proof. By the converse of Theorem 6.14 (cf. Exercise 8, p. 206), since $AD = DB$ and $BE = EC$, $DB/AB = BE/BC = \tfrac{1}{2}$ and therefore $\overline{DE} \parallel \overline{AC}$. Since $\overline{DE} \parallel \overline{AC}$, $\angle BDE \cong \angle BAC$ and $\angle BED \cong \angle BCA$. (Why?) Thus $\triangle ABC \sim \triangle DBE$ (why?) and $DB/AB = DE/AC = \tfrac{1}{2}$.

EXERCISES

⊙⊙ *Rigorous exercises*

Prove the following theorems analytically.

1. The midpoint of the hypotenuse of a right triangle is equidistant from all three vertices of the triangle.
2. The diagonals of a parallelogram bisect each other.
3. The diagonals of a rectangle are congruent.
4. The segments joining the midpoints of the sides of a quadrilateral form a parallelogram.
5. If P is a point on the perpendicular bisector of a segment $\overline{AB}$, then $PA = PB$.
6. The diagonals of a rhombus are perpendicular.
7. The medians of a triangle are concurrent at a point that divides each of the medians in the ratio 2 : 1. [*Hint:* Choose vertices with coordinates $(3a, 0)$, $(3b, 0)$, and $(0, 3c)$.]
8. The altitudes of a triangle are concurrent.
9. The perpendicular bisectors of the sides of a triangle are concurrent.

Networks

Which of the following figures can be traced without lifting the pencil, or crossing or retracing a segment?

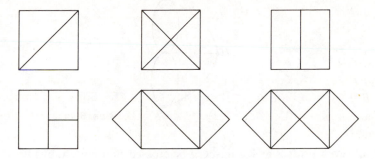

If a point is the endpoint of an odd number of segments (or curves), it is called an *odd vertex*; otherwise it is an *even vertex*.

Odd vertices   Even vertices

The famous Swiss mathematician, Euler, found a way to tell whether a figure could be traced as described above. He counted the number of odd vertices. What was his secret?

One of the most famous problems involving networks is The Seven Bridges at Königsberg. The problem is to cross all seven bridges in a continuous walk without recrossing any of them. (Illus. 8.43.)

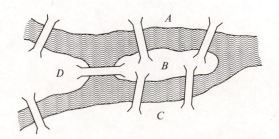

Illus. 8.43

Euler saw that this problem was like the one below (Illus. 8.44) where the four bodies of land are represented by the vertices of the figure.

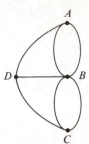

Illus. 8.44

Can all seven bridges at Königsberg be crossed in a continuous walk without recrossing any of them?

CHAPTER 9

TRANSFORMATIONS IN A PLANE

1. INTRODUCTION

When the notion of congruence was introduced in Chapter 4, it was pointed out that one might attempt a definition of congruence by using the idea of superposition. That is, two figures would be congruent if and only if they coincided when one was superimposed on the other. The difficulty with this approach is that geometric figures are abstractions with no physical characteristics and are therefore not capable of physical motion.

It is, however, easy to find geometric applications in the physical world that involve the concept of motion. For example, a dressmaker *places* a paper pattern on top of dress material in order to cut out pieces that are the same size and shape as the original pattern. And the reader may recall *turning* and *flipping* a piece of a puzzle to see if it fitted the outline of a missing piece. The fact is that we do move objects to see if they have the same size and shape as another, and we do assume, although tacitly perhaps, that motion has no effect on the geometric properties of the objects. More specifically, we assume that motions like those above do not alter the size or shape of an object, a phenomenon sometimes described as *rigid motion*. On the other hand, there are motions such as stretching which do alter the size and sometimes the shape of an object.

In the physical world, we think of moving an object from one location to another. In our geometry, we think of a *correspondence* between points. Instead of "picking up" $\triangle ABC$ and superimposing it on $\triangle A'B'C'$ in Illus. 9.1, we simply think of a matching or correspondence such that each point of $\triangle ABC$ is paired with a unique point of $\triangle A'B'C'$, and each point of $\triangle A'B'C'$ is paired with a unique point of $\triangle ABC$. What we have described here is, roughly speaking, a one-to-one correspondence. In geometry, one-to-one correspondences between sets of points are called *transformations*.

As we shall see, there are different types of transformations, as for example, transformations that represent the physical motion of sliding in a certain direction,

269

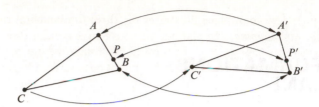

or of rotating about a fixed point, or of reflecting about a line. There are even transformations that represent the physical notion of stretching.

This chapter is only a brief introduction to the subject of transformations. The development is essentially informal and the reader is encouraged to participate in the exploratory activities provided in most sections of the chapter.

We shall begin our study of transformations by looking at some fundamental motions on a plane surface.

2. FUNDAMENTAL MOTIONS IN A PLANE

In Illus. 9.2 a puzzle piece is shown in three different positions. How would you move each piece to fit it into the puzzle space at the right? It is fairly easy to see that for (a) we need only to *slide* the piece to the right without any turning or twisting. For (b), we would need to *turn* the piece as well as slide it to the right, and for (c), we must *flip* the piece over.

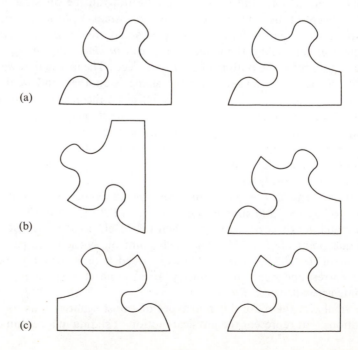

(a)

(b)

(c)

Fitting a puzzle piece in place illustrates the three fundamental rigid motions: *sliding*, *turning*, and *flipping*. In sliding, we move in a certain direction a given distance AB as shown in Illus. 9.3. In turning, we rotate the figure about a point through a certain angle as in Illus. 9.4. A flip might be described as turning a figure

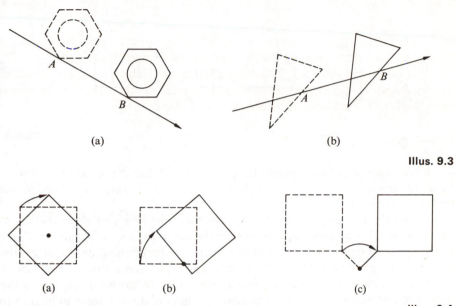

(a)   (b)

Illus. 9.3

(a)   (b)   (c)

Illus. 9.4

around a fixed line in the plane as in Illus. 9.5. Note that the figure is "lifted out of the plane" (except for any part on the fixed line).

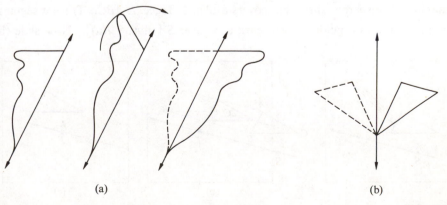

(a)   (b)

Illus. 9.5

In each part of Illus. 9.6, we have a pair of congruent figures. What motion or combination of motions could be used to superimpose one on the other? To help you decide, you might trace one of the figures on a piece of paper and move it to supermpose your figure on the second one. Is there more than one set of motions in each case?

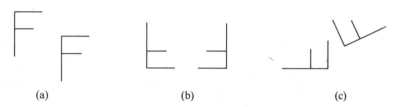

      (a)   (b)   (c)

Illus. 9.6

If you trace one of the figures on a piece of paper, are the figure and the traced figure congruent? If you move the paper around, does the size and shape of your traced figure change?

Intuitively speaking, if two figures are the same size and shape, we would expect to be able to superimpose one on the other by one or a combination of slides, turns, or flips. Conversely, if one figure is superimposed on the other, then sliding, turning, or flipping one of them to a new position should not change the fact that it is the same size and shape as the other. What we are saying is that, from physical experience, sliding, turning, or flipping do not seem to change the size or shape of an object; in other words, the object remains "rigid" throughout the motion.

Now let's explore some relationships between physical motions and geometric transformations.

Suppose we wish to "slide" $\triangle ABC$ a distance ST in the direction of ray $\overrightarrow{ST}$. Imagine a transparent sheet that covers a plane. Trace $\triangle ABC$. Trace a segment on $\overrightarrow{ST}$ for use as a guide. Then mark a $\times$ over S [Illus. 9.7(a)]. Now slide the

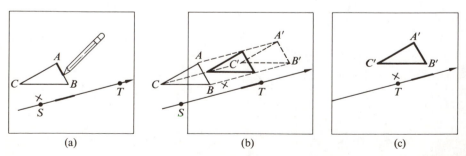

         (a)   (b)   (c)

Illus. 9.7

sheet along $\overrightarrow{ST}$ [Illus. 9.7(b)] until the × on the sheet is over the point T in the plane [Illus. 9.7(c)]. The traced triangle on the sheet is now superimposed on $\triangle A'B'C'$.

Note that none of the points in the plane has "moved." Sliding the sheet helps to illustrate how each point in the plane is paired with another point.

In sliding the sheet, the × moved from S to T. This shows that S is paired with T. We call S the *original* point and T the *image* point or simply the *image*. Thus we have an ordered pair which can be denoted as "(S, T)" or "$S \rightarrow T$."

EXERCISES

Exploratory exercises

Exercises 1–8 refer to Illus. 9.7.

1. What are the images for A, B, and C?

2. How do the distances BB' and ST compare? [See Illus. 9.7(b).]

3. What other distances are equal to ST?

4. Which segments are parallel?

5. Is $\triangle ABC \cong \triangle A'B'C'$? Why?

6. Suppose D is a point in the interior of $\triangle ABC$, and $D \rightarrow D'$. Where is D'? What can you say about the distance DD'? Is $\overline{DD'}$ parallel to any other segment?

7. If P is any point in the plane, how could you find its image?

8. Suppose we had begun by tracing $\triangle A'B'C'$ and then had slid our sheet along $\overleftrightarrow{ST}$ from T to S. What would be the images of A', B', and C'?

In the example above, each point P in the plane has exactly one image P'. Furthermore, each point P' in the plane is the image for exactly one original point P. This one-to-one correspondence between the set of all points in the plane with itself is an example of a *one-to-one transformation* in the plane.

A transformation determined by a slide such as the one above is called a *translation*.

Suppose we wish to "turn" $\triangle ABC$ θ degrees around a point O. Again let us imagine a transparent sheet over our plane. Trace $\triangle ABC$. Trace segment $\overline{OS}$ for use as a guide [Illus. 9.8(a)]. Now place the pencil point on O and turn the sheet, as in Illus. 9.8(b), until $\overline{OS}$ coincides with $\overline{OT}$ [Illus. 9.8(c)].

The triangle on the sheet is now superimposed on $\triangle A'B'C'$. Note again, no points in the plane have "moved." Turning the sheet illustrates how each point in the plane is paired with its image.

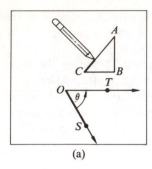
(a)

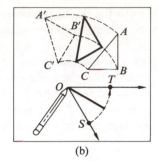
(b)

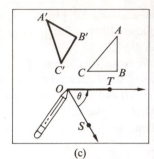
(c)

Illus. 9.8

EXERCISES

Exploratory exercises

1. What are the images of A, B, and C [Illus. 9.8]?
2. How do the central angles, $\angle COC'$ and $\angle TOS$, compare [Illus. 9.9]?

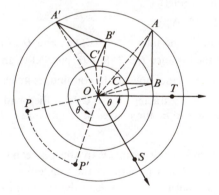

Illus. 9.9

3. What other angles are congruent to $\angle TOS$?
4. What other central angles are congruent to each other?
5. Is $\triangle A'OB' \cong \triangle AOB$? Why?
6. Is $\triangle ABC \cong \triangle A'B'C'$? Why?
7. Suppose P is any point in the plane. How could you find its image?
8. Suppose we turned our sheet around the point O from $\overrightarrow{OT}$ to $\overrightarrow{OS}$. What would be the images of A', B', and C'?

From Illus. 9.9 one can see that any point P in the plane determines a radius $\overline{OP}$. If we rotate $\overline{OP}$ θ degrees in the given direction, we obtain the image P'.

This correspondence is another example of a *one-to-one transformation* in the plane.

A transformation determined by a turn about a point such as that above is called a *rotation*.

Suppose we wish to "flip" $\triangle ABC$ about a line $\overleftrightarrow{ST}$. Trace segment $\overline{ST}$ and $\triangle ABC$ on the transparent sheet [Illus. 9.10(a)]. Now pick up the sheet and turn it over [Illus. 9.10(b)], superimposing the traced segment $\overline{ST}$ on the line $\overleftrightarrow{ST}$ [Illus. 9.10(c)]. The triangle on the sheet is now superimposed on $\triangle A'B'C'$ in the plane. Segments $\overline{AA'}$, $\overline{BB'}$, and $\overline{CC'}$ intersect $\overleftrightarrow{ST}$ at X, Y, and Z, respectively. Note again that no points in the plane have "moved." Flipping the sheet over illustrates how each point in the plane is paired with its image.

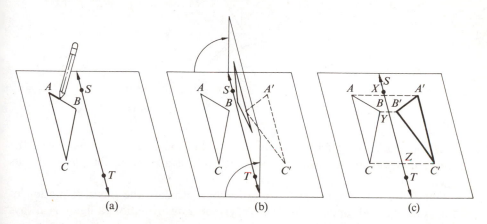

(a)   (b)   (c)

Illus. 9.10

EXERCISES

Exploratory exercises

Exercises 1–7 refer to Illus. 9.10.

1. What are the images of A, B, and C?

2. How is $\angle CZS$ related to $\angle C'ZS$? Since they are supplementary angles, what is the measure of each?

3. Compare the lengths of $\overline{CZ}$ and $\overline{C'Z}$.

4. How is $\overleftrightarrow{ST}$ related to $\overline{CC'}$?

5. How is $\overleftrightarrow{ST}$ related to $\overline{AA'}$? To $\overline{BB'}$?

6. If P is any point in the plane, how could you find its image?

7. If Q is on the line $\overleftrightarrow{ST}$, what is its image?

The activity outlined above illustrates a third example of a correspondence which is a one-to-one transformation in a plane.

A transformation determined by a flip about a line such as that above is called a *line reflection* (about a line $\overleftrightarrow{ST}$).

Line reflection is an appropriate description since the image of any point appears to be a reflection of the original point on the opposite side of the line. The line is often referred to as a *line of reflection*. This is illustrated in Illus. 9.11 where the set of image points is shown dashed.

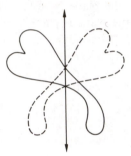

Illus. 9.11

As the reader may have already noted, the line in a line reflection is the perpendicular bisector of every segment which joins a point in the plane with its image.

Translations, rotations, and line reflections simulate the physical motion of sliding, turning, and flipping. With each type of transformation, a figure is always congruent to its image. More precisely, for any pair of points P, Q in a plane, if $P \rightarrow P'$ and $Q \rightarrow Q'$, under either a translation, rotation, or line reflection, then $PQ = P'Q'$. Because of this we say that translations, rotations, and line reflections are *distance-preserving* transformations. A distance-preserving transformation is called an *isometry*. The isometries of the plane—translations, rotations, and line reflections— are also referred to as "rigid motions" of the plane.

EXERCISES

1. Under which kind of transformation could one figure be the image of the other?

(a)   (b)   (c)   (d)

2. a) Given a quadrilateral $ABCD$ and a translation determined by $A \rightarrow A'$ (Illus. 9.12), construct the image of the quadrilateral.

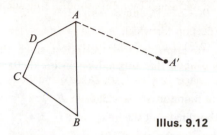

Illus. 9.12

b) Given the quadrilateral $ABCD$ and a line reflection determined by $A \rightarrow A'$ (Illus. 9.12), construct the image of the quadrilateral.

c) Find a rotation such that $A \rightarrow A'$, and construct the image of quadrilateral $ABCD$. [*Hint*: Use the line of reflection in (b).]

d) How many rotations are there for which $A \rightarrow A'$? Under each of these rotations, how would the images of quadrilateral $ABCD$ compare?

e) How do the images in (b) and (c) compare?

f) How do the images in (b) and (c) compare with the image in (a)?

3. Which of the isometries could fit the following descriptions?

a) For every point P in the plane, the image of P is different from P.

b) $P \rightarrow P'$, and every other point Q has an image that is different from Q.

c) $P \rightarrow P$, $Q \rightarrow Q$, and $R \rightarrow R'$, where $R \neq R'$.

4. a) Draw a circle and a line which does not intersect the circle. Then fold along the line and trace to get the image.

b) What type of isometry is illustrated here?

c) Could you get the same image with a translation? If so, describe the translation.

d) Could you get the same image with a rotation? If so, describe the rotation.

e) Does each point on the original circle have the same image under all three isometries? Explain.

5. Trace the simple closed curve below on your paper, draw a line, and then answer parts (a) through (d) of Exercise 4 for this figure.

6. Is it possible that two congruent figures could be positioned in a plane so that neither is the image of the other under a single isometry? If so, give some examples.

7. Consider a translation for which $A(0, 2) \rightarrow B(4, 5)$. What are the coordinates of the images of $C(0, 0)$, $D(-2, -1)$, and $E(-4, -1)$?

8. Consider a line reflection for which $A(0, 1) \rightarrow B(4, -1)$. What are the coordinates of the images of $C(2, 3)$, $D(-1, 2)$, and $E(6, -3)$?

9. Consider a rotation about the origin for which $A(2, 0) \rightarrow B(0, 2)$. What are the coordinates of the images of $C(-1, 0)$, $D(2, 3)$, and $E(-2, -4)$?

10. Consider a line reflection for which $A(3, 5) \rightarrow B(5, 3)$.

    a) What are the coordinates of the image of $C(-2, 4)$?

    b) Write an equation for the line of reflection.

3. COMBINING ISOMETRIES

Intuitively speaking, it is not difficult to see that translations, rotations, and line reflections may be combined to obtain other rigid motions. If, for example, we wish to associate the figures in Illus. 9.13, we might consider a translation, as indicated, followed by a reflection about $\overleftrightarrow{AB}$.

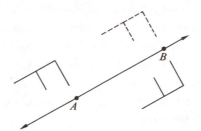

Illus. 9.13

For convenience, let us use the following abbreviations: T, translation; R, rotation; and S, line reflection (symmetry). The isometry in Illus. 9.13 might then be denoted as TS, which we read, "a translation *followed by* a line reflection." Note that we get the same result with the isometry ST (that is, the line reflection followed by the translation). Since this is true, TS and ST are really the same isometry and we write $TS = ST$. You may wish to conjecture at this point as to whether the order in combining isometries ever makes a difference.

The isometry TS is often referred to as a *glide reflection*. Although we shall not prove it here, it is a fact that every isometry of the plane is either a translation, a rotation, a line reflection, or a glide reflection.

As will be seen shortly, combinations of isometries may leave every point as its own image. That is, a transformation is obtained in which every point in the plane is associated with itself. This transformation is called the *identity transformation*. We denote it by I.

In the exploratory exercises below we shall experiment with combining isometries. We will write T_{AB} for a translation along $\overleftrightarrow{AB}$ from A to B, S_m for a reflection about a line m, and $R_{P(30)}$ for a counterclockwise rotation of 30 degrees about a point P.

EXERCISES

Exploratory exercises

1. A "flip" followed by another "flip" about the same line m has what effect (Illus. 9.14)? $S_m S_m$ is the same as what transformation?

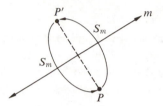

Illus. 9.14

2. a) What other translation would you combine with T_{AB} to obtain the identity transformation?

   b) Under what conditions would $R_{P(30)}R_{P(b)} = I$?

3. a) Copy Illus. 9.15 on a piece of paper. Fold along line l to find the first image. Then fold along line m to find the second image. The isometry $S_l S_m$ associates the original "F" with the second image.

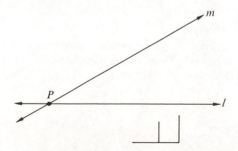

Illus. 9.15

   b) Copy Illus. 9.15 on a second paper and superimpose it over the first drawing Put your pencil on P and rotate the top paper. What happens?

   c) Measure the acute angle made by l and m. Measure the angle of your rotation.

   d) What rotation is equivalent to $S_l S_m$?

4. a) Copy Illus. 9.16 on a piece of paper. Fold along l and then along m to find the image of "F" under the isometry $S_l S_m$.

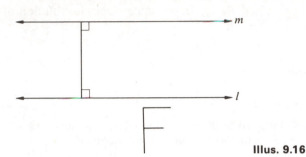

Illus. 9.16

b) How are lines l and m related?

c) Measure the distance between l and m. Measure the distance between corresponding points of "F" and its image.

d) Describe an isometry that is equal to $S_l S_m$.

5. Describe a single isometry that is equal to $T_{AB} T_{BC}$, where A, B, C are distinct points.

6. a) Trace Illus. 9.17 on a piece of paper.

A B **Illus. 9.17**

b) Find the image of "F" under the isometry $T_{AB} R_{B(180)}$.

c) Is $T_{AB} R_{B(180)} = R_{A(180)} T_{AB}$?

d) Is $T_{AB} S_{AB} = T_{AB} R_{B(180)}$?

7. a) Trace Illus. 9.18. Fold the paper so that P coincides with P'. What does the fold in the paper represent? One "F" is superimposed on the other by what type of rigid motion?

P P'

Illus. 9.18

b) Trace Illus. 9.19. Fold the paper so that P coincides with P', and trace the image you obtain. What additional fold should you make to obtain the same result as in (a) above?

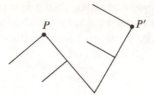

Illus. 9.19

c) Trace Illus. 9.20. Continue to fold the paper and trace images until one is co-incident with the figure at the right. How many line reflections were required to associate the two figures?

Illus. 9.20

d) Choose another figure and copy it in two places on a piece of paper. Note how many line reflections are required to associate one figure with the other.

In going through the exercises above, the reader probably noticed that line reflections seem to play an important role. In particular, a rotation can be ac-complished by a pair of line reflections, and a translation can be duplicated by two line reflections about parallel lines. These properties are summarized in the following theorems. We present them here without proof.

Theorem 9.1 If lines *l* and *m* intersect at a point *P*, then a line reflection about *l* followed by a line reflection about *m* is equal to a rotation about *P*, where the measure of the rotation is twice the measure of the angle formed by *l* and *m* (Illus. 9.21).

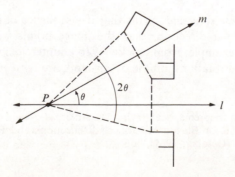

Illus. 9.21

Theorem 9.2 If lines *l* and *m* are parallel, then a line reflection about *l* followed by a line reflection about *m* is equal to a translation along a line perpendicular to *l* and *m*, and twice the distance between *l* and *m* (Illus. 9.22).

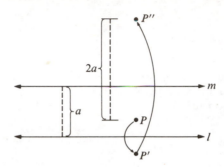

Illus. 9.22

Exercise 7 in the previous exercises points up a rather interesting and an important property of line reflections which is summarized in the theorem below. We omit the proof.*

Theorem 9.3 Every rigid motion (isometry) in the plane is equal to a combination of three or fewer line reflections.

Note that line reflections change the orientation. We say that line reflections are *sense reversing*. This is shown in Illus. 9.23. It follows, of course, that glide reflections are also *sense reversing*.

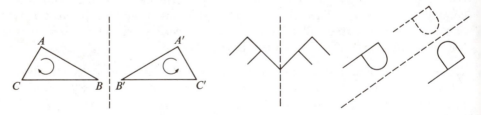

Illus. 9.23

Very simply, *sense reversing* means that if a sequence of points in a figure are ordered in a particular direction, then the image points will have a reverse or opposite order, for example, from clockwise to counterclockwise.

Translations and rotations, on the other hand, are *sense preserving* as shown in Illus. 9.24.

---

* For proofs of these theorems, see: Irving A. Adler, *A New Look at Geometry*, New York: John Day, 1966, or Burton W. Jones, "Reflections and Rotations," *The Mathematics Teacher*, *LIV* (October, 1961), 406–10.

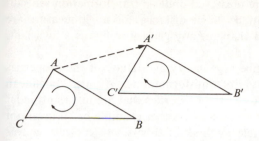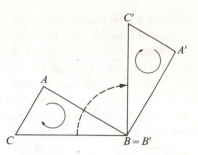

EXERCISES

1. Given a pair of translations T_{AB} and T_{CD}, is $T_{AB}T_{CD} = T_{CD}T_{AB}$ always true? Support your answer with reasons.

2. Consider a pair of rotations, $R_{P(a)}$ and $R_{P(b)}$.

   a) If the rotations are in the same direction, is $R_{P(a)}R_{P(b)} = R_{P(b)}R_{P(a)}$ always true?

   b) If the rotations are in *opposite* directions, what is the answer to (a) above?

3. Consider a pair of line reflections, S_l and S_m.

   a) If $l \parallel m$, is S_lS_m always equivalent to S_mS_l?

   b) If $l \nparallel m$, is $S_lS_m = S_mS_l$?

4. Construct an example to illustrate each of the following.

   a) A glide reflection followed by a translation is equal to a single glide reflection.

   b) A line reflection S_m followed by a glide reflection TS_m is equal to the translation T.

   c) A rotation R_P followed by a line reflection S_m is equal to a single glide reflection.

5. Which of the following are sense reversing? Which are sense preserving?

   a) RT b) SR

   c) $(TS)S$ (that is, a glide reflection followed by a line reflection).

6. Choose a geometric figure and copy it in two different positions on a piece of paper. By tracing and folding, find the successive line reflections that will superimpose one figure on the other. Repeat with three other figures.

4. INVARIANCE

Although we have not proved it, our exploratory exercises earlier in the chapter presented an intuitive argument that the isometries are distance preserving transformations. This means if $A \rightarrow A'$ and $B \rightarrow B'$ under any one of the isometries, then $AB = A'B'$. It is also true that the isometries preserve the measures of angles. If $C \rightarrow C'$ in the above example, then $m \angle ABC = m \angle A'B'C'$. Properties such as distance and angle measure that are preserved under a transformation are said to be *invariant* under the transformation. Since distance and angle measure are invariant, it follows that the size and shape of any geometric figure is invariant, hence the term "rigid motion."

Mathematicians sometimes describe a geometry as a study of the properties that are invariant under a certain group of transformations. As such, Euclidean geometry could be described as a study of the properties that are invariant under the isometries or rigid motions. Betweenness is another example of an invariant property and the reader should be able to think of others. In contrast, it was noted in the previous section that the sense or orientation of a figure is changed under a line reflection or a glide reflection and hence is not an invariant property.

Sometimes the image of a figure is the same as the figure itself. In Illus. 9.25, the image of $\triangle ABC$ under the line reflection S_{BD} is $\triangle ABC$. We may write $ABC \leftrightarrow CBA$ to show that $A \rightarrow C$, $B \rightarrow B$, and $C \rightarrow A$. In this case, we say that $\triangle ABC$ is *invariant* under the line reflection S_{BD}.

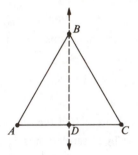

Illus. 9.25

Whenever the image of a figure is the same as the figure itself, we say that the figure is *invariant* under the transformation.

It will be noted also in the case above that B is its own image, as is D. That is, under S_{BD}, $B \rightarrow B$ and $D \rightarrow D$. In such cases, B and D are referred to as *fixed points*.

Now suppose that P is equidistant from the vertices of $\triangle ABC$. Illustration 9.26(a) shows a rotation of 90 degrees about P while Illus. 9.26(b) shows a rotation of 120 degrees about P. Under which rotation is $\triangle ABC$ invariant? Does $\triangle ABC$ have any fixed points under either rotation? Under which other rotations would $\triangle ABC$ be invariant?

Could any geometric figure be invariant under a translation? Illustration 9.27 provides one example where $\overrightarrow{AB}$ is invariant under T_{AB}. Are there any fixed points on $\overleftrightarrow{AB}$? Can you think of any other type of geometric figure that would be invariant under a translation?

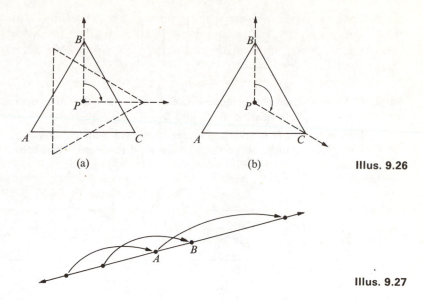

(a)   (b)   **Illus. 9.26**

Illus. 9.27

EXERCISES

1. Consider the equilateral triangle ABC in the above example. There are six rigid motions for which $\triangle ABC$ is invariant. Describe each one and identify any fixed points.

2. For each of the following, describe all rigid motions for which the figure is invariant.

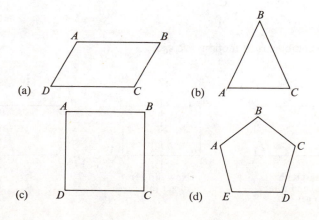

(a)   (b)

(c)   (d)

3. For each capital letter at the top of page 286, identify any line reflections or rotations for which it is invariant.

AEHINSX

4. a) Draw two congruent rectangles *ABCD*. Cut one out of the paper and label the corners, on both sides, *A*, *B*, *C*, and *D*.

   b) Rectangle *ABCD* is invariant under four rigid motions: R_0, R_{180}, S_v, and S_h. Use your rectangles to go through the rigid motions. Then copy and label the corners.

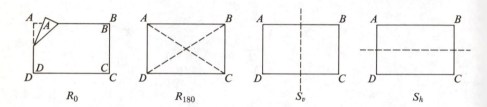

$$R_0 \qquad\qquad R_{180} \qquad\qquad S_v \qquad\qquad S_h$$

   c) We can describe each rigid motion by naming the images of the vertices. Complete this description.

R_0	R_{180}	S_v	S_h
$A \rightarrow A$	$A \rightarrow C$		
$B \rightarrow B$	B		
$C \rightarrow C$	C		
$D \rightarrow D$	D		

   d) We can combine rigid motions: $R_{180}S_v$.

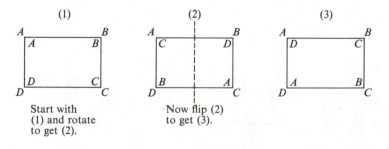

(1)	(2)	(3)
Start with (1) and rotate to get (2).	Now flip (2) to get (3).	

   What rigid motion did you get in (3)? $R_{180}S_v = $ _____?

   e) We can use (c) to find a combination.

Example. $S_v R_{180}$

$$S_v R_{180} \quad = \quad S_h$$
$$A \to B \to D \qquad A \to D$$
$$B \to A \to C \qquad B \to C$$
$$C \to D \to B \qquad C \to B$$
$$D \to C \to A \qquad D \to A$$

Complete the table below.

	R_0	R_{180}	S_v	S_h	
R_0					
R_{180}			S_h		
S_v		S_h			
S_h					

5. Follow the instructions in Exercise 4 for an equilateral triangle.

6. Follow the instructions in Exercise 4 for a square.

7. A line $\overleftrightarrow{AB}$ is invariant under T_{AB}. The line $\overleftrightarrow{AB}$ is also invariant under what other rigid motion(s)?

8. A ray $\overrightarrow{AB}$ is invariant under what rigid motion?

9. An angle $\angle ABC$ (nonstraight) is invariant under what rigid motion(s)?

10. A segment $\overline{AB}$ is invariant under what rigid motion(s)?

11. Where are the fixed points in each of the following isometries? (Disregard the identity transformation.)

  a) translation   b) rotation   c) line reflection

5. DEVELOPING GEOMETRY THROUGH TRANSFORMATIONS

There are many ways of developing Euclidean geometry. In this section we shall take a brief look at how one might develop a small part of Euclidean geometry using the concept of transformations. This, of course, does not affect the previous development. It merely illustrates an alternative approach to the development of a mathematical system—in this case Euclidean geometry.

We have accepted the fact that distance and angle measure are invariant properties under any isometry. We now turn our attention to some additional properties of translations and rotations of lines.

Consider a translation T_{AB} (Illus. 9.28). Trace $\overleftrightarrow{AC}$ on paper and slide it along $\overrightarrow{AB}$ (using $\overrightarrow{CD}$ as a guide) until $\overline{AC}$ is superimposed on $\overline{BD}$.

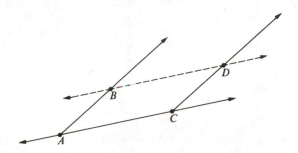

Illus. 9.28

How is the image line, $\overleftrightarrow{BD}$, related to $\overleftrightarrow{AC}$? Suppose $\overleftrightarrow{BD}$ did intersect $\overleftrightarrow{AC}$ at some point X. How would this contradict the meaning of a translation? This discussion points to the following theorem.

Theorem 9.4 For any line m and any translation T, either m is invariant (its own image) under T, or the image of m is another line that does not intersect m (Illus. 9.29).

Illus. 9.29

All of this suggests a way to define parallel lines using transformations as a fundamental concept.

Definition 9.1 Two lines are *parallel* if and only if one line is the image of the other under some translation.

Let us now consider the effect of a 180-degree rotation of a line. Try this experiment. Draw a line m and choose a point P not on m. Trace m and P on another paper and rotate 180 degrees about P. How is the image related to m? Repeat the experiment for several other lines and points. Next, choose a point P on the line m and rotate 180 degrees about P. How is the image related to m? The experiment above suggests the following theorem.

Theorem 9.5 The image of a line under a rotation $R_{P(180)}$ is (a) parallel to the line if P is not on the line, or (b) identical to the line if P is on the line (Illus. 9.30).
    As a consequence of the properties above, we have the following theorem.

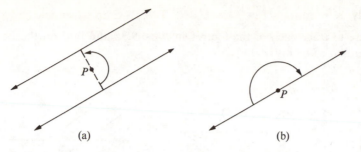

(a)   (b)

Illus. 9.30

Theorem 9.6 If two parallel lines are cut by a transversal, the corresponding angles are congruent.

Proof. Consider parallel lines *l* and *m* cut by transversal $\overleftrightarrow{BE}$ (Illus. 9.31). Since *l* ∥ *m* by the definition of parallel that we would be using, there is a translation *T* for which $\overleftrightarrow{BE}$ is invariant and $\overleftrightarrow{DE}$ is the image of $\overleftrightarrow{AB}$. In particular, ∠*DEF* is the image of ∠*ABC*. It follows that ∠*DEF* ≅ ∠*ABC*. By similar reasoning and with the same translation *T*, other corresponding pairs of angles are congruent.

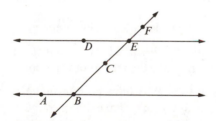

Illus. 9.31

EXERCISES

1. Supply the reasons for each of the statements in Theorem 9.6.

2. Use our generalization about 180-degree rotations to prove that vertical angles are congruent.

3. Prove that parallel lines cut by a transversal have congruent alternate interior angles.

4. Write a definition of perpendicular lines using the concept of invariance and a rigid motion.

5. Suppose you have a line *m* and a point *P* not on *m* on a piece of paper. Describe two ways to construct the perpendicular from *P* to *m*, using the properties of rigid motions.

6. Use paper folding to
    a) bisect a segment,   b) bisect an angle.

Symmetry

If you were to trace each of the figures in Illus. 9.32 and fold on the line, which would you find to be invariant?

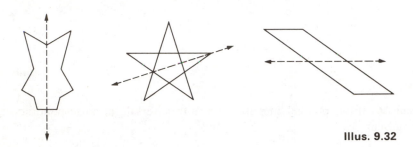

Illus. 9.32

Definition 9.2 A figure is *symmetrical about a line* if and only if it is invariant under a reflection about that line. The line is called a *line of symmetry*.

The first two figures in Illus. 9.32 are symmetrical about the line while the third figure is not symmetrical. Some figures have more than one line of symmetry as is the case with the star above. The first figure, of course, has only one line of symmetry.

A second kind of symmetry is called point symmetry.

Definition 9.3 A figure is *symmetrical about a point* if and only if it is invariant under some rotation less than 360 degrees about the point.

Each of the figures in Illus. 9.33 have point symmetry about *P*. As shown in the illustrations above, a figure may have line symmetry without having point symmetry, and vice versa. The star and equilateral triangle, of course, have both kinds of symmetry. The reader may wish to verify each point symmetry by tracing and rotating each figure.

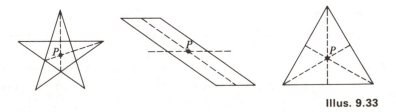

Illus. 9.33

EXERCISES

1. Copy the triangles in Illus. 9.34. Find all line symmetries. Find all point symmetries.

2. Define *isosceles triangle* using the concept of symmetry.

3. Define *equilateral triangle* using the concept of symmetry.

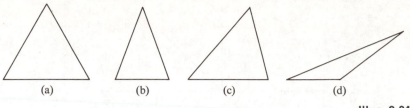

(a)   (b)   (c)   (d)

4. How is a point symmetry for a triangle dependent upon the line symmetries of the triangle?

5. Why is it impossible for a triangle to have exactly two lines of symmetry?

6. Draw as many different kinds of quadrilaterals as you can that have

   a) no lines of symmetry,

   b) exactly one line of symmetry,

   c) exactly two lines of symmetry,

   d) exactly three lines of symmetry,

   e) exactly four lines of symmetry.

7. Which kinds of quadrilaterals have a point of symmetry?

8. Write a definition of *parallelogram* using the concept of symmetry.

9. Write a definition of *square* using the concept of symmetry.

10. An octagon would be invariant under how many rotations? Under how many line reflections?

11. A circle has how many lines of symmetry?

12. Construct a figure different from those above that has

   a) a line of symmetry but no point of symmetry,

   b) a point of symmetry but no line of symmetry,

   c) four lines of symmetry.

6. SIMILARITY TRANSFORMATIONS

Thus far we have considered transformations of the plane that preserve distance and angle measure. Under these transformations every figure is congruent to its image. We turn now to another type of transformation that preserves only angle measure. Such transformations are called *similarity transformations*, since the image of a figure has the same shape, so to speak, but not necessarily the same size.

Some Rubber Sheet Experiments

Imagine a transparent sheet made of rubber so that it is easily stretched in any direction. A pair of coordinate axes are drawn on the sheet and are then superimposed on the axes in the plane. Finally we trace $\triangle ABC$ on our sheet, noting the coordinates of each vertex in the plane beneath (Illus. 9.35).

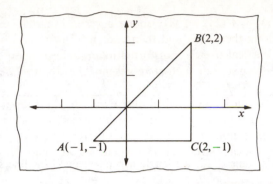

Illus. 9.35

Now suppose that the rubber sheet is firmly attached along the *y*-axis in the plane and we stretch the sheet as shown in Illus. 9.36 so that the distance between any point on our rubber sheet and the *y*-axis is doubled. Points on the *y*-axis remain fixed. Study this illustration and then do the exercises below.

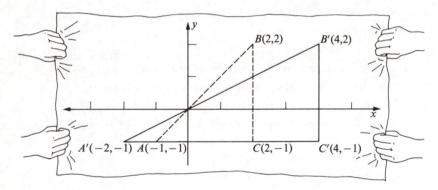

Illus. 9.36

EXERCISES

Exploratory exercises

1. Under this "horizontal stretch" transformation, what are the coordinates of the image points for *A*, *B*, and *C*?

2. With regard to $\triangle ABC$, what distance has been preserved? What distances have been changed?

3. Is the shape of $\triangle ABC$ the same as the shape of $\triangle A'B'C'$?

4. Is $m\angle BAC = m\angle B'A'C'$? Compare the measures of other corresponding angles. Has angle measure been preserved?

It is not difficult to see that under a horizontal stretch transformation, vertical distances are preserved while other distances are not, and with this particular stretch, horizontal distances are doubled. One might describe this stretch trans-

formation as $A(x, y) \rightarrow A'(2x, y)$. To obtain a vertical stretch, we would attach
our rubber sheet along the x-axis and then stretch it in the opposite direction to
that in Illus. 9.36. Vertical stretch transformations would be denoted as $A(x, y) \rightarrow$
$A'(x, ky)$, where k (the stretch factor) represents the amount of stretch in the
vertical direction.

While the transformations described above are not the similarity transformations we are looking for, they do provide us with a means of obtaining them. In
the meantime there are a number of interesting questions that may be raised about
the horizontal (or vertical) stretch transformations, such as: Is parallelism or
perpendicularity preserved under these transformations? Are any figures invariant,
and if so, under what conditions?

While we do not seek to establish rigorously the answers to any of the questions
above, the exercises below should help to provide the reader with some basis for
conjectures.

EXERCISES

1. For each of the figures below, construct the images under the stretch transformations
   $A(x, y) \rightarrow A'(2x, y)$ and $A(x, y) \rightarrow A'(3x, y)$ [that is, each image point is three times
   the distance from the y-axis as the original point].
   a) Triangle: (with vertices) $A(-2, -1)$, $B(2, 4)$, $C(1, 1)$
   b) Square: $A(0, 2)$, $B(2, 0)$, $C(0, -2)$, $D(-2, 0)$
   c) Parallelogram: $A(1, 1)$, $B(3, 1)$, $C(1, -3)$, $D(-1, -3)$
   d) Parallelogram: $A(0, 0)$, $B(2, -\frac{1}{2})$, $C(2\frac{1}{2}, -2\frac{1}{2})$, $D(\frac{1}{2}, -2)$
   e) Trapezoid: $A(0, 1\frac{1}{2})$, $B(0, 3)$, $C(5, 0)$, $D(2\frac{1}{2}, 0)$
   f) The line $y = 2$
   g) The line $x = 2$
   h) The line $y = -\frac{1}{2}x - 1$

2. Is parallelism preserved under either type of stretch transformation? Why or why not?

3. Is perpendicularity preserved under either type of stretch transformation? Why or
   why not?

4. If the stretch factor of a transformation is 1, what kind of transformation is it?

5. Assuming that $k \neq 1$, what type of figure is invariant under stretch transformations
   and under what conditions?

6. Under the rigid motions, the images of triangles are triangles, of parallelograms are
   parallelograms, etc. Is this also true for the stretch transformations? Discuss.

7. Suppose that $k = \frac{1}{2}$, that is $A(x, y) \rightarrow A'(\frac{1}{2}x, y)$. [It might be more descriptive to
   call this a "shrink" transformation, but for consistency we also call it a stretch
   transformation.] What is the image of $\triangle A'B'C'$ in Illus. 9.36?

8. Sketch the image of the circle below under the transformation $A(x, y) \rightarrow A'(2x, y)$.
   [*Hint:* Use a ruler to find several points in each quadrant.]

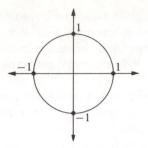

The Two-Way Stretch

We have seen that neither horizontal nor vertical stretch transformations preserve distance or angle measure. Because of this, perpendicularity is not preserved and rhombi, rectangles, and squares, being classified by angle measure or distance as well as by parallel sides, may not have the same type image.

On the other hand, the image of a line is a line, and parallel (or intersecting) lines have parallel (or intersecting) images. It also follows, of course, that triangles, trapezoids, and parallelograms have the same respective images.

Now let's see what happens if we put a horizontal and a vertical stretch together—sort of a two-way stretch. This time we attach our rubber sheet at the origin only and then stretch it as shown in Illus. 9.37.

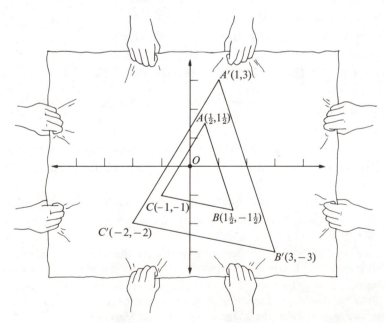

Illus. 9.37

We assume that the sheet is stretched uniformly, which means that the stretch factor, k, is the same for the horizontal and the vertical stretches. In this example, the origin, O, is the *center of stretch.** It is the only fixed point, unless $k = 1$ in which case we have the identity transformation and all points are fixed.

In Illus. 9.37 we began by tracing $\triangle ABC$ and then stretched the sheet to obtain the image $\triangle A'B'C'$. Note that each vertex and its image are specific examples of the correspondence

$$P(x, y) \rightarrow P'(2x, 2y).$$

That is, P' is twice as far as P is from the y-axis (due to the horizontal stretch) and also twice as far as P is from the x-axis (due to the vertical stretch), the only exceptions being those points on the axes. We shall refer to a correspondence like that above as a *two-way stretch transformation* about the origin since it simulates the physical movement of a two-way stretch.

More generally, we can describe any *two-way stretch transformation* about the origin by the correspondence

$$P(x, y) \rightarrow P'(kx, ky),$$

where $k > 0$ is the stretch factor.

After a careful look at Illus. 9.37, it would appear that two-way stretch transformations do preserve angle measure. If so, we have an example of a similarity transformation. Some other significant properties of the two-way stretch transformation in Illus. 9.37 are brought out below.

EXERCISES

Exploratory exercises

1. Use a ruler in Illus. 9.37 to help you find as many sets of collinear points as you can.

2. For each pair of segments, compare their measures.
   a) $\overline{OA}$ and $\overline{OA'}$ b) $\overline{OB}$ and $\overline{OB'}$ c) $\overline{OC}$ and $\overline{OC'}$
   d) $\overline{AB}$ and $\overline{A'B'}$ e) $\overline{BC}$ and $\overline{B'C'}$ f) $\overline{AC}$ and $\overline{A'C'}$

3. Which segments appear to be parallel?

4. a) Trace the axes and $\triangle ABC$ from Illus. 9.37 and then construct the image $\triangle A''B''C''$ under the two-way stretch transformation $P(x, y) \rightarrow P'(3x, 3y)$.
   b) Find sets of collinear points as in Exercise 1 above.
   c) Compare measures of segments as in Exercise 2 above.
   d) Which segments appear to be parallel?

---

* The center of stretch may be any point in the plane. For simplicity, we consider here only the origin as a center of stretch.

e) Choose any point P on $\triangle ABC$. Use a ruler to find a point Q on $\triangle A''B''C''$ such that O-P-Q.

f) What is the measure of $\overline{OP}$? Of $\overline{OQ}$?

g) How are points P and Q related?

The exercises above point to some important properties of two-way stretch transformations. They are summarized in the theorem below.

Theorem 9.7 Given a two-way stretch transformation $P(x, y) \to P'(kx, ky)$, $k > 0$, and Q, R any two points in the plane,

i) O-Q-Q' or O-Q'-Q, where O is the center of stretch,

ii) $OQ' = k(OQ)$,

iii) $Q'R' = k(QR)$,

iv) angle measure is preserved.

Outline of proof

i) We wish to show that any point and its image are collinear with O, the center of stretch, or that $O \in \overleftrightarrow{QQ'}$. Consider a point $Q(x_1, y_1)$ and its image $Q'(kx_1, ky_1)$. Using the two-point equation of a line (Theorem 8.6), we get

$$y - ky_1 = \frac{ky_1 - y_1}{kx_1 - x_1} (x - kx_1).$$

To show that $O \in \overleftrightarrow{QQ'}$, substitute the coordinates $(0, 0)$ for x and y in the equation above. Since this substitution satisfies the equation, Q and Q' are collinear with O, the center of stretch.

ii) Using the distance formula (Theorem 8.3), we get

$$OQ = \sqrt{x_1^2 + y_1^2} \quad \text{and} \quad \begin{aligned} OQ' &= \sqrt{(kx_1)^2 + (ky_1)^2} \\ &= \sqrt{k^2 x_1^2 + k^2 y_1^2} \\ &= k\sqrt{x_1^2 + y_1^2} \\ &= k(OQ). \end{aligned}$$

iii) We note that the coordinates are $Q(x_1, y_1)$, $Q'(kx_1, ky_1)$, $R(x_2, y_2)$, and $R'(kx_2, ky_2)$, and then use the distance formula as in (ii) above.

iv) It is sufficient to consider a triangle QRS and its image $\triangle Q'R'S'$ (Illus. 9.38). By (iii) above,

$$\frac{S'R'}{SR} = \frac{R'Q'}{RQ} = \frac{S'Q'}{SQ} = k.$$

It follows (Exercise 12, page 206) that $\triangle QRS \sim \triangle Q'R'S'$, which means by definition that corresponding angles have the same measure.

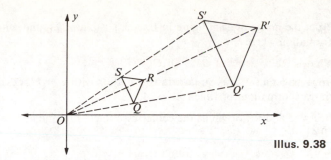

Illus. 9.38

The theorem above establishes the fact that two-way stretch transformations do preserve angle measure, and are therefore similarity transformations. The theorem also provides us with a method of constructing the image of a figure, given the stretch factor, k, and the center of stretch. In Illus. 9.38, for example, S', the image of S, is a point on $\overrightarrow{OS}$ such that $OS' = 2\frac{1}{2}(OS)$. Points R' and Q' are located in the same manner.

EXERCISES

1. Given O the center of stretch, find the images of A, B, and C for each of these stretch factors (Illus. 9.39).

   a) $k = 1\frac{1}{2}$ b) $k = \frac{1}{2}$

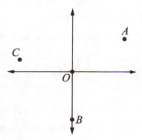

Illus. 9.39

2. Consider $A \to A'$ and $B \to B'$ under a two-way stretch transformation as shown in Illus. 9.40.

   a) Locate the center of stretch.

   b) Estimate the stretch factor k.

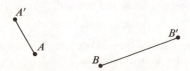

Illus. 9.40

3. Construct the image of each figure in Exercise 1, page 293, under a two-way stretch transformation where

   a) $k = 2$ b) $k = \frac{1}{2}$

4. If $A \rightarrow A'$ under a two-way stretch transformation with $k = 1\frac{1}{2}$, describe the two-way stretch transformation for which $A' \rightarrow A$.

5. Given $k = 2$ and the origin as the center of stretch, find the images of the circles in Illus. 9.41.

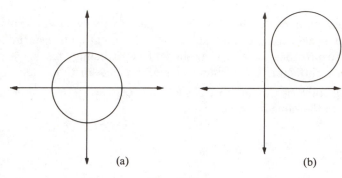

(a)   (b)

Illus. 9.41

6. Is perpendicularity preserved under two-way stretch transformations? Explain.

7. Assume that $k > 0$ and $k \neq 1$. What figures are invariant under two-way stretch transformations and under what conditions?

8. If there is exactly one fixed point under a two-way stretch transformation, what can be said about k?

9. If there is more than one fixed point under a two-way stretch transformation, what can be said about k?

10. Consider a two-way stretch transformation with stretch factor k.

    a) How does the area of a triangular region compare with the area of its image?

    b) How does the area of a circular region compare with the area of its image?

11. Consider a transformation $P(x, y) \rightarrow P'(2x, 3y)$ which has two different stretch factors.   For example, under this transformation $A(3, 5) \rightarrow A'(6, 15)$, $B(0, 2) \rightarrow B'(0, 6)$, and $C(-1, 0) \rightarrow C'(-2, 0)$. Construct the image of each figure in Exercise 1, p. 293, under this transformation.

12. On the basis of your observations in Exercise 11, which, if any, of the following are preserved under transformations of the form $P(x, y) \rightarrow P'(jx, ky)$, $j \neq k$?

    a) Distance   b) Angle measure   c) Parallelism

13. Are any figures invariant under transformations like those in Exercise 12 above? If so, which ones, and under what conditions?

⊙⊙ *Rigorous exercise*

14. Prove that parallelism is preserved under two-way stretch transformations ($k > 0$).

The Moebius Strip

Most surfaces have two sides. Around the middle of the 19th century, a German mathematician named Moebius discovered that there are surfaces with only *one* side. The simplest example of such a surface is called a *Moebius strip*.

To construct a Moebius strip, cut a strip of paper about 1 inch wide and 11 inches long. Turn one end over (that is, 180° twist) and then connect the ends with tape (Illus. 9.42).

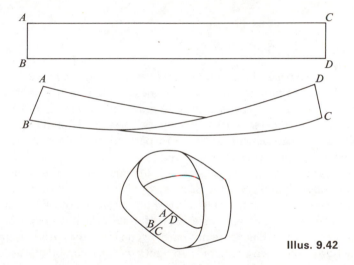

Illus. 9.42

EXERCISES

⊙⊙ 1. Make a Moebius strip. Draw a line along the center of the strip, continuing until you reach your starting point. Did you need to lift your pencil from the paper? What does this tell us about the number of sides of this surface?

2. Now cut the Moebius strip along the line drawn in Exercise 1 above. Describe the result.

3. Take another Moebius strip and make a mark one-third of the way in from an edge. Cut the strip parallel to the edge, continuing until you reach your starting point. [*Note:* Don't stop when you are *across* from your starting point.] Describe the result, comparing the widths and lengths of the loops.

4. Perform the same experiment as in Exercise 3 above, this time cutting along a line parallel to the edge and ¼ of the way in from the edge. Describe the result as in Exercise 3 above.

CHAPTER 10

SYSTEMS OF MEASURES

1. INTRODUCTION

The theoretical basis of measurement was established in Chapter 5. The application of that theory to practical problems of measurement is an important aspect of the daily activities of many people, including not only scientists and engineers but also housewives, plumbers, and storekeepers. An understanding of some of the problems of measurement is therefore particularly important for the teacher of mathematics or the teacher of science.

In this chapter we shall consider a bit of the history of measures and the problems of handling unit, or dimension, symbols. In the following chapter, errors in measurement will be considered. We shall consider not only measures of length, area, and volume, but also measures of time and mass, or weight. Although measures of time and mass are not purely geometric, they arise often in applications of measures and measurement. In fact, various combinations of measures of distance, area, volume, time, and mass are quite common in applications. We shall find that for applied problems of measures and measurement there are three kinds of measures that are fundamental, *length*, *time*, and *mass*. All of the other kinds are combinations of these.

2. EARLY UNITS OF MEASURE

In early civilizations accuracy in measurement was not too important and the choice of a unit was determined almost entirely on the basis of accessibility. Because of this, many of the earliest units used by man were related to parts of the human body.

Length

One of the earliest known units of length was the *cubit*, which was the distance from the elbow to the tip of the middle finger—about 18 inches. According to the Bible, Noah's Ark was supposed to have been 300 cubits long, 50 cubits wide, and 30 cubits high. One can only speculate as to the actual dimensions of such an ark in terms of today's units, since history does not record the size of Noah, the source of the unit.

The distance from a man's nose to the tip of his fingers is about two cubits and is the source of the unit *yard*. A *fathom* was originally designated as the distance from the fingertips of one hand to the fingertips of the other with the arms outstretched. Today, the fathom is still used as a unit of length and is officially designated as six feet. When a man's hand was spread, the distance between the thumb and the little finger was called a *span*, which is about 9 inches or half a cubit. The width of the hand itself was called a half-span or simply a *hand*. This unit of measure is still used today to measure the height of horses and is officially set at 4 inches.

The *inch* was originally thought of as the width of a man's thumb or the length of the index finger from the last joint to the tip of the finger. The expression "rule of thumb" originates from this early method of measuring with the thumb. The *foot* was, of course, the length of a man's foot, and the *yard*, as pointed out previously, was the distance from the tip of the nose to the tip of the fingers of one hand. It is interesting to note that these three familiar units originated independently of each other, and it was more or less by chance that the early units turned out to be approximately related as we think of them today. The word "inch" can be traced back to the Middle English word "inche," the Anglo Saxon "ynce," and finally to the Latin "uncia," meaning *twelfth part*. The word "yard" is derived from the Middle English "yerd" meaning *rod* or *staff*.

At the beginning of the fourteenth century, King Edward I of England attempted to standardize these units by having an iron bar made which he proclaimed to be the official yard. A foot was designated as $\frac{1}{3}$ of this length and an inch as $\frac{1}{36}$ of it. The stimulating effect this might have had on the commerce of the country was never realized, however, since the bar was inaccessible to the merchants and craftsmen who needed it. King Edward II recognized this and proclaimed the inch to be the length of "three barley corns, round and dry," placed end to end, thus sacrificing some uniformity for more accessibility.

Greater distances were also measured with units that were associated with the body. During the time of the Roman Empire, Roman soldiers used a *pace* or *double-step* as a unit of distance, and a large stone was placed at the end of a thousand paces to keep track of the distance they had traveled. Since a pace is approximately 5 feet, the distance between stones was approximately one mile, and the word "mile" is derived from the Latin "milia passuum," meaning "a thousand paces."

Weight

Early man also chose readily accessible objects to use as units of weight. One of the earliest units was a grain of wheat, since grains of wheat were of relatively uniform size. Although we no longer use the grain of wheat as a unit, the name persists and today we use the word "grain" to designate a unit of weight. For weighing gems or precious stones, a unit called a "carat" is usually used. This term very likely evolved from the *karob*, which was a small bean used by the Arabs as a unit of weight.

The Babylonians kept round polished stones as units of weight and today the word "stone" refers to a unit of weight most commonly used in England. A *stone* is officially set at 14 pounds. The word "pound" can be traced back to the Latin "pondo" which represented a unit of weight used by the Romans. The Roman pound was the predecessor of the *Troy pound*, which is equivalent to 12 ounces, and the word "ounce," like inch, is derived from the Latin "uncia," meaning *twelfth part*. The Troy pound was first used to check the value of gold and silver coins at the fairs in Troyes, France. Another pound unit was used during the same period and was referred to as *avoirdupois weight*. This pound is equivalent to 16 ounces and is the one most commonly used today.

Time

Due to the nature of the solar system, two of our fundamental units of time, the year and the day, are more or less naturally determined. The Chaldean astronomers in Babylonia were among the first to compute the length of a year in days, with surprising accuracy. Despite this, some early Roman calendars recorded only 304 days in a year until Julius Caesar changed it to 365 days with an additional day every fourth year. The sun is higher above the horizon at midday in the summer than it is in the winter. The ancient astronomers probably noted that the sun appeared to pass through a cycle, from a low point on the horizon to a high point and back again. The length of this complete cycle was taken as a *year*.

The word "month" is derived from the Middle English "moneth" or the Anglo Saxon "monath," both referring to the moon. Early man observed the phases of the moon and the time required for a complete cycle was called a *moon*, which later became a *month*. Since there were about 12 moons in a year, the year was divided into 12 months. Interestingly enough, the Egyptians divided the year into 11 months of 30 days each and one month of 35 days.

The beginning of the year varied considerably throughout Europe, and although Julius Caesar set the first of January as the beginning of the year, it was not until 1752 that England accepted this as a beginning date. The seven days of the week were derived from what the Romans considered to be the seven planets, Saturn, Jupiter, Mars, Venus, Mercury, the Sun, and the Moon.

The pattern of twelve months was carried on into the division of a day, and sunrise to sunset was divided into 12 "hours." Since the Babylonians used a numeration system with a base of 60, they divided an hour into 60 parts (minutes) and each of these parts into 60 smaller parts (seconds). The word "minute" is derived from the Latin "partes minuta prima," meaning "primary or first little parts," and "second" from "partes minuta secunda," meaning "second little parts."

3. STANDARDIZATION OF UNITS OF MEASURE

As we have seen, the primary requisite for a unit of measure in early times was its accessibility, and for this reason most of the early units evolved from parts of the human body. The obvious difficulty, however, was that the actual size of a unit varied from person to person, thus making commercial trade rather difficult. The kind of accuracy needed for careful scientific work was virtually impossible.

Although some attempts were made to maintain a set of standards, these were usually restricted to a small locality, and the amount of standardization of units of measure throughout a country was usually directly related to its political unity. As an example, in France, shortly before the French Revolution, there were over two hundred different sets of units of measure that had been established by feudal lords. England was less committed to a feudal system and her *Magna Carta* of 1215 stated in part that "There shall be one measure of wine and one of ale through our whole realm; and one measure of corn, that is to say the London quarter; and one breadth of dyed cloth . . . ; and it shall be of weights as it is of measures."

Shortly following the French Revolution, the National Assembly of France passed a law unifying the systems of measures, and a committee of scientists was entrusted with the task of establishing this system. The system they established was the *metric system* and the basic unit of length was the *meter*, which was defined to be one ten-millionth of the distance from the North Pole to the equator, along a meridian passing through Paris. Although their original measurement was wrong, it was the first step in establishing a system of measures which was later to be adopted on an international scale. The primary advantage of the metric system is that it is a decimal system in which subunits as well as multiple-units are related to the meter by powers of ten.

In 1872 the International Bureau of Standards was organized and the metric system was officially adopted by most of the civilized countries of the world. The United States adopted this as a legal (but not compulsory) system in 1866, and although the British yard is more commonly used, it is officially defined as 0.9143992 meters. The official meter is prescribed as the length between two scratches on a certain platinum-iridium bar (at a certain temperature) and is preserved by the International Bureau of Weights and Measures at Sèvres, France. A copy of this bar was sent to the United States in 1890 and is preserved by the United States Bureau of Standards in Washington, D.C. As spectroscopy was developed, it became possible to define length in terms of wavelengths of light, and the American physicist Michelson determined the length of one meter to be 1,553,164.13 wavelengths of a certain red radiation in the cadmium spectrum.

The most important units of mass now used in the United States are the British pound and the metric kilogram, both legal standards. The pound is designated as 0.453592428 kilograms and a copy of the standard kilogram is also preserved by the United States Bureau of Standards. The more common relations legalized by the United States are: 1 meter = 39.37 inches and 1 kilogram = 2.2046 pounds.

The standard unit of time is the *mean solar second* and there are 86,400 seconds in a *mean solar day*. The time interval between two successive transits of a fixed star, the second one being taken after the earth has rotated, is called a *sidereal day* and is 86,164.1 seconds. Standards of time in the United States are maintained by the United States Naval Observatory in Washington, D.C.

EXERCISES

1. See what you can find out about the following units of measure. (Dictionaries and encyclopedias may be good sources.)

   a) furlong   b) league   c) nautical mile   d) bolt (cloth)

   e) acre   f) em (or pica)   g) rod

2. a) What was the Julian calendar? The Gregorian calendar?

   b) What were the names of the 7th through 10th months in the old Roman calendar?

   c) When was the Gregorian calendar adopted in England? What happened when it was adopted?

3. What is a light-year?

REVIEW PRACTICE EXERCISES

1. Add.

   Example. $2x + 3x = (2 + 3)x = 5x$

   a) $7y + 2y$ b) $(2x^3 + 4) + (4x^3 + 5)$
   c) $(6x^2 + 3x) + 5x^2$ d) $3xy + 11xy$
   e) $4x^3 + 6x - 2 + 5x^3 - 6x$ f) $3x(y + 3) + 4x(y + 3)$

2. Multiply.

   Example. $(2x)(6x^2) = 2 \cdot 6 \cdot x \cdot x^2 = 12x^3$

   a) $(17y)(3y)$ b) $(3x)(2x^2)$
   c) $(5y^2)(8x)$ d) $(0x)(6xy)$
   e) $(-7yx^2)(3xy)(4y^3x)$ f) $(14xyz)(3y^2)(5zx)(-2xz)$

3. Perform the indicated operations and simplify.

   Example. $4x \cdot \dfrac{12y}{x} = \dfrac{4x \cdot 12y}{x} = 48y \cdot \dfrac{x}{x} = 48y$

   a) $93x \cdot \dfrac{y}{12x}$ b) $3y \cdot \dfrac{60x}{y} \cdot \dfrac{60z}{x}$

   c) $4x \cdot \dfrac{y}{12x} \cdot \dfrac{z}{3y}$ d) $\dfrac{3x}{y} \cdot \dfrac{r}{3x} \cdot \dfrac{60y}{s}$

   e) $7x^2 \cdot \dfrac{9y^2}{x^2}$ f) $9y^3 \cdot \dfrac{3x^3}{4z^3} \cdot \dfrac{7z^3}{6y^3}$

4. When algebraic expressions are associated with a number system such as the rational numbers, the operations performed on these expressions are based on properties similar to the number system.

   a) What properties are illustrated in Exercise 1?

   b) What properties are illustrated in Exercise 2?

   c) What properties are illustrated in Exercise 3?

4. THE BRITISH-AMERICAN SYSTEM OF MEASURES

The most commonly used system of measures in the United States today is taken from the British system and is referred to as the British-American System of Measures. The three fundamental units are the *yard*, *pound*, and *second*. All other units are defined in terms of these.

Length

The most common units of length used in this system and the relationships among them are as follows.

$$1 \text{ foot } = 12 \text{ inches}$$

$$1 \text{ yard } = 3 \text{ feet } = 36 \text{ inches}$$

$$1 \text{ mile } = 1760 \text{ yards } = 5280 \text{ feet.}$$

Conversion

In a theoretical development of measures we do not find it necessary to use more than one unit, whereas in experimental measurement it is convenient to have several, for example, inches, feet, yards, and miles. The existence and use of several different units poses the problem of converting from one to another. For example, if a measurement is given in feet we may on occasion wish to have it in inches, in which case we must make a conversion.

A simple conversion of units, such as changing from feet to inches, can be done by multiplying by 12, since there are 12 in. in 1 ft. To do the reverse, converting from inches to feet, one of course divides by 12. It is a common error in situations like these to divide rather than multiply or to multiply rather than divide. In more complex types of unit conversion, the problem of knowing whether to multiply or divide becomes a good deal more difficult, and finding the number by which to multiply or divide is more difficult still, unless some systematic procedure is used.

In the succeeding development a systematic procedure for accomplishing unit conversions will be developed. It is based on the notion that we may treat unit symbols as if they were numerals or variables, obeying the commutative, associative, and distributive laws, and manipulating them accordingly. The symbol

"3 in.," for example, will be treated in the same way that "$3x$" would be treated, as if "3 in." meant "3 multiplied by *in.*" No attempt will be made here to prove that this procedure is valid or to construct a mathematical system containing numbers and measures,* since it would take us beyond the scope of this book. The method outlined here, however, is in common use by scientists and engineers and, as will be seen, its use simplifies many problems that are difficult or virtually impossible under methods traditionally used in elementary and secondary schools. The first few examples involve algebraic-type substitutions.

Example 1. Convert 4 ft to inches.

Since

$$1 \text{ ft} = 12 \text{ in.}$$

$$4 \text{ ft} = 4 \cdot (12 \text{ in.}) \qquad \text{(substituting "12 in." for "1 ft")}$$

$$= 48 \text{ in.} \qquad \text{(simplifying).}$$

Example 2. Convert 93 in. to feet.

Since

$$1 \text{ ft} = 12 \text{ in.}$$

$$\tfrac{1}{12} \text{ ft} = 1 \text{ in.} \qquad \text{(dividing by 12)}$$

$$93 \text{ in.} = 93 \cdot \tfrac{1}{12} \text{ ft} \qquad \text{(substituting)}$$

$$= \tfrac{93}{12} \text{ ft} \qquad \text{(simplifying).}$$

Example 3. Convert 374 in. to yards.

$$374 \text{ in.} = 374 \cdot \tfrac{1}{36} \text{ yd} = \tfrac{374}{36} \text{ yd} = 10\tfrac{7}{18} \text{ yd.}$$

These same conversions may be accomplished by means of a somewhat different technique, involving the use of the "multiplicative identity." Since 1 ft = 12 in., the symbol

$$\frac{1 \text{ ft}}{12 \text{ in.}}$$

acts like the number 1, as does the symbol

$$\frac{12 \text{ in.}}{1 \text{ ft}}.$$

---

* It is possible to construct a formal mathematical system in which the real numbers and units of measure are combined. Such a system has properties very similar to those of a *field*.

Example 4. Convert 4 ft to in.

$$4 \text{ ft} = 4 \text{ ft} \cdot \frac{12 \text{ in.}}{1 \text{ ft}} \qquad \text{(``multiplying'' by the identity)}$$

$$= \frac{4 \cdot 12 \text{ in.-ft}}{1 \text{ ft}}$$

$$= \frac{4 \cdot 12 \text{ in.}}{1} \cdot \frac{\text{ft}}{\text{ft}}$$

$$= 48 \text{ in.}$$

Example 5. Convert 93 in. to feet.

$$93 \text{ in.} = 93 \text{ in.} \cdot \frac{1 \text{ ft}}{12 \text{ in.}}$$

$$= \frac{93 \cdot 1 \text{ ft-in.}}{12 \text{ in.}}$$

$$= \frac{93 \text{ ft}}{12} \cdot \frac{\text{in.}}{\text{in.}} = \frac{93}{12} \text{ ft.}$$

In applications one is often required to add measures. It is easy to determine valid ways of manipulating symbols if we remember that the numerals and unit symbols obey the distributive law. The following examples illustrate.

Example 6. Add 4 ft and 3 ft.

$$4 \text{ ft} + 3 \text{ ft} = (4 + 3) \text{ ft} \qquad \text{(applying the distributive law)}$$

$$= 7 \text{ ft.}$$

Example 7. Add 5 ft and 8 in.

We write

$$\text{``5 ft} + 8 \text{ in.''} \quad \text{or} \quad 5'8''. \tag{a}$$

If these symbols are not the kind most useful in the particular application, then the distributive law can be applied in one of two ways.

$$5 \text{ ft} \cdot \frac{12 \text{ in.}}{1 \text{ ft}} + 8 \text{ in.} = 60 \text{ in.} + 8 \text{ in.} \tag{b}$$

$$= (60 + 8) \text{ in.}$$

$$= 68 \text{ in.}$$

$$5 \text{ ft} + 8 \text{ in.} \cdot \frac{1 \text{ ft}}{12 \text{ in.}} = 5 \text{ ft} + \tfrac{2}{3} \text{ ft} \tag{c}$$

$$= (5 + \tfrac{2}{3}) \text{ ft}$$

$$= 5\tfrac{2}{3} \text{ ft.}$$

For many purposes one of the symbols "5 ft + 8 in." or 5′8″ is simpler and more useful than either of those obtained in parts (b) and (c) of the above example. In expressing a person's height, for example, 5′8″ is ordinarily preferred to either 68″ or $5\tfrac{2}{3}$ ft. A statement often heard in the schoolroom, to the effect that "one cannot add feet and inches," is completely false. When we write "5 ft + 8 in.," we have added, in the same way that when we write "$3 + \sqrt{2}$," we have added 3 and $\sqrt{2}$. It is not possible in the latter case to use the distributive law to simplify, but the addition has been performed. We can use the distributive law in the case of "5 ft + 8 in." if we first make a conversion of units. The statement "one cannot add feet and inches" might be considered valid, provided we agree that it means that we cannot do any combining by the use of the distributive law unless we first make a conversion of units.

EXERCISES

1. Perform the indicated conversions as in Example 1 above.
   a) 15 ft = ＿＿in.
   b) 62 yd = ＿＿ft
   c) 8 in. = ＿＿ft
   d) 17 yd = ＿＿in.
   e) 3 mi = ＿＿yd
   f) 440 yd = ＿＿mi
   g) 52 ft = ＿＿yd
   h) 144 in. = ＿＿yd

2. Perform the indicated conversions as in Example 4 above.
   a) 100,000 ft = ＿＿mi
   b) 10 mi = ＿＿ft
   c) 62.5 ft = ＿＿yd
   d) $14\tfrac{5}{6}$ yd = ＿＿in.
   e) $1\tfrac{1}{8}$ mi = ＿＿furlongs
   f) 728 in. = ＿＿yd
   g) 40 rods = ＿＿ft
   h) 1 mi = ＿＿in.

3. Add and simplify as in the above example, showing how the distributive law applies. Do not make any unit conversions.
   a) 7 yd and 5 yd
   b) 6 mi 10 ft and 3 mi
   c) 3 mi and 44 mi
   d) 6 ft and 37 ft

4. Add and simplify, making unit conversions so that the distributive law may be used.
   a) 4 yd and 12 ft
   b) 5 mi and 440 yd
   c) 35 yd and 72 in.
   d) 3 mi and 88 ft

5. If 1 cubit = 18 in., what were the dimensions of Noah's Ark?

6. Mount Everest is 29,028 ft above sea level. What is its height in miles?

Area

The measure of a two-dimensional region is, of course, called its *area*. The most commonly used type of area unit is a square region whose sides have linear measure 1. If the linear unit is an inch, then the corresponding area unit is called a square inch, abbreviated sq in. If the linear unit is a yard, then the corresponding area unit is called a *square yard*, abbreviated sq yd. These area units have other abbreviations that turn out to be more useful, as we shall see.

    Measures can be added, as we have seen. They can also be multiplied and, as for addition, the numerals and unit symbols can be manipulated as if they were all numerals or variables referring to numbers. Let us find the area of a square inch, remembering that the area of a rectangular region may be found by multiplying length by width. In this case (Illus. 10.1) length and width are both 1 in., so the area is found as follows:

$$A = l \cdot w = (1 \text{ in.}) \cdot (1 \text{ in.})$$
$$= (1 \cdot 1) \cdot (\text{in.} \cdot \text{in.}) \quad \text{(using commutativity and associativity)}$$
$$= 1 \text{ in.}^2.$$

The symbol in^2 can thus be used in place of sq in. and is generally more useful. Similarly the symbols ft^2 and yd^2 can be used in place of sq ft and sq yd, respectively.

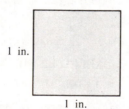

1 in.

1 in.   **Illus. 10.1**

    Let us now find some relationships between some of the commonly used units of area.

$$1 \text{ ft}^2 = (1 \text{ ft}) \cdot (1 \text{ ft}) \qquad\qquad 1 \text{ yd}^2 = (1 \text{ yd}) \cdot (1 \text{ yd})$$
$$= (12 \text{ in.}) \cdot (12 \text{ in.}) \qquad\qquad = (3 \text{ ft}) \cdot (3 \text{ ft})$$
$$= 144 \text{ in.}^2. \qquad\qquad\qquad = 9 \text{ ft}^2.$$

$$1 \text{ yd}^2 = (36 \text{ in.}) \cdot (36 \text{ in.})$$
$$= 1296 \text{ in.}^2.$$

There are two other units of area commonly used in the United States, the *acre* and the *township*. They are related as follows:

$$1 \text{ mi}^2 = 640 \text{ acres (abbreviated A)},$$
$$1 \text{ township (twp)} = 36 \text{ mi}^2.$$

To find areas, we may use available formulas, substituting measures in them.

Example 1. Find the area of a rectangle having length 8 ft and width 30 in.

$$A = l \cdot w = (8 \text{ ft}) \cdot (30 \text{ in.})$$
$$= (8 \cdot 30) \cdot (\text{ft-in.}) \qquad\qquad\qquad (a)$$
$$= 240 \text{ ft-in.}$$

In (a), different units were used for length and width, and the example shows that this is permissible. The only thing that may seem strange is that the unit of area is not a square region. The unit obtained would be called a "foot-inch" or an "inch-foot." This unit would be a rectangular region 1 in. wide and 1 ft long. If we wish to have the area in square feet, we may easily find it as follows.

$$A = l \cdot w = (8 \text{ ft}) \cdot (30 \text{ in.}) \qquad\qquad\qquad (b)$$
$$= (8 \text{ ft}) \cdot \left(30 \text{ in.} \cdot \frac{1\text{ft}}{12 \text{ in.}}\right)$$
$$= \frac{8 \cdot 30}{12} \text{ ft}^2$$
$$= 20 \text{ ft}^2.$$

Example 2. Find the area of a circle if a radius measures 4.2 yd.

$$A = \pi r^2 = \pi \cdot (4.2 \text{ yd})^2$$
$$= \pi \cdot (4.2 \text{ yd}) \cdot (4.2 \text{ yd})$$
$$\approx 3.14 \cdot (4.2 \text{ yd}) \cdot (4.2 \text{ yd})$$
$$\approx (3.14) \cdot (17.64) \text{ yd}^2.$$

In Example 2 the symbol $\approx$ means "is approximately equal to." This is appropriate since $\pi \neq 3.14$, and when 3.14 is used for π, an approximation is introduced.

Converting from one unit to another is often necessary when working with area units as it is with distance units. If one is not sure of the conversion factor, substitution is usually the best procedure, as illustrated in the following example.

Example 3. Convert 32 ft^2 to in^2.

$$32 \text{ ft}^2 = 32 \cdot (12 \text{ in.})^2 \qquad \text{(substituting "12 in." for "ft")}$$
$$= 32 \cdot (12 \text{ in.}) \cdot (12 \text{ in.})$$
$$= 32 \cdot 12 \cdot 12 \text{ in}^2$$
$$= 4608 \text{ in}^2.$$

EXERCISES

1. Perform the indicated conversions.
   a) 1008 in² = _____ft²
   b) 15 ft² = _____in²
   c) 3 yd² = _____in²
   d) 958 in² = _____yd²
   e) 17 yd² = _____ft²
   f) 163 ft² = _____yd²
   g) 1 mi² = _____yd²
   h) 1 mi² = _____ft²
   i) 644 in² = _____yd²
   j) 1 acre = _____ft²
   k) 1 acre = _____yd²
   l) 1 twp = _____A
   m) 200 A = _____mi²

2. A farmer owns a square region of land whose area is 160 A. What is the length, in miles, of each side of this square?

3. A realtor advertises a rectangular lot as having an area of 1 A. If the front of the lot is 150 ft long, how deep is the lot?

4. Suppose you wished to paper a wall which is 135 in. long and 93 in. high. What is the area of the wall in square feet? If a single roll is 20 in. wide and 24 ft long, how many rolls will you need? (Ignore excess needed for matching.)

5. A circular flowerbed has a radius of 6 ft. What is its area in square feet? (Use $\frac{22}{7}$ for π.)

6. Suppose a circular region concentric with that of Exercise 5 is marked off so that its area is one-half of the flowerbed. How long is a radius of the inner flowerbed?

7. What is the difference between the phrases "6 feet square" and "6 square feet"?

Volume

The measure of a three-dimensional region is, of course, called its *volume*. The most commonly used type of volume unit is a cubic region whose edges have linear measure 1. If such a unit has edges that measure 1 in., then the volume unit is called a *cubic inch*, abbreviated "cu in." or "in³." Let us now find some relationships between some of the commonly used volume units.

$$1 \text{ ft}^3 = 1 \text{ (12 in.)}^3 \qquad\qquad 1 \text{ yd}^3 = (3 \text{ ft})^3 = 27 \text{ ft}^3$$
$$= (12 \text{ in.}) \cdot (12 \text{ in.}) \cdot (12 \text{ in.}) \qquad = (36 \text{ in.})^3$$
$$= 12^3 \text{ in}^3 \qquad\qquad\qquad = 36^3 \text{ in}^3$$
$$= 1728 \text{ in}^3. \qquad\qquad\qquad = 46{,}656 \text{ in}^3.$$

Some of the other volume units used in commerce, possibly already familiar, are as follows.

Dry measure:
$$1 \text{ quart (qt)} = 2 \text{ pints (pt)},$$
$$1 \text{ peck (pk)} = 8 \text{ quarts},$$
$$1 \text{ bushel (bu)} = 4 \text{ pecks}.$$

Liquid measure:

$$1 \text{ pint (pt)} = 16 \text{ fluid ounces (fl oz)},$$
$$1 \text{ quart} = 2 \text{ pints},$$
$$1 \text{ gallon (gal)} = 4 \text{ quarts}.$$

It should be noted that a dry quart is *not* the same size as a liquid quart and, of course, a dry pint is different from a liquid pint. Both, however, are expressible in terms of cubic inches, 1 dry qt being 67.2 in^3 and 1 liquid qt being 58 in^3.

Other units of volume often used in the United States are the *board foot*, which is 1 ft by 1 ft by 1 in., the *barrel*, which is 31.5 gal, and the *dram* (fluid), which is $\frac{1}{8}$ of a fl oz.

Conversions can be accomplished as before, by manipulating unit names as if they were algebraic symbols.

EXERCISES

1. Perform the indicated conversions.

   a) $7 \text{ yd}^3 =$ _____ft^3 b) $3 \text{ ft}^3 =$ _____in^3

   c) $144 \text{ in}^3 =$ _____ft^3 d) $53 \text{ ft}^3 =$ _____yd^3

   e) $1 \text{ pt} =$ _____drams   f) $24 \text{ board ft} =$ _____in^3

   g) $6 \text{ oz} =$ _____qt   h) $7 \text{ qt} =$ _____pt

   i) $1 \text{ gal} =$ _____fl oz   j) $9 \text{ pk} =$ _____bu

   k) $2 \text{ barrels} =$ _____qt   l) $4\frac{1}{2} \text{ bu} =$ _____pk

   m) $17 \text{ gal} =$ _____pt   n) $37.5 \text{ bu} =$ _____qt

2. A cubic yard of gravel or dirt is often referred to as a "yard." How many yards of gravel would be needed to cover a driveway 3 in. thick which is 60 ft long and 9 ft wide? If a yard of gravel weighs $1\frac{1}{3}$ tons, how many tons would be needed?

3. A cardboard container is 22″ wide, 36″ long, and 33″ high. What is its capacity in cubic inches? in cubic feet? in cubic yards?

4. How many gallons of water would be required to fill a circular wading pool which is 18″ deep and has a radius of 30″?

Weight and Mass

The *mass* of an object refers to the amount of matter or material in the object, while the *weight* of the object refers to the force of gravity on it. Thus, if an astronaut weighs 150 pounds on the earth, he would weigh about 25 pounds on the moon, since the force of gravity on the moon is about $\frac{1}{6}$ of that on the earth. To further illustrate this concept, let us suppose our astronaut hangs on a spring scale as shown in Illus. 10.2. On earth, the gravitational force pulls him downward, stretching the spring, thus indicating a weight of 150 pounds. If he and the scale were transported to the moon, the gravitational force, being about $\frac{1}{6}$ that of the earth, would exert less force on the spring and his indicated weight would be about 25 pounds. And, of course, on the way to the moon, our astronaut (and his scale) would undergo a period of *weightlessness* when the gravitational force is zero.

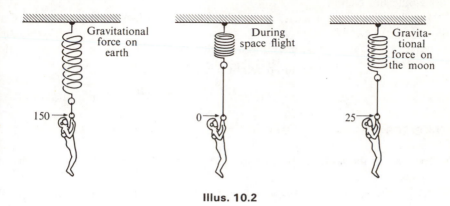

Illus. 10.2

The mass of the astronaut, on the other hand, remains essentially the same, whether on the earth, on the moon, or in space (Illus. 10.3). Thus, if we used a balance on earth to equate his mass with a multiple of standard units of mass, we would find that the scale also balanced on the moon.*

The basic unit of mass in the British-American system is the *pound*. Other units are ounces and tons. The basic unit of weight in this system is also called a *pound*, even though mass and weight are not the same thing. The pound of weight, or force, is defined to be the force of the gravity of the earth on a pound of mass at

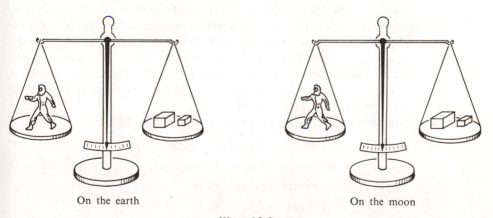

On the earth   On the moon

Illus. 10.3

---

* We are, of course, ignoring minor loss or gain of mass in the astronaut. Also, during the space flight it would be rather difficult to check the balance, since all masses would be weightless and would therefore exert no force on either arm of the balance.

standard sea level. As long as we remain reasonably near sea level on the earth, a pound of mass will weigh one pound. If we should move that mass onto a high mountain or to the moon, its weight would be less than a pound although its mass would not change.

The familiar units of mass, or weight, in the British-American system are as follows.

$$1 \text{ pound (lb)} = 16 \text{ oz}$$

$$1 \text{ hundredweight (cwt)} = 100 \text{ lb}$$

$$1 \text{ ton (T)} = 2000 \text{ lb.}$$

EXERCISES

1. The mass of the earth has been estimated to be 6×10^{21} tons. What is its mass in pounds?

2. The gravity on the moon is about $\frac{1}{6}$ that of the earth. How much would a 180-lb astronaut weigh on the moon?

3. The gravitational force of the earth is slightly less at the equator than it is at the poles. How would the weight of an object be affected as it is moved from the equator to the North Pole? How would the mass be affected?

5. THE METRIC SYSTEM

Relationships between various units of the British-American system are such that conversions are more difficult than they need be. The metric system, now in use in a large part of the world, and actually the legal system in the United States, is much more convenient in this regard since it is a decimal system. Units of different size have ratios that are powers of ten, and hence changing units amounts only to moving decimal points. It is possible that the metric system may become not only the legal U.S. system but the one actually used predominantly. This is one reason that it should be taught to every elementary school youngster today. Another reason is that in travel to foreign countries or in international commerce, a familiarity with the metric system is necessary.

The metric units of length and their relationships are as follows.

$$1 \textit{ kilo}\text{meter} \quad \text{(km)} = 10^3 \text{ meters (m)}$$

$$1 \textit{ hecto}\text{meter} \text{ (hm)} = 10^2 \text{ m}$$

$$1 \textit{ deka}\text{meter (dkm)} = 10 \text{ m}$$

$$1 \textit{ deci}\text{meter} \quad \text{(dm)} = \tfrac{1}{10} \text{ m}$$

$$1 \textit{ centi}\text{meter} \quad \text{(cm)} = \tfrac{1}{100} \text{ m}$$

$$1 \textit{ milli}\text{meter} \text{ (mm)} = \tfrac{1}{1000} \text{ m}$$

The metric prefixes, italicized above, should be memorized, since they are used for identifying other units, such as mass and time. The basic unit of mass is the *gram*, and a table like that above could be made for mass. One *kilogram*, for example, is 10^3 grams. Two other metric prefixes that find occasional use are *mega*, 10^6 or one million, and *micro*, 10^{-6} or one millionth. A chart like the following, which is similar to a place-value chart for numeration, may be of help in determining what conversion factor is to be used.

10^3	10^2	10		10^{-1}	10^{-2}	10^{-3}
km	hm	dkm	m	dm	cm	mm

Each position in the chart has a value ten times that to the right of it or one-tenth that to the left of it. If one were to convert from hectometers to decimeters, this would represent a move of three places to the right and, hence, a conversion factor of 10^3 would be used. That is, to change from hectometers to decimeters we multiply by 10^3. To change from centimeters to kilometers would represent a move of five places to the left. Hence we would multiply by 10^{-5}.

Another aid, which does not depend on the chart above, consists of multiplying successively by the identity, as in the following.

Example. Convert 3 dkm to centimeters.

$$3 \text{ dkm} = 3 \text{ dkm} \cdot \frac{10 \text{ m}}{1 \text{ dkm}} \cdot \frac{100 \text{ cm}}{1 \text{ m}}$$

$$= 3 \cdot 10 \cdot 100 \, \frac{\text{dkm}}{\text{dkm}} \cdot \frac{\text{m}}{\text{m}} \cdot \text{cm}$$

$$= 3 \times 10^3 \text{ cm}.$$

Note that in the above example, the first step was to convert to meters and the second to convert to cm. This allows us to use the prefix table on p. 314 directly for each step in the conversion.

Either of the procedures outlined above may be helpful to the novice, but neither is recommended for use routinely. As one gains experience, it is better not to rely on procedures as complex as this.

EXERCISES

1. Perform the indicated conversions.

a) 10 meters = ____dekameters   b) 100 meters = ____dekameters

c) 1 hectometer = ____dekameters   d) 1 kilometer = ____hectometers

e) ____dekameters = 1 kilometer   f) 247 dm = ____dkm

g) 105 mm = ____cm   h) 60dm = ____m

i) 10 mm = ____dm   j) 25 mm = ____cm
k) 54 dm = ____mm

2. Complete the following.

a) $1 \ m^2$ = ____dm^2 = ____cm^2 = ____mm^2

b) $1 \ km^2$ = ____hm^2 = ____dkm^2 = ____m^2

c) 1 cu m = ____cu dm = ____cu cm = ____cu mm

d) 1 mg = ____gm   e) 1 cg = ____gm   f) 1 dg = ____gm

g) 1 kg = ____hg   h) 1 hg = ____dkg   i) 1 dkg = ____gm

3. Perform the indicated conversions.

a) $4 \ m^2$ = ____cm^2 b) $650 \ m^2$ = ____km^2 c) $4 \ dkm^2$ = ____cm^2

d) $72 \ mm^2$ = ____dm^2 e) $42.07 \ cm^2$ = ____mm^2 f) $1 \ m^3$ = ____liter

g) 230 gm = ____kg   h) 22 mg = ____gm

6. RELATIONSHIPS BETWEEN METRIC AND BRITISH-AMERICAN UNITS

Since both systems of measures are used in the United States, it is often necessary to change from one system to the other. Here are the basic relationships between the two systems.

1 inch = 2.54 centimeters   1 liter (ℓ) = 1.06 quarts
1 meter = 39.37 inches   1 kilogram = 2.2 pounds
1 mile = 1.61 kilometers

While some of the relationships above are approximate, they are nevertheless sufficiently accurate for most purposes. It should also be noted that one linear relationship would have been sufficient; it is, however, more convenient to remember the three above and derive other relationships as they are needed.

Another unit of linear measure appropriate to mention here is the *nautical mile*. It is most commonly used by ships and is defined as 1 minute of longitude at the equator; that is, 1/21,600 the distance around the equator. The mile used on land is called the *statute mile*. The nautical mile is related to the statute mile and the kilometer as follows:

1 nautical mile = 1.1508 statute miles
1 nautical mile = 1.853 kilometers.

EXERCISES

1. Perform the indicated conversions.

a) 1 km = ____mi   b) 1 m = ____yd   c) 1 qt = ____ℓ

d) 1 lb = ____kg   e) ____ℓ = 1 gal   f) 1 gm = ____oz

g) 1 yd = ____m   h) 1 cm = ____in.   i) 1 oz = ____gm

j) 1 mi = ____nautical mi   k) 1 nautical mi = ____yd

2. Two artillery pieces used by the United States are the 105-mm and the 155-mm Howitzers. "105-mm" and "155-mm" refer to the diameters of their bores. Find the diameters of their bores in inches.

3. Some companies specify the net weight (mass) of an item in both the metric and British-American systems. If a can of soup weighs 298 grams, what is its weight in ounces?

4. Length and mass are given in metric units at the Olympic games. Convert each of the following to British-American units.

   a) 100-m dash—to yards

   b) 2.229-m high jump—to ft and in. (combined)

   c) 20-km long-distance run—to mi

   d) 151.4-kg press (weightlifting)—to lb

7. COMBINATIONS OF UNITS

Area and volume units have already been described as being combinations of linear units. There are many other kinds of combinations of units, and many of them are found in everyday use. Some examples are *miles per hour, dollars per day, bushels per acre, passenger-miles,* and *man-hours.* Units such as these arise from multiplications or divisions. Again, if the unit symbols are treated as though they were numerals or variables, ways of manipulating them can be found easily.

Example 1.

$$\frac{135 \text{ mi}}{3 \text{ hr}} = \frac{135}{3} \frac{\text{mi}}{\text{hr}} = 45 \frac{\text{mi}}{\text{hr}}.$$

Example 1 illustrates a means of finding *average speed*: dividing the distance by the time. The unit of speed obtained is the quotient of the distance unit and the time unit. Of course, there are many speed units that can be derived in this fashion. There is one that has a special name; the *knot* is a nautical mile per hour.

Example 2.

$$3 \text{ men} \cdot 8 \text{ hours} = 3 \cdot 8 \text{ man-hours}$$
$$= 24 \text{ man-hours}.$$

Example 2 illustrates how multiplication gives rise to a complex unit. If 3 men work together for 8 hours, we say that there have been 24 man-hours of work. Again, the unit name is obtained by algebraic manipulation of the unit symbols.

Conversions of Units

To convert from one complex unit to another is not difficult if we remember to multiply successively by the identity, as in the following example.

Example 3. Convert 60 mi/hr to ft/sec.

$$60 \frac{\text{mi}}{\text{hr}} = 60 \frac{\text{mi}}{\text{hr}} \cdot \frac{1 \text{ hr}}{60 \text{ min}} \cdot \frac{1 \text{ min}}{60 \text{ sec}} \cdot \frac{5280 \text{ ft}}{1 \text{ mi}}$$

$$= \frac{60 \cdot 5280}{60 \cdot 60} \frac{\text{mi}}{\text{mi}} \cdot \frac{\text{hr}}{\text{hr}} \cdot \frac{\text{min}}{\text{min}} \cdot \frac{\text{ft}}{\text{sec}}$$

$$= 88 \frac{\text{ft}}{\text{sec}} .$$

EXERCISES

1. A farmer planted 23 acres of corn and harvested 750 bushels. What was his yield in bushels per acre?

2. A TV network charges $276,000 for 46 one-minute commercial time slots. What is the advertiser's average cost in dollars per minute?

3. A small plane flew $5\frac{1}{2}$ hours and used 33 gallons of gasoline. What was the average consumption of gasoline in gallons per hour?

4. If a cubic decimeter of gold weighs 19.3 kg, what is its density in grams per cubic centimeter?

5. Perform the indicated conversions.

a) $40 \frac{\text{mi}}{\text{hr}} = $ _____ $\frac{\text{ft}}{\text{sec}}$

b) $3 \frac{\text{gal}}{\text{min}} = $ _____ $\frac{\text{gal}}{\text{day}}$

c) $103 \frac{\text{ft}}{\text{sec}} = $ _____ $\frac{\text{mi}}{\text{hr}}$

d) $4 \frac{\text{oz}}{\text{pt}} = $ _____ $\frac{\text{oz}}{\text{gal}}$

e) 28 knots $= $ _____ $\frac{\text{mi}}{\text{hr}}$

f) $4 \frac{\text{grams}}{\text{cm}^3} = $ _____ $\frac{\text{kg}}{\text{liter}}$

g) $3.7 \frac{\text{yd}}{\text{min}} = $ _____ $\frac{\text{yd}}{\text{day}}$

h) $56 \frac{\text{mi}}{\text{hr}} = $ _____ knots

i) $400 \frac{\text{ft}^2}{\text{gal}} = $ _____ $\frac{\text{yd}^2}{\text{gal}}$

j) $420 \frac{\text{lbs}}{\text{ft}^2} = $ _____ $\frac{\text{oz}}{\text{in}^2}$

6. One company sells an orange juice drink for $2.00 per gal, while another sells 48 oz of a comparable drink for $0.70. Which is the better buy?

7. A manufacturer recommends a mixture of 1 pt of sizing for every 10 lb of paste. How many fluid oz would be mixed with $\frac{1}{2}$ lb of paste?

8. INDIRECT MEASUREMENT

To find the area of a rectangular floor, one usually measures the length and the width and then multiplies these measures. The length and width are determined *directly* through the use of a ruler or other linear measuring device. The area is not determined directly, however, since it is obtained from a pair of linear measurements. To determine the area directly, one would have to cover the floor with square regions and then count them—a rather difficult and impractical process. In this example the length and width of the floor were determined by *direct measurement*, and the area by *indirect measurement*. In general, when a measure is determined by computing with other measures, it is called indirect measurement.

Average speed is sometimes determined by indirect measurement. During the time trials at the Indianapolis 500-mile race each year, the average speed is determined from the amount of time required to complete the $2\frac{1}{2}$ mile course. If, for example, a driver completes the $2\frac{1}{2}$ miles in 58.1 sec, his average speed would be determined as follows:

$$\frac{2.5 \text{ mi}}{58.1 \text{ sec}} \cdot \frac{3600 \text{ sec}}{1 \text{ hr}} \approx 155 \frac{\text{mi}}{\text{hr}}.$$

Speed may also be measured directly, using a speedometer, which determines speed without computation. Similarly, many other measuring instruments such as altimeters, thermometers, pressure gauges, etc., have been developed to determine measures directly which might otherwise be difficult or even impossible to determine by indirect measurement.

Occasionally it is necessary to determine length indirectly. For example, one may wish to determine the height of a building or the width of a river. This type of indirect measurement involves trigonometric concepts and was discussed in Chapter 6.

EXERCISES

1. How many cubic feet of air space are there in a room whose dimensions are 20 ft by 12 ft by 8 ft?

2. What is the volume of a cylindrical gas tank whose length is 25 ft and whose radius is 5 ft?

3. The interior of a freezer is 16″ by 30″ by 4′. What is its capacity in cubic feet?

4. How many cubic meters of helium are needed to fill a spherical balloon whose radius is 15 m?

5. A cylindrical sports arena is built with a hemisphere for a roof. If the radius of the cylinder and sphere is 100 ft and the height of the cylinder is 30 ft, what is the volume of the arena in cu ft?

6. If the diameter of a golf ball is 4 cm, what is its volume in cu cm?

7. A sphere is submerged in a cylinder of water and the water rises $\frac{1}{4}″$. If the radius of the cylinder is 7″, what is the radius of the sphere?

8. The Great Pyramid was originally 768 ft square and had a height of 482 ft. How many cubic feet of stone were required to build it?

9. Assuming the earth is a sphere whose radius is 3900 mi, what is its volume in cubic feet?

10. A manufacturer wishes to make a cylindrical container with marks to indicate the volume of liquid in the container. If the radius of the container is 4 mm, what should the distance between marks be for an increase of 1 cm³?

REVIEW PRACTICE EXERCISES

In Exercises 1 through 6, multiply and combine similar terms.

Example

$$
\begin{aligned}
(x + 3)(2x - 5) &= x(2x - 5) + 3(2x - 5) \\
&= 2x^2 - 5x + 6x - 15 \\
&= 2x^2 + x - 15
\end{aligned}
$$

1. $(x - 7)(x + 4)$

2. $(3x + 2)(x + 6)$

3. $(y - 9)(\frac{1}{2}y + 3)$

4. $(x + y)(x + 2y)$

5. $(3x^2 + 2)(x - 4)$

6. $(xy - 6)(xy + 3)$

Solve the equations in Exercises 7 through 12.

7. $3x - 6 = 0$

8. $2(x + 5) - 8 = 0$

9. $3y - 5 = 4y + 9$

10. $\frac{1}{2}y + 3y - 6 = 15$

11. $4 - x^2 = 0$

12. $ax + b = 0$

Solve for the given variable in Exercises 13 through 18.

13. $3x + 3y - 6 = 0; y$

14. $2 = ab + 3b; b$

15. $2 = ab + 3b; a$

16. $3c + 2a = 2ac - 3; a$

17. $a = bc + cd; d$

18. $a = bc + xy; c$

CHAPTER 11

ERRORS IN MEASUREMENT AND CALCULATION WITH APPROXIMATE DATA

In previous chapters we considered a theoretical development of measures. We also began to consider practical problems involved in the experimental determination of measures of real-world objects like, for example in Chapter 6, the determination of measures from trigonometric tables.

1. EXPERIMENTALLY DETERMINED MEASURES ARE APPROXIMATE

The use of trigonometric tables illustrates quite clearly that certain experimentally determined measures are approximate. Such tables are always rounded to a certain number of digits; hence any measure determined by the use of the tables is approximate, not exact. It is perhaps not quite so clear that *all* measures are inexact, whether determined with the help of tables or not. To illustrate, let us consider a very simple kind of experimental measurement—the determination of length with a ruler. In Illus. 11.1 a segment is being measured with a ruler, and it appears that its measure is between 3 and $3\frac{1}{4}$. In fact, from the drawing it appears that the measure might be very nearly $3\frac{1}{8}$. If the instrument had eighth-units marked on it, we would be better able to tell. In that event, the endpoint of the segment might be seen to be to the right or the left of the $3\frac{1}{8}$ mark. Then we would perhaps wish to have sixteenth-units marked on the instrument. If we should continue to mark smaller and smaller subunits on the ruler, eventually the endpoint of the segment being measured would appear to fall "exactly" on one of the calibration marks, say $3\frac{3}{16}$. We might be tempted to say that the measure of the segment is "exactly $3\frac{3}{16}$." A moment's reflection shows, however, that this is not correct. The marks on the ruler must have some thickness or they could not be seen. We should thus ask ourselves, "Just where on the calibration mark does the endpoint of the segment fall?" Is it in the middle of the mark, near one edge, or where? Using a magnifier might help to answer such a question, and if that were done, we might then be tempted to say that the measurement was exact. But the same problem would remain, because a more powerful magnifier could be used. The most one can ever hope to accomplish in experimental measurement is a narrowing of the limits. If we use a very precise instrument, we should be able to approximate a measure within quite narrow limits, but in no case can we find a measure exactly. All measurements are thus approximations.

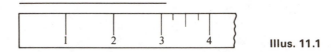

Illus. 11.1

It is not only impossible to find an exact measure of a physical object, no such exact measure even exists! Would a twig have an "exact" length, for example? We would have to decide which points of it were the endpoints, and our judgment in choosing these points would be open to question. At least, no two people would do it in exactly the same way. A similar dilemma exists with respect to the line segment of Illus. 11.1. If it is viewed with a microscope, it will be seen not to have a definite endpoint. Rather, the ink marks will be seen to have a jagged edge. Thus, exact measures do not exist for physical objects, and we could not find them if they did.

Since all measurements are approximations, every measurement involves some error. It is important not only to be aware of this but also to have some means of handling the approximate data sensibly. The remainder of this chapter is devoted to a consideration of the latter.

2. HOW GOOD IS A MEASUREMENT?

Whenever a measurement is made, some instrument must be used. There are, of course, many kinds of measuring instruments, and when we ask, "How good is a measurement?" the first thing we must do is consider the instrument. There are several things we might wish to know about it.

Least Count

An instrument must be calibrated. That is, it must have marks of some kind on it to indicate units and subunits. Some instruments have smaller subunits marked than others and, hence, would seem to be better instruments. Although this is not necessarily so, it is one criterion for the quality of an instrument. The instrument with the smaller subunits marked on it is better, provided the instruments are comparable in other respects.

The smallest subunit marked on a measuring instrument is known as its *least count*. The least count of the ruler in Illus. 11.1 is $\frac{1}{4}$ in. If eighth-units were marked on it, then its least count would be $\frac{1}{8}$ in.

Greatest Possible Error

Most instruments can be read to the nearest calibration mark at least. In Illus. 11.2, for example, the endpoint of a segment being measured is nearest the calibration mark for $1\frac{1}{4}$. If the endpoint were nearer the calibration mark for 1 or for $1\frac{1}{2}$, it would be apparent. Thus we can be sure that the endpoint lies in the interval between $1\frac{1}{8}$ and $1\frac{3}{8}$.

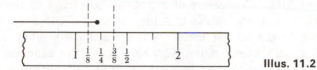

Illus. 11.2

From the situation shown in Illus. 11.2 it is apparent that the maximum error that should occur in the measurement is $\frac{1}{8}$ in. Note that the least count of the instrument is $\frac{1}{4}$ in. The maximum error is half the least count of the instrument. This is true of many measuring instruments and gives rise to the definition of the purely mathematical term *greatest possible error* (g.p.e.). As we shall see, this is something of a misnomer. We shall define it as follows: The "greatest possible error" of a measurement is half the least count of the instrument used.*

Precision

If one measuring instrument has a smaller *least count* than another, then it will usually give us finer measurements. If it is also a good instrument, giving consistent results, then we would say that the instrument with the smaller least count is more *precise*.

The word precision is used with various and sometimes contradictory meanings. We define it here in a manner to conform with the most general scientific usage. The *precision* of a measurement refers to the narrowness of the limits between which the measurement can be made consistently. If one measurement is made to the nearest tenth of an inch and another to the nearest hundredth of an inch, then the second measurement is more precise, *provided* it can be repeated to consistently obtain a result within the same narrow limits. If, on the other hand, the instrument that measures to the nearest hundredth of an inch is not consistent, perhaps giving different readings as the temperature of it or the force on it changes, then the second instrument might be less precise than the first. The ability of the measurer is also a factor here. It may be that an expert using impeccable technique would get more precise results with a less precise instrument than would a novice using a better instrument.

---

* For some measurements the "greatest possible error" is not half the least count of the instrument. We use the g.p.e. unless some better information is available. Our definition is thus really an oversimplification. It will serve the present purpose, however.

An instrument with a small least count, however, is usually more precise than an instrument with a large least count. If the instruments are comparable in other respects, and they are used with comparable technique, then the least count is a sufficient criterion for deciding which instrument is more precise.

The precision of a measurement may be indicated in several ways. Whenever subunits are used, this is fairly easy. A measurement of 12.43, for example, is understood to be precise to the nearest hundredth, the g.p.e. being half a hundredth, or 0.005. A measure of 215.60 is also precise to the nearest hundredth. The final "0" indicates this fact. A measurement of 215.6 would be understood to be precise to the nearest tenth. When fractional symbols are used, a similar device is often (but not universally) employed. For example, a measurement of $8\frac{1}{2}$ would be understood to be to the nearest half-unit, whereas a measurement of $8\frac{5}{10}$ is precise to the nearest tenth-unit. If subunits are not used, it is sometimes difficult to tell about the precision. A measurement of 12,000, for example, is precise to the nearest thousand, at least, since the thousands digit is not zero. If this measure is the population of a city, it may, in fact, be no more precise than this. On the other hand, the measure might be more precise than the nearest thousand. If, for example, the measure is the number of seats in a stadium, it might be precise to the nearest ten. When 0's occur as they do here, one may have no way of knowing how precise the measurement is. Another often-used device can be illustrated as follows: A measurement of 18.35 ± 0.001 indicates that the measurement was, and can consistently be, obtained to be a number between 18.351 and 18.349. The 0.001 in this case may be the g.p.e., as defined here, or some statistically determined number.

EXERCISES

1. Find the least count and the g.p.e. of the following.
   a) A ruler marked with half-units, quarter-units, eighth-units, and sixteenth-units.
   b) A ruler marked in centimeters and millimeters.
   c) A protractor marked in degrees.

2. Find the g.p.e. of each of these measurements.
   a) 4.01 ft   b) 14.0 cm   c) 14 cm
   d) 2.1000 yd   e) 2.10 yd   f) 2.1 yd

3. Within what limits would each of these measurements lie?
   a) 34.12 ± 0.01 lb   b) 34.12 ± 0.006 lb   c) 4.01 km
   d) 14.0 mi   e) 14 mi   f) 263 ft

4. A ruler marked in centimeters and millimeters has 2 mm worn off its end but is otherwise correct.
   a) What is the g.p.e. of a measurement made with it?

   b) What is the actual maximum possible error, assuming the measurer is not aware of the shortened end and uses the instrument correctly otherwise?

5. A stopwatch with a least count of 0.01 sec gains 3 sec/min. Another watch with a least count of 1 sec keeps correct time.

   a) If a measurement of 3 min 41.02 sec is made with the first watch, about what reading would be shown on the second watch, making the same measurement?

   b) Find the g.p.e. of both measurements.

   c) What is the actual maximum error in the two measurements, assuming the technique of measurement introduced no error?

   d) Assuming both watches give consistent results, which measurement is more precise?

   e) Which measurement is more correct?

Accuracy

The word *accuracy* refers to the correctness of a measurement. Although this word is used with various meanings, the meaning used here is that which conforms best to its scientific usage. A measurement may be very precise and yet be inaccurate. For example, an electric meter that gives consistent readings and has a small least count would be precise. If, however, it has been bent or otherwise damaged, its precise readings may be incorrect. The instrument is not very *accurate*, even though it is quite precise.

Relative Error

An error of 1 ft is not usually a serious error in a measurement of several miles. An error of 1 ft in a measurement of several yards, however, is apt to be an important error. In most situations we strive for an error that is small relative to the measurement itself, and this gives rise to the mathematical concept of relative error, defined as follows:

$$\text{Relative error} = \frac{\text{g.p.e.}}{\text{measurement}}.$$

Relative error is thus the ratio of the g.p.e. to the measurement itself and is often expressed in percent of error.

Example. If there is a maximum error of 0.1 m in a measurement of 3.75 m, what is the relative error in percent?

$$\text{Relative error} = \frac{0.1}{3.75} = 2\tfrac{2}{3}\%.$$

   The concept of relative error is, of course, a mathematical one and tells us nothing about the correctness of the instruments used or the validity of the technique used. However, if instruments are not faulty and they are used with good technique, then the relative error is a good measure of accuracy.* In this case a measurement with a small relative error is more accurate than a measurement with a large relative error.

EXERCISES

1. Find the relative error of each measurement of Exercise 2, page 324.

2. Find the relative error of the two measurements of Exercise 5, page 325.

3. Suppose that two measurements of the same quantity have different precision but the instruments and techniques are comparably valid. Is the more precise measurement also more accurate? Why?

3. CALCULATING WITH APPROXIMATE DATA

It is now clear that numbers resulting from measurement are approximations, never exact. So far we have not concerned ourselves, however, with indirect measurements, in which the final result is obtained as a result of some direct measurement and some calculation. Most measurements are actually indirect in this sense. For example, the finding of an area ordinarily involves calculating and is thus indirect. We measure directly the length and the width of a rectangle and then obtain the area by multiplying.

Since in indirect measurement we calculate with numbers that are approximations, we obtain approximate results. It is important to have some idea of how good a measurement is, be it direct or indirect, and thus it is important to consider how good an approximation is when it results from a calculation that uses approximate data.

Not all approximations result from measurement, of course. We often use a rational-number approximation to the irrational number π or the irrational number $\sqrt{2}$. While these approximations do not result from measurement, the result of a calculation with such approximations is also approximate, just as though the approximations did result from measurement. Thus, in the next sections we shall discuss calculating with approximate data, and what we say will apply to measurements as well as to any other approximate data.

---

* In many elementary books the words *precision* and *accuracy* are not used as they are here. They are both given purely mathematical meanings. In fact, the word *accuracy* is often defined to mean what we here call *relative error*. The authors of this book have chosen to use words this way, not only because it is prevalent among scientists, but also in order that the vocabulary force some attention to the possibility that mistakes can be made in measuring or that instruments can be faulty. When all of the words used are given purely mathematical definitions, it is difficult even to talk about errors due to faulty technique or faulty instruments and the student may (and often does) miss the important fact that there is something to measurement besides reading instruments and calculating.

Addition

Suppose we wish to add two numbers that have resulted from approximations. In the first place it is important to have some idea of the size of the error in each addend, and then we shall wish to know something about the error that consequently occurs in the sum.

    Let us represent two numbers to be added as "a" and "b," and the error of each "Δa" and "Δb" (read "delta a" and "delta b"), respectively. The "true"* numbers are, respectively, $(a + \Delta a)$ and $(b + \Delta b)$. (Note that if a is in error by being too small, then Δa is a positive number. If a is too large, then Δa is negative.) The "true" sum is $(a + \Delta a) + (b + \Delta b)$, or

$$(a + b) + (\Delta a + \Delta b).$$

When we merely add the two original numbers, a and b, we obtain $(a + b)$. The "true" sum differs from this by $(\Delta a + \Delta b)$. Hence, the error in the sum may be found by adding the errors of the addends.

EXERCISES

1. a) If an approximation a is 13.6 and is too small by 0.1, what is Δa? Is it positive or negative?

   b) If an approximation b is 214.8 and is too large by 0.2, what is Δb? Is it positive or negative?

   c) Calculate $a + b$.

   d) What is the error in $a + b$? Is it positive or negative?

2. a) If in Exercise 1, Δa were $+0.1$ and Δb were $+0.2$, what would be the error in $a + b$? Is it positive or negative?

   b) If Δa were -0.1 and Δb were -0.2, what would be the error in $a + b$? Is it positive or negative?

3. Two correctly determined measures are 3162 ± 1.5 and 413 ± 2.5.

   a) Find the sum and express it as ＿＿ $\pm$ ＿＿.

   b) What is the most the sum could be?

   c) What is the least the sum could be?

   d) What is the most the error of the sum could be?

   e) What is the least the error of the sum could be?

4. One measure is 6.123 and its g.p.e. is 0.0005. Another measure is 14.5 and its g.p.e. is 0.05. Both of them are correctly determined.

   a) What is the most each of these measures could be?

---

* Quotation marks are used here because in the case of measurement there is no exact measure.

b) What is the least each of these measures could be?

c) Add the measures.

d) Add the maxima, from part (a). What is the most the sum could be?

e) Add the minima from part (b). What is the least the sum could be?

f) In view of your answers to parts (d) and (e), how many digits is it sensible to keep in your answer to part (c)? Why?

Multiplication

We have found a simple way to calculate the error of a sum. We simply add the errors of the addends. We now wish to find a way to determine the error of a product. We shall use a geometric argument.

The area of a rectangular region is found by multiplying two numbers, so the error in a product can be represented geometrically as the error of the area of a rectangular region, based on the errors of the measures of the sides. Suppose the measures of the sides are a and b, with errors Δa and Δb, respectively. Then the "true" measures are $a + \Delta a$ and $b + \Delta b$, and the "true" area, as shown in Illus. 11.3 (if Δa and Δb are both positive), is $(a + \Delta a)(b + \Delta b)$. A little algebra shows that this is also

$$ab + b \cdot \Delta a + a \cdot \Delta b + \Delta a \cdot \Delta b.$$

When we find the product ab, then, the error is

$$b \cdot \Delta a + a \cdot \Delta b + \Delta a \cdot \Delta b.$$

Let us now examine the geometric interpretation of this expression for the error. Region I has area $b \cdot \Delta a$, region II has area $a \cdot \Delta b$, and region III has area $\Delta a \cdot \Delta b$. Since region III is so small, relatively speaking, we may ignore it. Thus we obtain a fairly simple expression for the error of the product ab.

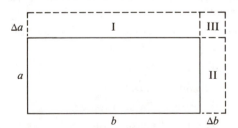

Illus. 11.3

The error of a product ab is approximately

$$a \cdot \Delta b + b \cdot \Delta a.$$

In other words, the error of ab is a times the error of b plus b times the error of a.

Example. Correctly determined measures of 11.1 and 4.2 each have a g.p.e. of

0.05. What is the maximum error in their product, and if the result is to be rounded, how should it be done?

$$
\begin{array}{r}
11.1 \\
4.2 \\
\hline
2\ 22 \\
44\ 4 \\
\hline
46.62
\end{array}
$$

The maximum error is $a \cdot \Delta b + b \cdot \Delta a$, when both terms are positive or both terms are negative. For simplicity we may assume they are positive, obtaining $11.1 \times 0.05 + 4.2 \times 0.05$, or 0.765.

The product is thus 46.62 ± 0.765. This means that the product is between 45.875 and 47.385. If the product is to be rounded, it should then be rounded to 46 or 47,* since it is in the units digit that the uncertainty first occurs.

The method of finding the error of a product can easily be extended to cases in which there are more than two factors. Let us consider a product of three factors, abc. We shall associate these factors as $(ab)c$, in which case we have two factors, as before. The error of the product is then

$$(ab)\ \Delta c + c\ \Delta(ab). \tag{11.1}$$

Now $\Delta(ab)$ is, again by the previous rule,

$$a\ \Delta b + b\ \Delta a.$$

Thus we find, upon substituting this in (11.1) and simplifying, that the error of abc is

$$ab\ \Delta c + ac\ \Delta b + bc\ \Delta a. \tag{11.2}$$

The expressions (11.1) and (11.2) both provide a rule, or procedure, for finding the error of a product of three factors. Either procedure is very much like that illustrated in the preceding example, except that for three factors the amount of calculation is greater than for two.

EXERCISES

1. Consider the following correctly determined measures:

    28.2 with g.p.e. 0.05 and 0.31 with g.p.e. 0.005.

    a) Find the product of the measures.
    b) Find the maximum error of the product, and express the product as ____ ± ____.
    c) What is the most the product could be? The least?
    d) Round the product to the most sensible number of digits.

---

* There is a well-known rule of rounding which says that if the first digit dropped is more than 5, the last digit retained should be increased by 1. This example shows that this rule is rather arbitrary.

2. Consider the following approximations: 312 and 431, where each is correct to within 0.5.
   a) Find the product.
   b) Find the maximum error of the product.
   c) What is the most the true product could be? The least?
   d) Round the product to the most sensible number of digits.
3. Consider the following approximations: 99.8, which is correct to within 0.05, and 4.61, which is correct to within 0.005.
   a) Find the product, and the maximum error of the product.
   b) What is the most the true product could be? The least?
   c) Round the product to the most sensible number of digits.
4. Consider the correctly determined measures 8.17 ± 0.05 and 1.02 ± 0.005.
   a) Find the product and the maximum error of the product.
   b) What is the most the product could be? The least?
   c) Round the product to the most sensible number of digits.
5. Consider the correctly determined measures 385 ± 0.5, 211 ± 0.05, and 42.3 ± 0.05.
   a) Find the product and the maximum error of the product.
   b) What is the most the product could be? The least?
   c) Round the product to the most sensible number of digits.
6. Make a drawing like that of Illus. 11.3 for the case where
   a) Δa and Δb are both negative,
   b) one of Δa, Δb is positive and the other negative.

Division

To find a means of calculating the error of a quotient, we shall recall that division and multiplication are opposite operations and shall use the result already developed for multiplication.

We wish to find the error of a quotient a/b. If we also represent this quotient by "c," then we have $a/b = c$, which is equivalent to

$$a = bc.$$

Now the error of a must be the same as the error of bc, since a and bc are the same. Thus

$$\Delta a = \Delta(bc) \approx b\,\Delta c + c\,\Delta b.$$

We are interested in the error of a/b, which is the same as Δc. We thus solve the above equation for Δc:

$$\Delta c = \Delta\left(\frac{a}{b}\right) \approx \frac{\Delta a - c\,\Delta b}{b}.$$

Remembering that $c = a/b$, we may substitute "a/b" for "c" in the numerator above. When we do this and simplify, we obtain the desired formula for the error of the quotient:

$$\Delta \left(\frac{a}{b}\right) \approx \frac{b\,\Delta a - a\,\Delta b}{b^2} .$$

To aid in remembering this formula one might verbalize it as follows: The error of a quotient is approximately *the denominator times the error of the numerator minus the numerator times the error of the denominator, all over the denominator squared.* The use of this formula in calculating an error is similar to the use of the formulas we have considered previously.

Example. Find the quotient of $41.3/2.21$. If the maximum error of the numerator is 0.05 and that of the denominator is 0.005, what is the maximum error of the quotient?

Dividing, we find that $41.3/2.21 = 18.6878$. To find the maximum error we use the formula above, obtaining

$$\frac{2.21 \times 0.05 - 41.3 \times 0.005}{(2.21)^2} .$$

At this point, however, we should recall that the errors of the numerator and denominator might be positive or negative. As we look at the formula we can easily see that the maximum error of the quotient will occur when one error is positive and the other negative. Let us suppose that the error of the denominator is negative and that of the numerator positive. We then obtain

$$\frac{2.21 \times 0.05 + 41.3 \times 0.005}{(2.21)^2} ,$$

which is about 0.06. The "true" quotient thus lies between 18.747 and 18.627. If the result is to be rounded, it should be rounded to 18.6 or 18.7.

EXERCISES

In Exercises 1 through 4, find the quotient, the maximum error, the most and least the quotient could be, and then round the quotient sensibly.

1. $418/314$, g.p.e. of numerator and denominator 0.5.

2. $16.21/4.32$, g.p.e. of numerator and denominator 0.005.

3. $953/0.102$, g.p.e. of numerator 0.5, g.p.e. of denominator 0.0005.

4. $0.075/521$, g.p.e. of numerator 0.0005, g.p.e. of denominator 0.5.

Rigorous exercise

5. Develop a formula for finding the error that occurs in finding a square root. [*Hint:* Consider the equation $a = b^2$.]

4. ROUNDING AND SIGNIFICANT DIGITS

In the preceding section a theory of errors was developed that is satisfactory for ordinary calculations except for the fact that finding maximum error is often time-consuming and tedious. In this section we shall consider a method of rounding the result of a calculation with approximations which is less precise but is more satisfactory in the sense that it is less time-consuming.

In a calculation with numbers that result from approximation, certain digits may occur which are not appropriate. For example, consider the following addition.

$$15.2367$$
$$\underline{3.02}$$
$$18.2567$$

Let us assume the g.p.e. of the larger addend to be 0.00005 and the g.p.e. of the other addend to be 0.005. The maximum error of the sum is then 0.00505, and the "true" sum is between 18.26175 and 18.25165. We cannot approximate the sum any more closely than this. The last two digits of the sums obtained by direct addition are therefore superfluous. When we round, we should drop these digits, thus obtaining 18.25 or 18.26. The digits retained are said to be *significant*. Those that were dropped are not. In the rounded sum, 18.25 or 18.26, the last digit is not known exactly. Nonetheless, it contributes to our knowledge of what the sum is. This is true of the final digit of any approximation. The final digit, although not known exactly, is still *significant*, because it contributes to our knowledge of how good the approximation is. In general, a digit is significant if it contributes to our knowledge of the degree of approximation. Superfluous digits as in the above example are not significant. All other nonzero digits are significant. Zero digits may or may not be significant. If their purpose is merely to show where a decimal point belongs, then they are not significant. For example, the zero digits in 0.0012 m are not significant. This measurement is the same as 0.12 cm, and moving the decimal point to change units does not affect our knowledge of how good the measurement is. Zero digits as in 13,000 may or may not be significant. If this number is the population of a city to the nearest thousand, then the zeros are not significant. If it is given to the nearest hundred, then one of the zeros is significant and the last two are not. On the other hand, if the number refers to a distance measured to the nearest foot then all the zeros are significant. Any zero occurring between two nonzero significant digits is significant, of course. Note that the addends of the preceding example contain 6 and 3 significant digits, respectively.

Addition

In the preceding example the last significant digit of the sum was the hundredths digit. The last significant digit of the addend 3.02 is also the hundredths digit.

This example illustrates a "rule of thumb" for rounding the result of an addition (or subtraction).

Look at the least precise addend and note the place value of its last digit. The digit of this place value in the sum is the last significant digit of the sum.

In practice, using this rule, one rounds the sum to correspond to the least precise addend.

Examples

1. 47.803
   12.3
   ─────
   60.103
      ⊥

   Not significant;
   round to 60.1

2. 1.0348
   0.001
   ─────
   1.0358
      ⊥

   Not significant;
   round to 1.036

Multiplication and Division

The rule of thumb most often used for multiplication is quite simple and is as follows.

Look at the factor with the least number of significant digits. The product has this same number of significant digits.

Examples

1.   4.02 (3 significant digits)
       3.1 (2 significant digits)
   ──────
      402
   12 06
   ──────
   12.462
   ⊥

   2 significant digits; round to 12

2.   0.004152 (4 significant digits)
          1.23 (3 significant digits)
   ──────
     12456
     8304
     4152
   ──────
   0.00510696
       ⊥

   3 significant digits;
   round to 0.00511

3.   2800 (4 significant digits)
      3.12 (3 significant digits)
   ──────
     5 600
   28 00
   840 0
   ──────
   873 600
     ⊥

   3 significant digits; round to 874

4.   43,000 (2 significant digits)
        4.2 (2 significant digits)
   ──────
     8600 0
   172000
   ──────
   180600.0
     ⊥

   2 significant digits;
   round to 180,000

In the preceding examples we note that the position of a decimal point is not considered when the significant digits are counted. The last two examples illustrate again that when a numeral contains terminal "0"s, those "0"s may or may not be significant. One cannot determine the number of significant digits by inspection in such cases, but must have some knowledge of the kind of approximation that is involved.

The rule of thumb for division is almost the same as that for multiplication.

Look at the divisor and dividend. Determine which has the least number of significant digits. The quotient has this same number of significant digits.

These rules of thumb work fairly well in a large number of cases, but there are a great many situations in which they fail to give correct results. Some caution in using them is therefore appropriate. In case of doubt in any situation for which maximum precision is important, one should resort to the procedures developed in the previous section.

EXERCISES

1. Determine the number of significant digits of each of the following, assuming they are rounded correctly.

   a) 1.304   b) 0.000125   c) 1.0044

   d) 417   e) 41006   f) 0.001001

   g) 17,200   h) 18.10   i) 0.01000

   j) 300   k) 47.0   l) 13,000,000

2. Add and round to the last significant digit, according to the rule of thumb in the preceding text.

   a) 403.12   b) 6.031   c) 0.000304   d) 38.402
      12.4076   42.58   17.5   14.256

3. Multiply and round to the last significant digit of the product according to the rule of thumb in the preceding text.

   a) 28.2×31 b) 312×431

   c) 99.8×4.61 d) 817.5×1.02

4. Divide and round to the last significant digit of the product according to the rule of thumb in the preceding text.

   a) $418/314$ b) $16.21/4.32$

   c) $953/0.102$ d) $0.075/521$

5. Compare the results of Exercise 3 with those of Exercises 1–4, pages 329–330. In which cases does the rule of thumb give a valid result?

6. Compare the results of Exercise 4 with those of page 331. In which cases does the rule of thumb give a valid result?

APPENDIX

This book was written with the assumption that the reader already knows certain things about mathematical symbolism, proof, and the properties of number systems. The authors realize that not every reader has been exposed to the prerequisite concepts and that the reader who has been exposed to them may need a bit of review. For these reasons, an outline of the essential prerequisite concepts is included here.

1. NUMBER SYSTEMS

A. Natural Numbers

The natural numbers are the counting numbers, 1, 2, 3, 4, 5, 6... They constitute a number system in which there are two fundamental operations, addition and multiplication. Subtraction is the opposite of addition, $a - b$ being defined as that number (if it exists) which when added to b gives a. Similarly, division is the opposite of multiplication, $a \div b$ or a/b being defined as the number (if it exists) which when multiplied by b gives a.

The fundamental properties of this system are as follows.

Closure. The set of natural numbers is closed under addition. This means that *for any* natural numbers a and b, the sum $a + b$ is in the set of natural numbers.

The set of natural numbers is also closed under multiplication.

Commutative laws. Addition is a commutative operation. This means that *for any* natural numbers a, b, $a + b = b + a$. In other words, the order in which numbers are added does not affect the sum.

Multiplication also is commutative.

Associative laws. Addition is an associative operation. This means that *for any*

335

natural numbers a, b, c, $a + (b + c) = (a + b) + c$. In other words, the grouping of numbers in addition does not affect the sum.

Multiplication also is associative.

Distributive law. Multiplication of natural numbers is distributive over addition. This means that *for any* natural numbers a, b, c,

$$a \cdot (b + c) = (a \cdot b) + (a \cdot c).$$

A multiplicative identity. The number 1 is a multiplicative identity. This means that when a number n is multiplied by 1, the result is n.

Order. The natural numbers can be arranged in order, beginning with the smallest number 1. This arrangement can be pictured on a number line as shown in Illus. A.1.

Illus. A.1

The sentence $a > b$ is read "a is greater than b," while $a < b$ is read "a is less than b." The relation $>$ is defined as follows.

$a > b$ *means that* $a - b$ *is a natural number.*

Thus $5 > 3$ because $5 - 3 = 2$ and 2 is a natural number. It is not true that $2 > 7$ because the difference $2 - 7$ does not exist in the set of natural numbers. The sentence $a < b < c$ means both that $a < b$ and $b < c$. If $a < b < c$ (or $a > b > c$), we say that b "is between" a and c.

B. Whole Numbers

When the number 0 is included with the natural numbers, the resulting set is called the set of *whole numbers*. The properties of the system of whole numbers are the same as those for the system of natural numbers, with the following additions.

Zero, an additive identity. The number 0 is an additive identity. This means that for any whole number a, $a + 0 = a$.

Multiplication by zero. When any whole number is multiplied by 0, the result is 0.

Order. The order of the whole numbers may be pictured as shown in Illus. A.2.

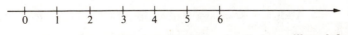

Illus. A.2

The sentence $a > b$ for whole numbers is defined to mean that $a - b$ is a natural number.

Division by zero. Division by zero is impossible!

C. Integers

The integers consist of the whole numbers and, in addition, the additive inverses of the whole numbers:

$$\ldots -7, -6, -5, -4, -3, -2, -1, 0, 1, 2, 3, 4, 5, \ldots$$

The properties of this system are as follows.

1. The set of integers is closed under both multiplication and addition. It is closed under subtraction also.

2. Both operations, addition and multiplication, are commutative and associative.

3. Multiplication is distributive over addition. It is also distributive over subtraction.

4. The numbers 0 and 1 are additive and multiplicative identities, respectively.

5. Every integer X has an additive inverse $-X$. This means that for any integer X, there is an integer $-X$ such that the sum $X + (-X)$ is 0, the additive identity.

   The additive inverse of a positive integer (natural number) is negative, and the additive inverse of a negative integer is positive. For example, the additive inverse of 5 is -5 and the additive inverse of -7 is 7. The additive inverse of 0 is 0.

6. To subtract $a - b$, one may add the inverse of the subtrahend; in other words, $a + (-b)$. For example, $-3 - 7 = -3 + (-7) = -10$.

Order. The order of the integers may be pictured as in Illus. A.3. The sentence $a > b$ is defined to mean that $a - b$ is positive. Thus $5 > 3$ because $5 - 3$ is positive; $-3 > -7$ because $-3 - (-7) = -3 + 7 = 4$, which is positive.

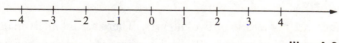

Illus. A.3

Absolute value. The absolute value of an integer n symbolized $|n|$, tells how far it is from 0. It is defined as

$$|n| = n \text{ when } n \text{ is positive or zero}$$

and

$$|n| = -n \text{ (the additive inverse of } n) \text{ when } n \text{ is negative.}$$

In other words, the absolute value of any nonnegative integer is the integer itself. The absolute value of a negative integer (its additive inverse) is positive. Thus $|5| = 5$ and $|-5| = 5$.

D. Rational Numbers

The rational numbers include the integers as well as other numbers (fractions) such as $\frac{2}{3}$, $-\frac{12}{5}$, etc. The quotient (ratio) of any two integers a/b, where $b \neq 0$, is a rational number. Moreover, every rational number is the ratio, or quotient, of two integers; in fact, the word *rational* stems from the word *ratio*.

The system of rational numbers has all of the properties listed for integers, as well as the following.

1. *Multiplicative inverses.* Every nonzero rational number has a multiplicative inverse, or *reciprocal*. The reciprocal of a number n is a number which when multiplied by n gives 1, the multiplicative identity.

The reciprocal of a number a/b is the number b/a. The reciprocal of a number x can also be named $1/x$.

2. The *order* of the rational numbers may be considered as superimposed on that of the integers, as is shown in Illus. A.4.

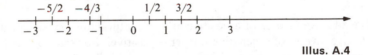

Illus. A.4

The relation $>$ is defined exactly as for integers. Absolute value is also defined in the same way.

The order of the rational numbers is *dense*. This means that between any two rational numbers there is a third. It then follows that between any two numbers there is an infinite set of numbers.

E. Real Numbers

Certain numbers are not a ratio of two integers. There are infinitely many of them. Some examples are $\sqrt{2}$, $\sqrt{5}$, and π. There are points of a number line corresponding to these numbers, however. The set of real numbers contains a number for every point of a line. This includes the rational numbers. The real numbers that are not also rational numbers are called *irrational* numbers.

The system of real numbers has all of the properties listed for rational numbers. Absolute value and $>$ are defined in the same way, and the order can be pictured on a line in the same way. There is, however, a real number for each point of the line, while this is not true of the rational numbers.

2. NUMERATION

Numbers are abstract ideas. We write certain symbols to represent them. Any symbol used to represent (or *name*) a number is called a *numeral*.

The scheme, or system, in use today is the *decimal* system. The word *decimal* is the adjective meaning "based on ten."

The meaning of a decimal numeral can be shown by expanded notation as follows:

$$5142.6153 = 5 \times 10^3 + 1 \times 10^2 + 4 \times 10 + 2 + 6 \times 10^{-1}$$
$$+ 1 \times 10^{-2} + 5 \times 10^{-3} + 3 \times 10^{-4}.$$

Place values are powers of ten, as shown in the following place-value chart.

10^3	10^2	10^1	10^0	10^{-1}	10^{-2}	10^{-3}
1000	100	10	1	$\frac{1}{10}$	$\frac{1}{100}$	$\frac{1}{1000}$

Nondecimal numeration. Numeration systems can be based on a number other than ten. If the base is b, the place values are powers of b as follows.

b^4	b^3	b^2	b	1	b^{-1}	b^{-2}	b^{-3}

For example, if 2 is used as a base, the place values are the following.

32	16	8	4	2	1	$\frac{1}{2}$	$\frac{1}{4}$	$\frac{1}{8}$	$\frac{1}{16}$

The binary numeral 10110_2 thus represents $1 \times 16 + 0 \times 8 + 1 \times 4 + 1 \times 2 + 0 \times 1$, or 22_{10}. The binary numeral 1.1101_2 represents

$$1 + 1 \times \tfrac{1}{2} + 1 \times \tfrac{1}{4} + 0 \times \tfrac{1}{8} + 1 \times \tfrac{1}{16} \quad \text{or} \quad 1\tfrac{13}{16}.$$

Numerals for rational numbers. Every rational number is a quotient of two integers, and can therefore be named by a fractional symbol a/b. Every rational number can also be named by a decimal numeral. For example,

$$\tfrac{12}{5} = 2.4 \quad \text{and} \quad \tfrac{2}{3} = 0.666\overline{6}.$$

The decimal numeral for a rational number either terminates, as in 2.4, or it repeats endlessly as in $0.66\overline{6}$.

The decimal numeral for a number a/b can be found by dividing a by b.

When a repeating decimal numeral is given, a fractional numeral can be found as in the following example.

Example. Find a fractional numeral for 0.414141 . . .

$$n = \ \ \ 0.414141$$
$$100n = 41.414141$$

$$99n = 41 \qquad \text{(subtracting } n \text{ from } 100n)$$
$$n = \tfrac{41}{99}$$

If rational numbers are named by nondecimal numerals, say numerals of base b, then those numerals either terminate or repeat, as is true of decimal numerals. For a given rational number, its base-b numeral may terminate while its numeral in some other base may repeat, but for any base, one of these things must happen.

If a base-b numeral does not end or repeat, then it names an irrational number and, conversely, every irrational number can be named by a nonterminating, nonrepeating numeral.

3. SET NOTATION

Sets are named by use of braces { }. The set of numbers one, two, and three can be named by writing numerals for these numbers, separated by commas, within braces, as

$$\{1, 2, 3\}.$$

Sets may also be named by so-called *set-builder* notation, where conditions under which an element belongs to the set are given. For example, the set of all numbers less than two hundred may be named as follows:

$$\{x \mid x < 200\}.$$

This is read "the set of all x such that x is less than 200."

The empty set. A set without any members is called the *empty set*, or the *null set*, or the *void set*, and is often named $\emptyset$.

Intersections. The intersection of two or more sets consists of those elements common to them. The symbol $\cap$ is used to denote set intersection. For example,

$$\{A, B, C, D\} \cap \{C, D, E\} = \{C, D\}.$$

This sentence says that the intersection of the sets $\{A, B, C, D\}$ and $\{C, D, E\}$ is the set $\{C, D\}$. The sentence

$$\{1, 2, 3, 4, 5\} \cap \{12, 15, 20\} = \emptyset$$

says that the intersection of the two sets named is empty.

If two or more sets have an empty intersection, they are said to be *disjoint*.

Unions. The *union* of two or more sets is the set obtained by combining them. The symbol $\cup$ is used to denote set union. For example,

$$\{A, B, E, G\} \cup \{B, G, T, V\} = \{A, B, E, G, T, V\}.$$

Complements. When one works with sets, some universal set is usually kept in mind, depending on the context. In algebra, for instance, the universal set might be the set of real numbers, while in geometry the universal set might be the set of all points in space. For a given set S, the *complement* of S, denoted $\bar{S}$, consists of all of the universal set except S. For example, if the universal set is $\{1, 2, 3, 4, 5, 6\}$ and $S = \{3, 5\}$, then $\bar{S} = \{1, 2, 4, 6\}$.

Set membership. To say that an element belongs to, or is a member of a set, we use the symbol $\in$ as a verb. The sentence $a \in S$ is read "a belongs to S."

Set inclusion. It may happen that one set is contained entirely within another. In this case the first set is said to be a *subset* of the other. The symbol $\subset$ is used as a verb for this context. The sentence $A \subset B$ says, for example, that set A *is included in* or *is a subset of* set B. The definition is as follows: $A \subset B$ if and only if *every* element of A is a member of B.

4. SENTENCES

The only kinds of sentences we shall consider are those which are used to express ideas in mathematics, namely declarative sentences.

Statements and open sentences. If a sentence is true, or if it is false, it is said to be a statement. No statement can be both true and false. Certain sentences contain variables, for which substitutions can be made.

Example.

$$3x = 5.$$

This sentence is neither true nor false until a replacement has been made for the variable "x." It is an *open sentence*.

Equality. The verb symbol $=$ is used with a single meaning throughout this book. A sentence $a = b$ says that the symbols "a" and "b" represent (or *name*) the same thing.

Example. "$\triangle ABC = \triangle DEF$" says that $\triangle ABC$ and $\triangle DEF$ are the same triangle.

It should be noted that every sentence with $=$ for the verb is reversible, that is, $a = b$ and $b = a$ say the same thing.

If $a = b$ is true, then "a" and "b" are names of the same thing. Hence, either name may be used in place of the other at any time. Replacing one name by another is often called *substitution*.

Conjunctions. When two or more sentences are jointed by the word *and*, a new sentence is formed, called the *conjunction* of the sentences. A conjunction of sentences is true when all of the individual sentences are true. It is false if one or more of them is false. The conjunction of sentences A and B is often denoted $A \wedge B$.

Disjunctions. When two or more sentences are joined by the word *or*, a new sentence is formed, called the *disjunction* of the sentences. A disjunction of sentences is false when all of the individual sentences are false. It is true if one or more of them is true. The disjunction of two sentences A and B is often denoted $A \vee B$.

Conditional sentences. A sentence of the form "If $A \ldots$, then $B \ldots$" is called a conditional sentence and is often denoted $A \rightarrow B$. This may be read "A *implies* B." The sentence A is called the *antecedent* and B is called the *consequent*. A conditional sentence $A \rightarrow B$ is false when A is true while B is false. It is true otherwise. It should be noted that $A \rightarrow B$ is true whenever A is false, regardless of the truth or falsity of B.

Negations. The negation of a sentence is also a sentence. The negation of a sentence A is "not A" and is often denoted $\sim A$. If A is true, then $\sim A$ is false, and if A is false, then $\sim A$ is true. The negation of "It is raining" is "It is not raining."

Negating a conjunction. The negation of a conjunction is the disjunction of the negations of the individual sentences; that is, $\sim(A \wedge B)$ is $(\sim A) \vee (\sim B)$. For example, the negation of "It is raining and 3 is an odd number" is the sentence "It is not raining or 3 is not an odd number."

Negating a disjunction. The negation of a disjunction is the conjunction of the negations of the individual sentences; that is, $\sim(A \vee B)$ is $(\sim A) \wedge (\sim B)$. For example, the negation of "It is snowing or I will stay home" is the sentence "It is not snowing and I will not stay home."

Negating a negation. The negation of the negation of a sentence is the original sentence; that is, $\sim(\sim A)$ is A.

Negating a conditional. The negation of a conditional sentence is a conjunction, as follows: $\sim(A \rightarrow B)$ is $A \wedge (\sim B)$. For example, the negation of "If x is odd, then x^3 is odd" is the sentence "x is odd and x^3 is not odd." It should be noted that the negation of a conditional sentence is *never* conditional.

Equivalent sentences. If two sentences, A and B, imply each other, that is, $A \rightarrow B$ and also $B \rightarrow A$, then the sentences A and B are said to be *equivalent*. The conjunction $(A \rightarrow B) \wedge (B \rightarrow A)$ is often abbreviated $A \leftrightarrow B$. This is read "A implies B and conversely," or "A if and only if B." If two sentences are equivalent, then they are true together and false together. Since $A \rightarrow B$ is the negation of $A \wedge (\sim B)$, the sentences $A \rightarrow B$ and $(\sim A) \vee B$ are equivalent.

Converses. The converse of a conditional $A \rightarrow B$ is the sentence obtained by interchanging antecedent and consequent. In other words, the converse of $A \rightarrow B$ is $B \rightarrow A$.

If a conditional $A \rightarrow B$ is true, its converse $B \rightarrow A$ may be true or it may be false.

Contrapositives. The contrapositive of a sentence $A \rightarrow B$ is obtained by negating antecedent and consequent, and then interchanging them. In other words, the contrapositive of $A \rightarrow B$ is $\sim B \rightarrow \sim A$.

Any conditional $A \rightarrow B$ is equivalent to its contrapositive $\sim B \rightarrow \sim A$, because both are equivalent to $(\sim A) \wedge B$. A conditional and its contrapositive are true together and false together, and one may be substituted for the other.

5. SENTENCES AND SETS

There is a close relationship between sentences and sets.

Replacement sets. Any open sentence contains one or more variables, for which replacements may be made. The set of elements from which the replacements come is called the *replacement set* for the sentence. On occasion, the replacement set for a sentence may be a set of numbers, or it may be a set of points, or it may be a set of geometric figures.

Solution sets. When a replacement is made for the variable(s) of an open sentence, sometimes a true sentence results and sometimes a false sentence results. The set of replacements that result in a true sentence is known as the *solution set* of the sentence. Any member of that set is called a *solution* of the sentence. With any open sentence, there is naturally associated its solution set.

Solution sets of conjunctions. The solution set of a conjunction $A \wedge B$ is the intersection of the solution sets of A and B. Thus sentence conjunction corresponds to set intersection.

Solution sets of disjunctions. The solution set of a disjunction $A \vee B$ is the union of the solution sets of A and B. Thus sentence disjunction corresponds to set union.

Solution sets of negations. The solution set of the negation of a sentence A is the complement of the solution set of A. Thus sentence negation corresponds to set complementation.

Solution sets and conditionals. If a conditional sentence $A \rightarrow B$ is true, then the solution set of A is a subset of the solution set of B. If the converse $B \rightarrow A$ should also be true, then of course the solution set of B would be a subset of the solution set of A. In this case the sentences A and B would have the same solution set. This is always true of equivalent sentences.

6. RELATIONS

The elements of a set may bear many kinds of relationships to each other. We often use the symbolism xRy to denote the fact that x bears a certain relation (called "R") to y. There are many kinds of relations, and we shall list and name some of the more important kinds.

Reflexive relations. If an element is in relation with itself, we might denote this as xRx. If every x involved in a relation is in relation with itself, then we say that the relation is *reflexive*. The relation "is the same color as" is a reflexive relation, for example.

Symmetric relations. It may happen that whenever an element x is in a certain relation with y, then y also bears that relation to x. For example, consider the relation "is equal to." If x is equal to y, then also y is equal to x." Such a relation, in which xRy always implies that yRx, is called a *symmetric* relation.

Transitive relations. Certain relations have the property that when xRy and yRz, then it follows that xRz. If this happens for any x, y, and z that are involved, then the relation is called *transitive*. The relation "is older than" is an example of a transitive relation. Whenever x is older than y and y is older than z, it follows that x is older than z.

Equivalence relations. Some relations have all three of the properties mentioned above. This kind of relation is an important one and is called an *equivalence* relation. An equivalence relation, then, is a relation that is reflexive, symmetric, and transitive.

An important property of any equivalence relation is that it *partitions* the set on which it is defined. This means that the set can be separated into disjoint subsets by using the relation in question. This would be done as follows. Choose any element x. Then find all the members of the set that are in relation with it. These members constitute one *cell* of the partition. Then choose another element of the original set. Find all the elements that are in relation with it. These constitute another cell. Continue in this manner until all members of the original set have been placed in some cell. As an example we might take the relation "has a birthday in the same month as" among the members of a mathematics class. We would partition the set into cells in which all members of a cell have birthdays in the same month. These cells would be nonoverlapping, or *disjoint*.

Order relations. An order relation is defined to be any relation that is *antisymmetric* and *transitive* and *complete*. If a relation is antisymmetric, this means that whenever xRy, then it is *not* true that yRx. To say that a relation is complete means that when we choose any two elements x and y in the set on which the relation is defined, either xRy or yRx. In other words, one of them bears the given relation to the other.

Any order relation (one which is antisymmetric, transitive, and complete) orders the set on which it is defined. Therefore by means of the relation concerned, it is possible to place the elements of the set in order. The relation $<$ (is less than) for numbers is an example. This relation is antisymmetric, transitive, and complete. We can place a set of numbers in order by means of this relation, from smallest to largest.

7. SOME RULES OF PROOF

There are no hard-and-fast rules for proving theorems, but there are a few general principles that may be of some help.

We might mention at the outset that a proof is essentially a series of assertions, where a supporting reason is given for each. The supporting reasons may be axioms, or *postulates* (these words are synonyms), which are nothing more than assumptions that underlie the particular mathematical discourse in question. The reasons may also be previously proved theorems or, in some cases, basic logical principles. We ordinarily try to make the last assertion in a proof a sentence which asserts the fact to be proved.

An assumption of a permanent nature is called an *axiom*, or *postulate*, but we often make assumptions of a temporary nature and call them *hypotheses*. A proof often begins with the assertion of some hypothesis, but this is by no means universally the case. A hypothesis may or may not be true. We never know whether or not a hypothesis is true. If we knew it to be true, then we would not call it an hypothesis.

Rule of conditional proof. Many theorems are stated by conditional sentences $A \to B$. When a theorem is stated in this way, we generally prove it by assuming A as hypothesis and showing that B follows. We then conclude that $A \to B$. This is known as the rule of *conditional proof*. We can state this as follows: From an argument in which B is shown to follow from a hypothesis A, we may infer that $A \to B$.

If a theorem is not stated as a conditional sentence, we often proceed in much the same manner, except that we must find a statement known to be true for the opening statement in the argument. There is no rule that can be given for finding an opening statement that will serve the purpose. The prover must search until he finds one.

Rule of modus ponens. A conditional sentence $A \to B$ tells us nothing about the truth of B, even though $A \to B$ itself is true. This is because $A \to B$ may be true when B is false, provided A is also false. However, if we know that A is true and $A \to B$ is true, we may conclude that B is true. When we reason in this manner, we are using a rule known as the rule of *modus ponens*, stated as follows: From A and $A \to B$, we may infer B.

Rule of substitution. Whenever two sentences, A and B, are equivalent, they have the same solution set and, hence, one may be used in place of the other without altering the truth or falsity of any sentence in which such substitution is made. Thus we have another rule, called the *rule of substitution*, as follows: If two sentences, A and B, are equivalent ($A \to B$ and also $B \to A$), then either sentence may be substituted for the other.

Proof by contradiction. Many proofs are made by contradiction. To do this, we assume the negation of what we are trying to show. Then we proceed to deduce something known to be false. We then conclude that our assumption was false and hence its negation (the theorem to be proved) is true. If a theorem $A \rightarrow B$ is to be proved by contradiction, then its negation $A \wedge \sim B$ is assumed at the beginning. When a contradiction is shown to follow, the negation of $A \wedge \sim B$, or $A \rightarrow B$, is then inferred to be true.

8. DEFINITIONS

A definition is usually stated in one of two ways:

a) It can be stated as an equivalence, "A if and only if B." An example is, "A figure is a triangle if and only if it is a polygon of three sides." By the rule of substitution, then, either "a figure is a triangle" or "a figure is a polygon of three sides" may be substituted for the other without changing the truth or falsity of anything.

b) It can be stated by using the *is* of logical identity. We may say, for example, that "a triangle *is* a polygon of three sides." In this case, we are asserting that the names "polygon of three sides" and "triangle" are two names of the same thing. In this case, it is clear—we can substitute one name for another.

It is a habit of mathematical writers to use the word "if" in a definition in place of the words "if and only if." (We do not do so in this book, however.) The reader must learn to supply the missing words.

LIST OF AXIOMS, DEFINITIONS, AND THEOREMS

Chapter 2

Undefined terms *Point* and *between*.

Definition 2.1 A *geometric figure* (or simply *figure*) is a nonempty set of points.

Axiom 2.1 There exist at least two points.

Axiom 2.2 For any points A, B, C, if A-B-C, then A, B, and C are three distinct (different) points.

Axiom 2.3 For any distinct points A and B, there is at least one point between them.

Axiom 2.4 If a point B is between points A and C, then it is also between C and A. (If A-B-C, then also C-B-A.)

Axiom 2.5 For any distinct points, A, B, C, and D, if A-B-C and A-D-B, then A-D-C and D-B-C.

Definition 2.2 For any two distinct points A and B, the set consisting of A, B, and all the points between A and B is called a *segment*. It may be named "$\overline{AB}$."

Theorem 2.1 For any two distinct points A and B, $\overline{AB} = \overline{BA}$.

Theorem 2.2 Every segment is an infinite set of points.

Definition 2.3 For any two distinct points A and B, the figure $\{X \mid A\text{-}X\text{-}B\}$ is called an *open segment* and may be named "$\overset{\circ\!-\!\circ}{AB}$." The figure $\{A\} \cup \{X \mid A\text{-}X\text{-}B\}$ is called a *half-open segment*. It may be named "AB."

Axiom 2.6 For any two distinct points A and B, there exists at least one point P such that A-B-P.

Definition 2.4 For any two distinct points A and B, $\overline{AB} \cup \{X \mid A\text{-}B\text{-}X\}$ is called a *ray*. It may be named "$\overrightarrow{AB}$."

Definition 2.5 For any two distinct points A and B, $\overrightarrow{AB} \cup \overrightarrow{BA}$ is called a *line*. It may be named "$\overleftrightarrow{AB}$."

Theorem 2.3. For any two distinct points A and B, $\overleftrightarrow{AB} = \overleftrightarrow{BA}$.

Definition 2.6 For any two distinct points A and B, $\overset{\circ}{\overrightarrow{AB}} \cup \{X \mid A\text{-}B\text{-}X\}$ is called a *half-line*. It may be named "$\overset{\circ}{\overrightarrow{AB}}$."

Axiom 2.7 For any three distinct points of a line, exactly one is between the other two.

Theorem 2.4 For any distinct points A and B, and for any two distinct points X and Y between them, one and only one of the following holds: $A\text{-}X\text{-}Y$ or $A\text{-}Y\text{-}X$.

Theorem 2.5 For any points A, X, and B, if $A\text{-}X\text{-}B$, then $\overline{AB} = \overline{AX} \cup \overline{XB}$.

Theorem 2.6 For any points A, B, X, and Y, if $A\text{-}X\text{-}B$ and $X\text{-}B\text{-}Y$, then $A\text{-}X\text{-}Y$ and also $A\text{-}B\text{-}Y$.

Theorem 2.7 For any ray $\overrightarrow{AB}$ and any point P on the ray except the endpoint, $\overrightarrow{AB} = \overrightarrow{AP}$.

Theorem 2.8 If C and D are any two distinct points on a line $\overleftrightarrow{AB}$, then $\overleftrightarrow{CD} = \overleftrightarrow{AB}$.

Theorem 2.9 For any two distinct points, there is one and only one line that contains them.

Theorem 2.10 Any two distinct lines meet in at most one point.

Axiom 2.8 For any line l there is at least one point P such that $P \notin l$.

Definition 2.7 For any three distinct, noncollinear points A, B, and C, the union $\{P \mid P \in \overrightarrow{AX}, \text{ where } X \in \overline{BC}\} \cup \{P \mid P \in \overrightarrow{BY}, \text{ where } Y \in \overline{AC}\} \cup \{P \mid P \in \overrightarrow{CZ}, \text{ where } Z \in \overline{AB}\}$ is called a *plane*.

Axiom 2.9 For any three distinct, noncollinear points, there is no more than one plane that contains them.

Theorem 2.11 If P and Q are any two points on a plane α, then $\overleftrightarrow{PQ} \subset \alpha$.

Theorem 2.12 For any line and any point not on that line, there is one and only one plane that they lie within.

Theorem 2.13 For any two distinct lines l and m, if $l \cap m \neq \emptyset$, then the two lines are contained in one and only one plane.

Axiom 2.10 For any plane α there is at least one point P not on α.

Theorem 2.14 For any line l and any plane α, if l is not on α, then $l \cap \alpha$ contains at most one point.

Axiom 2.11 If two planes intersect, their intersection contains at least two distinct points.

Theorem 2.15 If two distinct planes have a nonempty intersection, then that intersection is a line.

Definition 2.8 Two lines are said to be *parallel* if and only if they are in the same plane and have an empty intersection.

Definition 2.9 Any two of the following kinds of figures (segments, rays, half-lines, lines, and open segments) are said to be *parallel* if and only if they are subsets of lines that are respectively parallel.

Definition 2.10 Two lines are said to be *skew lines* if and only if they are not in the same plane.

Definition 2.11 Two planes are said to be *parallel* if and only if their intersection is empty.

Definition 2.12 A line and a plane are *parallel* if and only if their intersection is empty.

Chapter 3

Theorem 3.1 Given a point P on a line l, if two points A and B are on the same side of P, then the segment $\overline{AB}$ is contained in (is a subset of) that side of P and does not contain P. If A and B are on opposite sides of P, then the segment $\overline{AB}$ intersects both half-lines and contains the point P.

Definition 3.1 A set S is a *convex set* if and only if, for any two distinct points A, B in S, the segment $\overline{AB}$ is also contained in S.

Theorem 3.2 The intersection of any two convex sets is also a convex set.

Axiom 3.1 (Plane Separation Axiom) Given a line l in a plane, the set of all points in the plane not on l consists of two disjoint sets such that (1) each set is a convex set, and (2) if A is in one set and B is in the other, then $\overline{AB}$ intersects l.

Theorem 3.3 If P is a point on a line l and Q is a point not on this line, then the half-line $\overset{\circ}{PQ}$ is contained entirely in one side of the line l.

Definition 3.2 An *angle* is a figure consisting of two distinct rays with a common endpoint. If the rays are opposite rays of a line, the angle is called a *straight angle*. The rays are called *sides* and the common endpoint is called the *vertex*.

Definition 3.3 A point Q is in the *interior* of an angle $\angle ABC$ if and only if Q is in the intersection of half-planes $\overleftrightarrow{BC}/A \cap \overleftrightarrow{BA}/C$. A point Q is in the *exterior* of an angle $\angle ABC$ if and only if Q is not on the angle and is not in the interior of the angle.

Theorem 3.4 Given any nonstraight angle, the set of all points in its plane and not on the angle consists of two disjoint sets such that:

    a) One of the sets is convex and the other is not.

    b) If P is in one set and Q is in the other, then $\overleftrightarrow{PQ}$ meets the angle.

Definition 3.4 For any three distinct noncollinear points; A, B, and C, the union of the segments $\overline{AB}$, $\overline{BC}$, and $\overline{CA}$ ($\overline{AB} \cup \overline{BC} \cup \overline{CA}$) is called a *triangle*. The segments are called *sides*, and the points A, B, and C are called *vertices* (sing., *vertex*). Triangle ABC is denoted "$\triangle ABC$."

Definition 3.5 The *interior* of a triangle $\triangle ABC$ is the intersection of half-planes $\overleftrightarrow{BC}/A \cap \overleftrightarrow{AC}/B \cap \overleftrightarrow{BA}/C$. The *exterior* of a triangle $\triangle ABC$ is the set of all points in the plane of the triangle which are not on the triangle and are not in the interior of the triangle.

Theorem 3.5 For any triangle:

a) The interior is a convex set.

b) If P is in the interior and Q is in the exterior, then $\overline{PQ}$ meets the triangle.

Undefined terms *Curve, closed curve, polygon, simple polygon.*

Definition 3.6 A quadrilateral that has *at least one* pair of parallel sides is called a *trapezoid*. A quadrilateral that has two pairs of parallel sides is called a *parallelogram*.

Axiom 3.2 For any simple polygon in a plane, the set of all points in the plane not on the polygon consists of two sets such that (1) for each set, if two points, A and B, are in the set, there exists at least one polygonal curve in the plane that joins the points and does not intersect the polygon, (2) if A is in one set and B is in the other, every polygonal curve in the plane that joins these points intersects the polygon.

Definition 3.7 A simple polygon is a *convex polygon* if and only if each side lies on a line determining a half-plane which contains the remainder of the polygon. Any simple polygon which is not convex is a *concave polygon*.

Definition 3.8 The *interior* of a convex polygon is the intersection of all half-planes H with the following property: The edge of H contains a side of the polygon and the remainder of the polygon is contained in H.

Theorem 3.6 The interior of every convex polygon is a convex set.

Definition 3.9 A diagonal of a polygon is a segment that joins two vertices and is not a side of the polygon.

Jordan Curve Theorem For any simple closed curve in a plane, the set of all points in the plane not on the curve consists of two sets such that (1) for each set, if two points, A and B, are in the set, there exists at least one polygonal curve in the plane that joins the points and does not intersect the curve; (2) if A is in one set and B is in the other, every polygonal curve in the plane that joins these points intersects the curve; and (3) one of these sets contains a line (the exterior); the other contains no line (the interior).

Definition 3.10 If C is any plane curve and m is a line not on the plane but containing a point C, then the union of m and all lines parallel to m and containing a point of C is called a *cylindrical surface*. The line m is called a *directrix* and the curve C is called the *generatrix*. Any of the lines parallel to m are called *elements* of the surface.

Definition 3.11 Consider any cylindrical surface for which the generatrix C is a simple closed curve and two parallel planes, P_1 and P_2, which intersect all elements of the cylindrical surface, forming curves C_1 in P_1 and C_2 in P_2. The figure consisting of that portion of the cylindrical surface between P_1 and P_2 together with C_1 and C_2 and their interiors is called a *cylinder*. If C is a polygon, then the cylinder is also called a *prism*. The curves C_1 and C_2 and their respective interiors are called the *bases* and the portion between the bases is called the *lateral surface*. The polygons that constitute the lateral surface of a prism are called its *lateral faces*.

Definition 3.12 If C is a curve in a plane α and P a point not on the plane, then the union of all lines that contain P and a point of C is called a *conical surface*. Any of the lines contained in the surface is called an *element* of the surface. The curve C is called the *generatrix*.

Given a plane β which contains P' and is parallel to α, each portion of the conical surface in a half-space determined by β, together with the point P, is called a *nappe* of the surface.

Definition 3.13 Consider any conical surface for which the *generatrix* C is a simple closed curve. The figure consisting of C and its interior, together with that portion of the conical surface between C and the point P, together with point P, is called a *cone*.

The curve C with its interior is called the *base*. The rest of the figure is called the *lateral surface*. The point P is called the *vertex*. If C is a polygon, then the cone is also called a *pyramid*. The polygons that constitute the lateral surface of a pyramid are called its *lateral faces*.

Definition 3.14 A *polyhedron* is a closed surface consisting entirely of polygonal regions. The polygonal regions are called *faces*.

Euler's Formula If V, E, and F represent the number of vertices, edges, and faces, respectively, of any simple polyhedron, then $V - E + F = 2$.

Chapter 4

Congruence for segments is *undefined*.
Congruence for angles is *undefined*.

Axiom 4.1 For any segments, $\overline{AB}$, $\overline{CD}$, and $\overline{EF}$

  a) $\overline{AB} \cong \overline{AB}$;

  b) if $\overline{AB} \cong \overline{CD}$, then $\overline{CD} \cong \overline{AB}$;

  c) if $\overline{AB} \cong \overline{CD}$ and $\overline{CD} \cong \overline{EF}$, then $\overline{AB} \cong \overline{EF}$.

  (That is, congruence for segments is an equivalence relation.)

Axiom 4.2 Given a segment $\overline{AB}$ and a ray $\overrightarrow{CD}$, there is exactly one (a unique) point E on $\overrightarrow{CD}$ such that $\overline{AB} \cong \overline{CE}$.

  (This is sometimes referred to as the "compass axiom.")

Axiom 4.3 (Segment "Addition"). If A-B-C, A'-B'-C', $\overline{AB} \cong \overline{A'B'}$, and $\overline{BC} \cong \overline{B'C'}$, then $\overline{AC} \cong \overline{A'C'}$.

Axiom 4.4 (Segment "Subtraction") If A-B-C, A'-B'-C', $\overline{AB} \cong \overline{A'B'}$, and $\overline{AC} \cong \overline{A'C'}$, then $\overline{BC} \cong \overline{B'C'}$.

Definition 4.1

   a) A segment $\overline{AB}$ is *less than* a segment $\overline{CD}$ if and only if the point E on $\overrightarrow{CD}$ for which $\overline{AB} \cong \overline{CE}$ is between C and D.

   b) $\overline{AB}$ is *greater than* $\overline{CD}$ if and only if D is between C and E.

   c) To say that "$\overline{AB}$ is less than $\overline{CD}$" we may write "$\overline{AB} < \overline{CD}$." To say that "$\overline{AB}$ is greater than $\overline{CD}$" we may write "$\overline{AB} > \overline{CD}$."

Theorem 4.1 For any segments $\overline{AB}$ and $\overline{CD}$, exactly one of the following is true: $\overline{AB} \cong \overline{CD}$, $\overline{AB} < \overline{CD}$, or $\overline{AB} > \overline{CD}$.

Axiom 4.5 For any angles $\angle A$, $\angle B$, and $\angle C$

   1) $\angle A \cong \angle A$;

   2) if $\angle A \cong \angle B$, then $\angle B \cong \angle A$;

   3) if $\angle A \cong \angle B$ and $\angle B \cong \angle C$, then $\angle A \cong \angle C$.

   (That is, congruence for angles is an equivalence relation.)

Axiom 4.6

   a) For any nonstraight angle $\angle ABC$ and ray $\overrightarrow{B'C'}$ on the edge of a half-plane h_1, there is a unique ray $\overrightarrow{B'A'}$, with A' in h_1, such that $\angle ABC \cong \angle A'B'C'$.

   b) For any angles $\angle A$ and $\angle B$, if $\angle A$ is a straight angle, then $\angle A \cong \angle B$ if and only if $\angle B$ is a straight angle.

   (That is, all straight angles are congruent and no nonstraight angle is congruent to a straight angle.)

Axiom 4.7 (Angle "Addition") If D is interior to $\angle ABC$ (nonstraight), D' is interior to $\angle A'B'C'$ (nonstraight), $\angle ABD \cong \angle A'B'D'$, and $\angle DBC \cong \angle D'B'C'$, then $\angle ABC \cong \angle A'B'C'$.

Axiom 4.8 (Angle "Subtraction")

   a) If D is interior to $\angle ABC$ (nonstraight), D' is interior to $\angle A'B'C'$ (nonstraight), $\angle ABD \cong \angle A'B'D'$, and $\angle ABC \cong \angle A'B'C'$, then $\angle DBC \cong \angle D'B'C'$.

   b) For any straight angles, $\angle ABC$ and $\angle A'B'C'$, if D is any point not on $\angle ABC$, D' is any point not on $\angle A'B'C'$, and $\angle ABD \cong \angle A'B'D'$, then $\angle DBC \cong \angle D'B'C'$.

Definition 4.2 Suppose that $\angle ABC$ and $\angle DEF$ are any nonstraight angles, and consider $\angle GEF$ where $\overrightarrow{EG}$ is in the half-plane $\overleftrightarrow{EF}/D$, and $\angle GEF \cong \angle ABC$.

   a) $\angle ABC$ is *less than* $\angle DEF$ if and only if $\overrightarrow{EG}$ is in the interior of $\angle DEF$.

b) $\angle ABC$ is *greater than* $\angle DEF$ if and only if $\overset{\circ\longrightarrow}{EG}$ is in the exterior of $\angle DEF$.

c) If $\angle ABC$ is not straight and $\angle DEF$ is straight, then $\angle ABC$ is less than $\angle DEF$. As for *less than* and *greater than* for segments, we will denote "$\angle A$ less than $\angle B$" as "$\angle A \prec \angle B$" and "$\angle A$ greater than $\angle B$" as "$\angle A \succ \angle B$."

Theorem 4.2 For any angles $\angle ABC$ and $\angle A'B'C'$, exactly one of the following is true: $\angle ABC \cong \angle A'B'C'$, $\angle ABC \prec \angle A'B'C'$, or $\angle ABC \succ \angle A'B'C'$.

Definition 4.3 Two angles are said to be *adjacent* if and only if they have a common side and the remaining sides are on opposite sides of the common side.

Definition 4.4 Two angles, $\angle ABC$ and $\angle DBE$ are *vertical angles* if and only if C-B-D and A-B-E or A-B-D and C-B-E.

Theorem 4.3 Vertical angles are congruent.

Definition 4.5 $\angle A$ is *supplementary* to $\angle BCD$ if and only if $\angle A \cong \angle DCE$, where E is a point such that B-C-E.

Theorem 4.4 If two angles are congruent, then their supplements are also congruent.

Definition 4.6 A *right* angle is an angle that is congruent to every angle that is supplementary to it.

Definition 4.7 Two lines are *perpendicular* to each other if and only if they intersect and at least one angle determined by the intersection is a right angle.

Definition 4.8 Any two figures of the following kinds, segments, rays, half-lines, and lines, are *perpendicular* if and only if... (the lines containing them are perpendicular).

Definition 4.9 Two triangles are *congruent* if and only if they can be labeled ABC and $A'B'C'$ such that $\overline{AB} \cong \overline{A'B'}$, $\overline{BC} \cong \overline{B'C'}$, $\overline{CA} \cong \overline{C'A'}$, $\angle A \cong \angle A'$, $\angle B \cong \angle B'$, and $\angle C \cong \angle C'$.

We write "$\triangle ABC \cong \triangle A'B'C'$" to say that $\triangle ABC$ is congruent to $\triangle A'B'C'$.

Theorem 4.5 Congruence for triangles is an equivalence relation.

Axiom 4.9 (SAS) Given $\triangle ABC$ and $\triangle A'B'C'$, if $\overline{AB} \cong \overline{A'B'}$, $\overline{AC} \cong \overline{A'C'}$, and $\angle A \cong \angle A'$, then $\triangle ABC \cong \triangle A'B'C'$.

Theorem 4.6 (ASA) If, for $\triangle ABC$ and $\triangle A'B'C'$, $\angle A \cong \angle A'$, $\angle C \cong \angle C'$, and $\overline{AC} \cong \overline{A'C'}$, then $\triangle ABC \cong \triangle A'B'C'$.

Definition 4.10 An *isosceles* triangle is a triangle with at least two congruent sides. Given any two congruent sides, the remaining side is called the *base*, and the angles opposite the congruent sides are called the *base angles*.

Theorem 4.7 The base angles of an isosceles triangle are congruent.

Theorem 4.8 (SSS) If, for $\triangle ABC$ and $\triangle A'B'C'$, $\overline{AB} \cong \overline{A'B'}$, $\overline{AC} \cong \overline{A'C'}$, and $BC \cong \overline{B'C'}$, then $\triangle ABC \cong \triangle A'B'C'$.

Theorem 4.9 (AAS) If, for $\triangle ABC$ and $\triangle A'B'C'$, $\angle BAC \cong \angle B'A'C'$, $\angle ABC \cong \angle A'B'C'$, and $\overline{BC} \cong \overline{B'C'}$, then $\triangle ABC \cong \triangle A'B'C'$.

Theorem 4.10 Right angles exist.

Theorem 4.11 All right angles are congruent; no nonright angle is congruent to a right angle.

Theorem 4.12 (the "Crossbar Theorem") If a point D is in the interior of $\angle BAC$, then $\overset{\circ}{\overrightarrow{AD}}$ intersects $\overline{BC}$.

Theorem 4.13 Given a line l on a plane α and a point P on l, there is one and only one line m on α which contains P and is perpendicular to l.

Definition 4.11 An angle that is adjacent and supplementary to an angle of a triangle is called an *exterior angle* of the triangle. The two angles of the triangle not adjacent to the exterior angle are called *opposite interior* angles.

Theorem 4.14 (the Exterior Angle Theorem) Any exterior angle of a triangle is greater than either of the opposite interior angles.

Definition 4.12 A point C of a segment $\overline{AB}$ such that $\overline{AC} \cong \overline{CB}$ is called a *midpoint* (or *bisector*) of $\overline{AB}$.

Definition 4.13 A half-line $\overset{\circ}{\overrightarrow{BD}}$ interior to an angle $\angle ABC$ such that $\angle ABD \cong \angle DBC$ is called a *bisector* of $\angle ABC$.

Theorem 4.15 Every segment has one and only one midpoint.

Theorem 4.16 Every nonstraight angle has exactly one bisector.

Definition 4.14 An angle that is less than a right angle is called an *acute* angle.

An angle that is greater than a right angle but is less than a straight angle is called an *obtuse* angle.

Definition 4.15 An *acute* triangle is a triangle whose angles are all acute. A *right* triangle is a triangle with one right angle. The side opposite the right angle is called the *hypotenuse*. The other two sides are called the *legs* of the triangle. An *obtuse* triangle is a triangle with one obtuse angle.

Definition 4.16 A *scalene* triangle is a triangle with no pairs of congruent sides.

An *isosceles* triangle is a triangle with two or more congruent sides.

An *equilateral* triangle is a triangle whose three sides are all congruent.

Definition 4.17 A parallelogram whose sides are all congruent is called a *rhombus* (plural, *rhombi*). A parallelogram containing at least one right angle is called a *rectangle*.

A rhombus with at least one right angle is called a *square*.

Definition 4.18 Two figures are said to be *congruent* if and only if there exists a one-to-one correspondence between their points such that the segment determined by any two points of one figure is congruent to the segment determined by the corresponding points of the other figure.

Definition 4.19 Given a point O in a plane α and a segment $\overline{AB}$, the set of all points X on α such that $\overline{OX} \cong \overline{AB}$ is called a *circle*. Point O is called the *center* of the circle and any segment $\overline{OX}$, where X is on the circle, is called a *radius* of the circle.

Definition 4.20 A segment having both endpoints on a circle is called a *chord* of the circle.

A chord that contains the center of a circle is called a *diameter* of the circle.

Definition 4.21 Circles in the same plane having the same center are called *concentric* circles.

Definition 4.22 A line in the plane of a circle that contains one and only one point of the circle is said to be a *tangent* to the circle.

The point of intersection is called the *point of tangency*.

Theorem 4.17 Two circles are congruent if they have congruent radii.

Definition 4.23 Given a point O and a segment $\overline{AB}$, the set of all points X in space such that $\overline{OX} \cong \overline{AB}$ is called a *sphere*. The point O is called the *center* of the sphere and for any point X on the sphere, $\overline{OX}$ is called a *radius* of the sphere.

Theorem 4.18 If a plane contains the center of a sphere whose radii are congruent to $\overline{AB}$, then the intersection of the plane and the sphere is a circle having the same center and radii also congruent to $\overline{AB}$.

Definition 4.24 The intersection of a sphere and a plane that contains the center of the sphere is called a *great circle*.

Theorem 4.19 If, for two right triangles, the hypotenuse and one leg of one triangle are congruent to the hypotenuse and the corresponding leg of the other triangle, then the triangles are congruent.

Theorem 4.20 For any line l and any point P not on l there is one and only one line m containing P and a point of l, and perpendicular to l.

Definition 4.25 A line is *perpendicular* to a plane α at a point O if and only if the line intersects the plane at O and is perpendicular to every line in the plane that lies on the point O.

Theorem 4.21 If a plane α intersects a sphere S in more than one point, then the intersection of the plane and the sphere is a circle.

Definition 4.26

a) The intersection of a right circular cone and a plane not on the axis of the cone that intersects both nappes is called an *hyperbola*.

b) The intersection of a right circular cone and a plane that is parallel to an element of the cone is called a *parabola*.

c) The intersection of a right circular cone and a plane that:
   i) intersects only one nappe,
   ii) is not parallel to an element of the cone, and
   iii) is not perpendicular to the axis of the cone is called an *ellipse*.

Chapter 5

Axiom 5.1 (Archimedes' Axiom) If $\overline{AB}$ and $\overline{CD}$ are any segments, and on $\overrightarrow{CD}$ we choose points Q_1, Q_2, Q_3, and so on, such that $\overline{CQ_1}$, $\overline{Q_1Q_2}$, $\overline{Q_2Q_3}$, and so on, are all congruent to $\overline{AB}$, then eventually we shall find points Q_n and Q_{n+1} for which $D = Q_n$ or D is between Q_n and Q_{n+1}.

Theorem 5.1 For any segments $\overline{AB}$ and $\overline{CD}$, $\overline{AB} \cong \overline{CD}$ if and only if $m(\overline{AB}) = m(\overline{CD})$.

Theorem 5.2 For any segment $\overline{AB}$ and $\overline{CD}$, $\overline{AB} \prec \overline{CD}$ if and only if $m(\overline{AB}) < m(\overline{CD})$.

Theorem 5.3 For any collinear segments $\overline{AB}$ and $\overline{CD}$ for which $\overline{AB} \cup \overline{CD}$ is a segment, $m(\overline{AB} \cup \overline{CD}) = m(\overline{AB}) + m(\overline{CD}) - m(\overline{AB} \cap \overline{CD})$.

Theorem 5.4 For any angles $\angle A$ and $\angle B$, $\angle A \cong \angle B$ if and only if $m(\angle A) = m(\angle B)$.

Theorem 5.5 For any angles $\angle A$ and $\angle B$, $\angle A \prec \angle B$ if and only if $m(\angle A) < m(\angle B)$.

Cavalieri's Principle Consider two three-dimensional figures and some fixed plane. Suppose that the two figures can be oriented so that a plane parallel to the given plane which intersects one of the figures also intersects the other. If for each such plane the cross sections of the figures determined by that plane have the same area, then the two figures have the same volume.

Theorem 5.6 If two pyramids have the same height and the same base area, then they have the same volume.

Theorem 5.7 The volume of a triangular pyramid is one-third the product of the height and the base area.

Theorem 5.8 The volume of a sphere with radius r is $\frac{4}{3}\pi r^3$.

Theorem 5.9 The surface area of a sphere with radius r is $4\pi r^2$.

Chapter 6

Definition 6.1 Given a line n which meets two distinct lines l and m in a plane at distinct points E and F, respectively, and points A on l and D on m on opposite half-planes relative to n, line n is called a *transversal* with respect to lines l and m, and $\angle AEF$ and $\angle DFE$ are called *alternate interior angles*.

Theorem 6.1 If lines in a plane cut by a transversal are such that alternate interior angles are congruent, then the lines are parallel.

Theorem 6.2 For any line l and any point P not on l, there is (at least) one line m in the plane of l and P which contains P and does not meet l.

Axiom 6.1 (The Parallel Axiom) For any line l and a point P not on l, there is no more than one line parallel to l and containing the point P.

Theorem 6.3 If two parallel lines are cut by a transversal, the alternate interior angles are congruent.

Theorem 6.4 In any parallelogram $\square ABCD$, $\angle A \cong \angle C$ and $\angle B \cong \angle D$.

Corollary 6.5 All of the angles of a rectangle are right angles.

Theorem 6.6 In any parallelogram $\square ABCD$, $\overline{AB} \cong \overline{CD}$ and $\overline{AD} \cong \overline{BC}$.

Theorem 6.7 The sum of the measures of the angles of any triangle is 180°.

Theorem 6.8 (The Pythagorean Theorem) In any right triangle $\triangle ABC$, where $\angle C$ is a right angle, $[m(AB)]^2 = [m(BC)]^2 + [m(AC)]^2$. Or, if the measures of the sides opposite angles $\angle A$, $\angle B$, and $\angle C$ are respectively a, b, and c, then $a^2 + b^2 = c^2$.

Theorem 6.9 (Converse of Pythagorean Theorem) Consider any triangle $\triangle ABC$ in which the measures of the sides opposite angles $\angle A$, $\angle B$, and $\angle C$, respectively, are a, b, and c. If $a^2 + b^2 = c^2$, then $\angle C$ is a right angle.

Definition 6.2 Two segments are said to be *commensurable* if and only if there is a segment such that when it is used as a unit, each segment has a natural number for its measure.

Theorem 6.10 Two segments with measures M and N are commensurable if and only if M/N is a rational number.

Axiom 6.2 (Completeness) For any unit segment and any positive real number p, there exists a segment having p for its measure.

Theorem 6.11 For any line and for any segment as a unit, the line can be *co-ordinatized* in such a way that: (1) there is a one-to-one correspondence between the real numbers and the points of the line; (2) the length of any segment on the line is $|p - q|$, where p and q are the coordinates of the endpoints; (3) the points having coordinates 0 and 1 may be chosen arbitrarily, and the half-lines to be designated positive can be chosen arbitrarily.

Theorem 6.12 For any plane and any two distinct lines in the plane which intersect, the plane can be *coordinatized* in such a way that there is a one-to-one correspondence between the points in the plane and the set of all ordered pairs of real numbers.

Definition 6.3 Two triangles are *similar* if and only if they can be labeled ABC and $A'B'C'$ in such a way that $\angle A \cong \angle A'$, $\angle B \cong \angle B'$, and $\angle C \cong \angle C'$, and also such that

$$\frac{AB}{A'B'} = \frac{BC}{B'C'} = \frac{CA}{C'A'}.$$

We write "$\triangle ABC \sim \triangle A'B'C'$" to assert that the triangles are similar.

Theorem 6.13 If three parallel lines cut off congruent segments on one transversal, then they cut off congruent segments on any transversal.

Theorem 6.14 If a line meeting two sides of a triangle is parallel to the third side, then it divides the two sides proportionally. In other words, if line l meets $\triangle ABC$ at interior points of $\overline{AC}$ and $\overline{BC}$, at points D and E, respectively, and if $l \parallel \overline{AB}$, then $AC/DC = BC/EC$.

Theorem 6.15 If two triangles have two pairs of congruent angles, then they are similar, that is, if in $\triangle ABC$ and $\triangle A'B'C'$ we have $\angle A \cong \angle A'$ and $\angle C \cong \angle C'$, then $\triangle ABC \sim \triangle A'B'C'$.

Definition 6.4 Two figures are said to be *similar* if and only if there exists a one-to-one correspondence between their points such that the ratio of the distance between any pair of points of one figure and the distance between the corresponding pair of points of the other figure is constant.

Definition 6.5 In a right triangle $\triangle ABC$, with an acute angle $\angle A$, *sine A* (abbreviated "sin A") is defined to be

$$\frac{a}{c} \quad \text{or} \quad \frac{\text{length of side opposite } \angle A}{\text{length of hypotenuse}}.$$

Definition 6.6 In a right triangle $\triangle ABC$, with an acute angle $\angle A$, *cosine A* (abbreviated "cos A") is defined to be

$$\frac{b}{c} \quad \text{or} \quad \frac{\text{length of leg adjacent to } \angle A}{\text{length of hypotenuse}}.$$

Definition 6.7 In a right triangle $\triangle ABC$, with an acute angle $\angle A$, *tangent A* (abbreviated "tan A") is defined to be

$$\frac{a}{b} \quad \text{or} \quad \frac{\text{length of leg opposite } \angle A}{\text{length of leg adjacent to } \angle A}.$$

Chapter 7

Axiom 7.1 (Euclid's Fifth Postulate) If two lines are cut by a transversal such that the sum of the measures of the interior angles on one side of the transversal is less than 180°, then the two lines meet on that side of the transversal.

Axiom 7.2 For any distinct lines l and m and n, if $l \parallel n$ and $m \parallel n$, then $l \parallel m$.

Chapter 8

Theorem 8.1 Two points have the same first (second) coordinate if and only if the line containing them is parallel to or the same as the second (first) axis.

Theorem 8.2 If two points have coordinates (x, y_1) and (x, y_2), then the distance between these points is $|y_1 - y_2|$. If two points have coordinates (x_1, y) and (x_2, y), then the distance between these points is $|x_1 - x_2|$.

Theorem 8.3 (The Distance Formula) For any two points $A(x_1, y_1)$ and $B(x_2, y_2)$,

$$m(\overline{AB}) = \sqrt{(x_1 - x_2)^2 + (y_1 - y_2)^2}.$$

Theorem 8.4 Any line parallel to the x-axis may be described by the set $\{(x, y) \mid y \text{ is constant}\}$. The first axis is $\{(x, y) \mid y = 0\}$. Any line parallel to the y-axis may be described by the set $\{(x, y) \mid x \text{ is constant}\}$. The second axis is $\{(x, y) \mid x = 0\}$.

Definition 8.1 Given a line l, different from and not parallel to the y-axis, and two distinct points $A(x_1, y_1)$ and $B(x_2, y_2)$ on l, the real number

$$\frac{y_2 - y_1}{x_2 - x_1}$$

is called the *slope* of the line l.

Theorem 8.5 If a line has a slope, it is unique.

Theorem 8.6 Given the points $A(x_1, y_1)$ and $B(x_2, y_2)$, where $x_1 \neq x_2$, the line $\overleftrightarrow{AB}$ is

$$\left\{(x, y) \mid y - y_2 = \frac{y_2 - y_1}{x_2 - x_1}(x - x_2)\right\},$$

or

$$\{(x, y) \mid y - y_2 = m(x - x_2)\}$$

where m is the slope of the line.

Theorem 8.7 Two distinct nonvertical lines are parallel if and only if they have the same slope.

Theorem 8.8 Two nonvertical lines l_1 and l_2 with slopes m_1 and m_2 are perpendicular if and only if $m_1 m_2 = -1$.

Theorem 8.9 Any line in a plane can be defined by a first degree equation.

Theorem 8.10 (Converse of Theorem 8.9) The graph of every first-degree equation is a line.

Because of this correspondence, first-degree equations are often referred to as *linear* equations.

Theorem 8.11 For any real number a, the two half-planes determined by the line $x = a$ are the sets $\{(x, y) \mid x > a\}$ and $\{(x, y) \mid x < a\}$. For any real number b, the two half-planes determined by the line $y = b$ are the sets $\{(x, y) \mid y > b\}$ and $\{(x, y) \mid y < b\}$.

Theorem 8.12 For any real numbers m and b, the two half-planes determined by the line $y = mx + b$ are the sets $\{(x, y) \mid y > mx + b\}$ and $\{(x, y) \mid y < mx + b\}$.

Theorem 8.13 The midpoint of a segment having endpoints (x_1, y_1) and (x_2, y_2) is the point

$$\left(\frac{x_1 + x_2}{2}, \frac{y_1 + y_2}{2}\right).$$

Chapter 9

Theorem 9.1 If lines l and m intersect at a point P, then a line reflection about l followed by a line reflection about m is equal to a rotation about P, where the measure of the rotation is twice the measure of the angle formed by l and m.

Theorem 9.2 If lines l and m are parallel, then a line reflection about l followed by a line reflection about m is equal to a translation along a line perpendicular to l and m, and twice the distance between l and m.

Theorem 9.3 Every rigid motion (isometry) in the plane is equal to a combination of three or fewer line reflections.

Theorem 9.4 For any line m and any translation T, either m is invariant (its own image) under T, or the image of m is another line that does not intersect m.

Theorem 9.5 The image of a line under a rotation $R_{P(180)}$ is

a) parallel to the line if P is not on the line, or

b) identical to the line if P is on the line.

Theorem 9.6 If two parallel lines are cut by a transversal, the corresponding angles are congruent.

Definition 9.1 Two lines are *parallel* if and only if one line is the image of the other under some translation.

Definition 9.2 A figure is *symmetrical about a line* if and only if it is invariant under a reflection about that line. The line is called a *line of symmetry*.

Definition 9.3 A figure is *symmetrical about a point* if and only if it is invariant under some rotation less than 360 degrees about the point.

Theorem 9.7 Given a two-way stretch transformation $P(x, y) \rightarrow P'(kx, ky), k > 0$, and Q, R any two points in the plane,

i) $O\text{-}Q\text{-}Q'$ or $O\text{-}Q'\text{-}Q$, where O is the center of stretch,

ii) $OQ' = k(OQ)$,

iii) $Q'R' = k(QR)$,

iv) angle measure is preserved.

BIBLIOGRAPHY

Adler, Claire F., *Modern Geometry: An Integrated First Course*. McGraw-Hill, New York, 1958.

Adler, Irving, *A New Look at Geometry*. John Day, New York, 1966.

Arnold, B. H., *Intuitive Concepts in Elementary Topology*. Prentice Hall, Englewood Cliffs, N.J., 1962.

Birkhoff, George D. and Ralph Beatley, *Basic Geometry*. Chelsea, New York, 1959.

Blumenthal, Leonard M., *A Modern View of Geometry*. W. M. Freeman, San Francisco, 1961.

Copeland, Arthur H., *Geometry, Algebra and Trigonometry by Vector Methods*. Macmillan, New York, 1962.

Courant, Richard and H. E. Robbins, *What is Mathematics?* Oxford University Press, New York, 1941.

Coxeter, Harold S., *Introduction to Geometry*. John Wiley and Sons, New York, 1961.

Eaves, J. C. and A. J. Robinson, *An Introduction to Euclidean Geometry*. Addison-Wesley, Reading, Mass., 1957.

Euclid, *The Thirteen Books of Euclid's Elements*. Translated with introduction and commentary by Sir Thomas L. Heath. Dover Publications, New York, 1956.

Eves, Howard, *An Introduction to the History of Mathematics*. Rinehart, New York, 1960.

Eves, Howard, *A Survey of Geometry*. Allyn and Bacon, Boston, 1963.

Forder, Henry G., *Geometry, An Introduction*. Harper and Brothers, New York, 1962.

Hall, Dick Wick and Steven Szabo, *Plane Geometry, An Approach Through Isometries*. Prentice Hall, Englewood Cliffs, N.J., 1971.

Hilbert, David, *The Foundations of Geometry*. Translated by E. J. Townsend. Open Court, Chicago, 1921.

Keedy, Mervin L., *Number Systems: A Modern Introduction*. Addison-Wesley, Reading, Mass., 1969.

Keedy, Mervin L., R. E. Jameson, S. A. Smith, and E. Mould, *Exploring Geometry*. Holt, Rinehart, and Winston, New York, 1967.

Levi, Howard, *Foundations of Geometry and Trigonometry*. Prentice-Hall, New York, 1960.

Meserve, Bruce E. and Joseph A. Izzo, *Fundamentals of Geometry*. Addison-Wesley, Reading, Mass., 1969.

Moise, Edwin E., *Elementary Geometry from an Advanced Standpoint*. Addison-Wesley, Reading, Mass., 1963.

Moise, Edwin E. and Floyd Downs, *Geometry*. Addison-Wesley, Reading, Mass., 1964.

National Council of Teachers of Mathematics, *Insights Into Modern Mathematics, Twenty-third Yearbook*. The National Council of Teachers of Mathematics, Washington, D.C., 1957.

National Council of Teachers of Mathematics, *The Growth of Mathematical Ideas Grades K-12, Twenty-fourth Yearbook*. The National Council of Teachers of Mathematics, Washington, D.C., 1959.

Newman, James R., *The World of Mathematics*. Simon and Schuster, New York, 1956.

Ohmer, Merlin M., *Elementary Geometry for Teachers*. Addison-Wesley, Reading, Mass., 1969.

Piaget, Jean, Barbel Inhelder, and Alina Szeminska, *The Child's Conception of Geometry*. Translated from French by E. A. Lunzer. Routledge and Paul, London, 1960.

Rainich, George Y. and S. M. Dowdy, *Geometry For Teachers; An Introduction to Geometrical Theories*. John Wiley and Sons, New York, 1968.

Rosskopf, Myron F., Joan L. Levine, and Bruce R. Vogeli, *Geometry, A Perspective View*. McGraw-Hill, New York, 1969.

School Mathematics Study Group, *Geometry With Coordinates*. Yale University Press, New Haven, Conn., 1963.

Stabler, E. R., *An Introduction to Mathematical Thought*. Addison-Wesley, Reading, Mass., 1948.

Wolfe, Harold E., *Introduction to Non-Euclidean Geometry*. The Dryden Press, New York, 1945.

INDEX